# Genetic Prehistory in Selective Breeding:

a prelude to Mendel

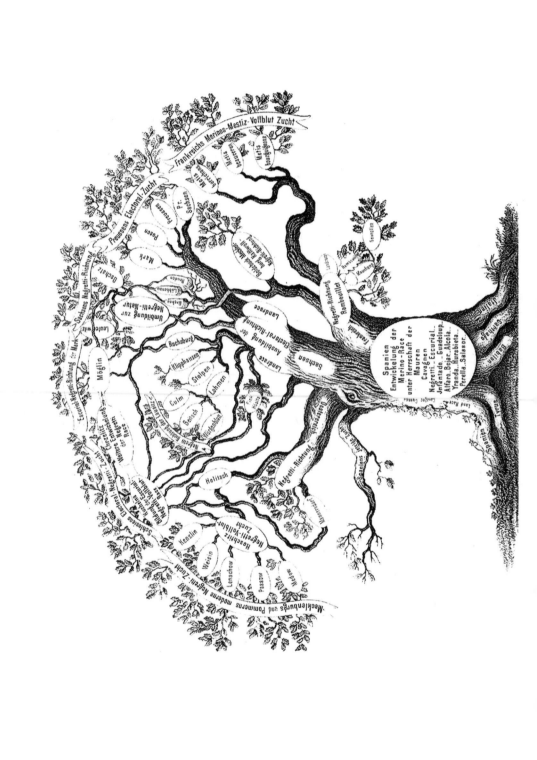

# Genetic Prehistory in Selective Breeding:

## a prelude to Mendel

**Roger J. Wood**

*University of Manchester*

**Vítězslav Orel**

*Emeritus Head of the Mendeleanum*

*Brno*

To my dear cousin Lena
With love and best wishes
Roger
5. 1. 07

OXFORD
UNIVERSITY PRESS

# OXFORD

UNIVERSITY PRESS

Great Clarendon Street, Oxford OX2 6DP

Oxford University Press is a department of the University of Oxford.
It furthers the University's objective of excellence in research, scholarship,
and education by publishing worldwide in

Oxford New York

Athens Auckland Bangkok Bogotá Buenos Aires
Cape Town Chennai Dar es Salaam Delhi Florence Hong Kong Istanbul
Karachi Kolkata Kuala Lumpur Madrid Melbourne Mexico City Mumbai
Nairobi Paris São Paulo Shanghai Singapore Taipei Tokyo Toronto Warsaw

with associated companies in Berlin Ibadan

Oxford is a registered trade mark of Oxford University Press
in the UK and in certain other countries

Published in the United States
by Oxford University Press Inc., New York

Library of Congress Cataloging in Publication Data
Wood, Roger J.
Genetic prehistory in selective breeding : a prelude to Mendel
/ Roger J. Wood, Vítězslav Orel.
Includes bibliographical references (p. ).
1. Sheep–Breeding–Europe–Histroy. 2. Merino sheep–Breeding–Europe–History.
I. Orel, Vítězslav. II. Title.
SF376.2W66 2001    636.3′082′094–dc21    2001021144

019 850584 1 (Hbk)

Typeset by J&L Composition Ltd, Filey, North Yorkshire
Printed in Great Britain on acid-free paper by
T. J. International Ltd., Padstow, Cornwall

*To Valerie and Olga*

# *Preface*

It was from experimental breeding that the principles of genetics were discovered[1]

*L.C. Dunn, 1965*

The origin of genetics, its 'prehistory' if we count Gregor Mendel (1822–84) as the first geneticist, is not only to be found in the clever writings of philosophers. More fundamental is the accumulated evidence of centuries of controlled breeding of plants and animals, and even human beings, from royal dynasties kept pure by incest to slaves selectively bred by their masters. Practical information and the breeding lore derived from it, refined by use and transmitted by word of mouth, were shared within families or between trusted friends in the same line of activity.

In this narrow world of secrets, it was not unusual to find breeders concentrating their skills upon a single species or even a single breed or variety. Of all domesticated animals, the one with a uniquely valuable coat attracted the most attention. Not for nothing did the insignia of the noble Order of the Golden Fleece, created to encourage knightly endeavour, come to represent economic power. Bruges, the city of the Order's foundation in 1431, had grown vastly rich as the centre of the textile industry and western European finance. Nor was it a matter of small moment that an English merchant two centuries later should describe the precious fibre that supplied the industry as 'the Flower and Strength, Revenue and Blood of England'.[2] For a substantial period of western European history, nothing compared with wool as a generator of wealth.

I thank God and ever shall

It was the sheep that payed for all[3]

From this much-valued economic base, the practitioners of the sheep-breeding art had every incentive to become more efficient. Their efforts to

---

[1] Dunn 1965, p. 25    [2] Gent 1656 (reprinted by Smith 1747, i, p. 197)
[3] Mediaeval couplet mentioned by Quennel and Quennel (1938, p. 199) as originating in Suffolk, and quoted by Power 1941, p. 17

improve the quality of their sheep in various respects led them to develop ideas about breeding and its theoretical basis in heredity. Differences among them encouraged the participation of naturalists and other experts in lively debate, particularly from the eighteenth century onwards. It is the purpose of this book to recount the phenomenon for the period 1700–1860, prior to the publication of Mendel's experiments. In so doing we have not only to review what was written by the acknowledged or self-styled experts of the day, but also to attempt an interpretation of the oral tradition behind the successes reported. Quite often this was reflected in remarks by farmers to journalists and 'economic tourists', a rich source of matter-of-fact information not only about the farmer's aims and achievements but also his preconceptions and prejudices. Breeding was an endeavour in which ideas arising out of observation-based logic could often be ahead of scholarly writing.

Whatever the subject of a breeder's attention, animal or plant, his overwhelming priority was to determine the conditions of breeding that affected resemblances between related individuals—the process of heredity. Although the word itself was not much used until towards the end of the period under consideration, the idea was gradually defined. This took place not only in the context of agriculture and horticulture but also in the work of writers representing other sections of society, most notably in medicine, veterinary as well as human, and in field sports. In situations where experiments were possible, theories were increasingly exposed to investigation. It was an activity that reached its most productive in communities in which talented individuals had freedom to act on their own initiative, making decisions on an empirical basis.

In the period under review, many good breeders were active, although one in particular, Robert Bakewell (1725–95) of Dishley in Leicestershire, served as a standard against which many others were to judge themselves. He came to epitomise breeding success at its highest level. His talent, which was directed mainly at increasing animal growth rates and maximising edible tissue proportions on the basis of minimum food intake, inspired a massive burst of experimentation in breeding on an international scale, setting a new agenda for an exploration of the truth about heredity. The words of those writing about the subject in this context have an enthusiastic, down-to-earth freshness about them, reflecting the fact that farmers and horticulturalists who have to live by their ideas show little patience with wild speculative theories. Although the language they used might sometimes seem old-fashioned, carrying a flavour of the ancient classics or the scriptures, it also had the capacity to embrace mechanistic ideas in the spirit of the scientific revolution. The fact that the proof of any theory, or justification for any particular technique, could affect a breeder's economic survival excluded all but the most able and innovative. Among such exceptional individuals, the drive to succeed encouraged indifference to certain pressures of convention, social, religious or even intellectual. Leading breeders set standards of their own, based on a comfortable, even complacent, certainty that they were furthering the cause of improvement, not only for their personal benefit but for the general good, a patriotic endeavour to be shared with others of similar vision. Working hypotheses to support their actions would be formulated on the basis of accumulated experience and spoken about with caution outside the chosen circles.

Bakewell's loyal disciple, George Culley (1735–1813), referred to the theory behind their own breeding activities as Bakewell's 'sheepish doctrine'. The expression is to be found in a letter to the Secretary of the Board of Agriculture in London, Arthur Young.[4] Of all domesticated species, sheep had revealed the most to Bakewell. The idea of learning from sheep was at the same time an in-joke and a shared revelation within the circle to which the secret was being revealed. The doctrine was both mysterious and controversial, and reaction to it could be empassioned, either for or against the revolutionary methods being employed. Those in opposition viewed the radical ideas that seemed to underlie Bakewell's actions as heresy, while the many who admired or envied his success tried to follow his example. As the doctrine was modified by successive agrarian reformers to meet local needs, it became adopted even beyond the shores of Britain, where there could be no greater honour than to be likened to him. On the other side of Europe, the distinguished Baron Ferdinand Geisslern (1751–1824), of Hoštice in Moravia, was pleased to acknowledge his reputation throughout Austria and Hungary as the 'Moravian Bakewell'. Geisslern, like Bakewell, is a name associated most strongly with the origin of pre-genetic ideas of heredity. In neither case had it much to do with what they wrote, rather more to do with what others reported of what they had said, but most of all to do with what they achieved. Both 'Bakewells' will figure largely in this book.

According to circumstances and economic pressures, individual sheep breeders would direct their activities towards different primary ends: meat production in one country or region, as in the case of the original Bakewell, and wool in another, as with his Moravian counterpart. In Moravia, the priority was to produce the highest quality fine wool, which meant breeding the best possible sheep for this purpose, the Spanish Merino. Whatever the priority, serious breeders were united in the conviction that success depended on their skill in applying techniques that had been tested, or could be tested, by experience. The most pressing practical problems were those that raised the most interesting theoretical questions. How, for example, does a farmer deal with a newly introduced race? Should it be kept pure? If so, how may it be guarded against degeneration as it becomes progressively inbred in an

[4] Wykes c. 1995 (quoting Northumberland Record Office ZCU9, 4 November 1784)

environment to which it is unaccustomed? If it is not to be inbred, should it be crossed to a locally adapted breed in order to spread its valuable traits more widely? If so, what is the role of artificial selection in fixing the desired type? Arising from the central dilemma, of whether to cross or not to cross, came a number of related issues: How do different traits interact? What is the relative importance of male and female in a cross? How does mere chance influence the course of events? In different countries breeders were facing similar problems and exchanging information with those around them. Critical experiments were required to answer the questions raised, not only in animals but also in cultivated plants.

It was Geisslern's exceptional success with imported Merino sheep on his Moravian estate during the early nineteenth century that stimulated the writing of this book. We took our lead from Johann Karl Nestler (1783–1841), Professor of Natural History and Agriculture at the Moravian University of Olomouc, who in 1831 wrote that 'The history of science forces us to think more than science alone can reveal.' Nestler had developed a particular interest in sheep breeding. Looking back to the previous century, he was thinking both about the background to Geisslern's achievements and about what followed. After Geisslern and his fellows had set about exploiting heredity in the cause of fine-wool production, using techniques derived from various sources, including Bakewell, the success of their activities inspired other commercial breeding projects, the most successful of which were directed towards improving fruit trees and vines. In this connection, important work was carried out by members of the monastic community of St Thomas in Brno, a tradition that would eventually encompass Mendel's own breeding experiments.[5] In the light of that history, we have felt bound to view the prolonged period of sheep-breeding activity centred in Mendel's territory with particular interest, as a potential link in a chain of events. Not, it must be stated at once, that we are claiming any tangible connection between the sheep breeders and Mendel. Nor would we ever expect to do so, because, in Mendel's time, sheep were yesterday's object of breeding interest. On the basis of the evidence presented in this book, we are prepared to claim (i) that the search for rules of heredity in sheep created an atmosphere of enquiry about heredity in general; (ii) that one of the major locations where the search took place was in Brno; and (iii) that certain public figures whom Mendel respected were deeply committed to the search.

In Moravia, our study is largely confined within a period of 50 years. Before 1790 there was little sign to the outside world of anything unusual about sheep in Moravia and by 1840 the interest in sheep had lost its impetus, for reasons which will be defined. But our search for information on the

---

[5] Orel 1996

contribution of sheep breeding in uncovering heredity's rules extends far beyond the borders of Moravia or even of the Hapsburg Monarchy. In attempting to trace the origin of the successful Moravian Merino breed, with such a widespread influence during just a few years, we are led to ideas and techniques developed well before 1790, as well as after 1840. Our researches in Brno have caused us to look westwards, to German lands and beyond (Chapters 2, 6 and 7) and, above all, to England, the home of the original Bakewell (Chapters 3, 4 and 5). It was no secret that heredity in sheep occupied the attention of a number of clever and original minds. Geisslern and his colleagues in Moravia made a successful synthesis of ideas and techniques drawn from several different western sources (Chapter 8). Information we have gathered on this body of knowledge, on its nature, its transmission through the written and spoken word, and its use on the breeding farm, has formed itself into a story of high endeavour and the pursuit of excellence, an inspiration for questions about heredity and a fitting prelude to Mendel's experiments (Chapters 1, 9 and 10).

# Contents

# Illustrations

# Acknowledgements

After Mendel's work was finally recognised for its true significance in 1900 (the so-called 'rediscovery'), fellow members of his religious order, the Augustinians, began to organise a Mendel Museum at the monastery. It was opened in 1922 for the benefit for geneticists celebrating the centenary of Mendel's birth. Then, in 1948, shortly after the Communist takeover in Czechoslovakia, the Government forced the monasteries to close. With the Augustinians expelled, the Museum ceased to function. Genetics was stigmatised as a reactionary science and the new ideological teaching of 'Lamarckian' heredity was introduced from the Soviet Union, in a campaign led by the notorious Trofim Lysenko (1898–1976). But great scientific discoveries in genetics after 1950 led slowly towards a different attitude in Communist countries. It was in the changing atmosphere that, in 1963, the Brno geneticist Jaroslav Kříženecký (1896–1964), who had served a period in prison as a victim of Lysenkoism, was entrusted with collecting materials for a new Mendel Museum (Mendelianum). Set up as a department of the Moravian Museum in Brno, it was housed in the former refectory of the monastery, immediately adjacent to the experimental garden where the scholar carried out his experiments. In 1965 the new museum organised, with international scientific support through UNESCO, an International Memorial Symposium, commemorating the one hundredth anniversary of Mendel's discovery. Since then, the staff have been collecting and examining ever-growing numbers of documents relating to Mendel, the origin of his experiments and his influence on the development of genetics. Without this body of systematic research, carried out in co-operation with other national and foreign scholars, our book could not have been written.

Many others could also be thanked, among whom we should like to name the following for particular mention: Dr J. Harwood and Mrs V.E. Wood MA for critical readings of the book in manuscript; Mr J.M. Visanji for his expert help with the references, enhanced by his knowledge of German; Dr D. Farnie for research into letters written by Count Berchtold; Prof. John Pickstone and Mr F. Hirschelmann, the latter based at the Bibliographisches

Institut AG in Mannheim, for information on Hans Moritz Graf von Brühl; Dr D.L. Wykes for his generosity in allowing us to quote from an unpublished manuscript on Bakewell; Mr Fred Hartley, Keeper of Harborough Museum, for copies of unpublished documentation collected by him for an exhibition on Bakewell in 1995; Mr M. Scafe MA for the gift of an unpublished resumé of information on the Order of the Golden Fleece; Mr James Beechey for a copy of Robert Bakewell's will and unpublished information on his family tree and connections; Dr C.O. Carter for copies of unique documentation on the Merino and other information; Mr K.-H. Suneson, Ms R. Edberg and Ms C. Wijkström for helping us to interpret the Swedish Merino story; Professor Donald R. Wood of the Department of Agronomy, Colorado State University, Fort Collins, for undertaking research for us on a drawing of a Hoštice Merino ram made by C.C. Fleischmann in 1847; Dr S.J.G. Hall for an interesting personal communication on Spanish sheep in England; Dr A. Matalová for her generous assistance in uncovering and interpreting information in the Mendelianum; Dr and Mrs J. Benedík for their most willing help in locating and visiting former sheep breeding estates in Moravia, and acting as interpreters; Mr P.C. Caswell MA for help with an aspect of German translation; Mr R.A. Neave for presenting us with his interpretative sketch of the face of Bakewell; and Dr Isabel Phillips for her unpublished translation of Rudolf André's book of 1816, 'Introduction to the Improvement of Sheep on Principles deriving from Nature and Experience', which she included as an annexe to her Ph.D. thesis (1989), 'submitted with the thesis for the use of other scholars in the field'. Finally we thank Miss Lisa Monks, whose patient and heroic effort in typing this book in its successive drafts, with an ever cheerful smile, has been deeply appreciated.

For permission to use illustrations for which they hold the copyright, we are indebted to the Leicestershire Museum Arts and Records Service, The Natural History Museum Trading Company Ltd (London), Albert Bonniers Förlag and the Svenskt Biografiskt Lexicon , Mr R. Edberg and Foreningen Gutefåret and Gerald Duckworth and Co. Ltd.

Except where otherwise stated, all translations from other languages into English were made by the authors. Except when translated, quotations are given as they appeared, including the author's punctuation, use of italics, underlining or upper case.

# 1

## *The elusive law*

That, so far, no generally applicable law governing the formation and development of hybrids has been successfully defined can hardly be wondered at by anyone who is acquainted with the extent of the task.

*G. Mendel, 1866*

*(original in German; translation by Bateson, 1909)*

Close to the heart of the industrial city of Brno, in a walled garden of the monastery of St Thomas, can still be seen the plot where Fr. Gregor Mendel grew his peas. Here, between 1856 and 1863, the famous Augustinian friar discovered heredity's fundamental secret, expressed in the formation and development of hybrids, which led to the universal concept of the gene. Few can visit the garden and the Mendel Museum which now adjoins it without a sense of excitement and curiosity. A discovery so profound, far-reaching and seemingly ahead of its time encourages speculation upon its origin. What factors, scientific, cultural or social, could have stimulated it into existence in this particular place, at that early date, in those particular hands? The answer has proved complex. Research into the origin of the discovery reveals a web of probable influences upon Mendel, uniquely integrated and still, in many respects, mysteriously obscure.[1]

Unconnected with any mainstream scientific grouping, Mendel is difficult even to categorise. As Bronowski has put it, Mendel 'remained all his life a farm boy in the way he went about his work, not a professor nor a gentleman naturalist like his contemporaries in England; he was a kitchen-garden

---

[1] Iltis 1924, 1932; Olby 1985, 1990; Orel 1996; Orel and Wood 2000a, 2000b

naturalist'.[2] At the same time it has to be acknowledged that Mendel was unusually broadly educated and that he found that university examinations in physics were much easier than those in certain aspects of biology. By some happy blend of these strengths and weaknesses, directed with curiosity, he came to ask some pointed questions bearing upon a mysterious subject, and to devise experiments sufficiently well designed to answer those questions precisely.

In this book we do not dwell much upon Mendel, although he is our ultimate inspiration for writing it. Here we are concerned with a particular aspect of farming practice, one that counted highly in Mendel's country because it formed the basis of national prosperity, namely the breeding of fine-wooled sheep. It might seem strange to introduce Mendel at all into an account of an activity with which, as far as we know, he was never personally associated. The reason centres upon our conviction that knowledge of this area of breeding can help in reaching a clearer understanding of the circumstances that led to his experiments. For several decades before Mendel arrived in Brno, the sheep breeders of Moravia had been deeply preoccupied with trying to manipulate the process of heredity and understand its basis. Sheer economics dictated that more was written then about sheep than about any other species, whether plant or animal. Intimately involved in the continuing discussions about heredity, which arose from efforts to improve wool quality, were a number of experts destined to have a direct influence on Mendel's career and intellectual development. Nobody enters this category with greater right than the man whom he would later succeed as abbot, Prelate Cyrill Franz Napp (1792–1867) (Figure 1.1), and it is with him that we start, by discussing his role as a promoter of ideas on the basis of reliable practice and sketching out the influence he was able to exert upon his successor. We also begin to explore from where he developed his ideas, a theme to be elaborated in greater detail later.

From the day Mendel entered the monastery in 1843, at the age of 21, until Napp's death in 1867, this powerful personality was to play a principal hand in determining his destiny. Not, it must be stressed, that Napp would have been at all inclined to force his interests and ideas upon the younger man. The dignified scholar kept his distance from junior members of the community, rarely if ever speaking directly with them. Yet his personal concern for each member of his community was gratefully acknowledged. Hugo Iltis, Mendel's first biographer, wrote that Napp 'was zealous in promoting the brethren of the order, being quick to develop talent among them and, where possible, further its development'.[3] In Mendel's case, Napp arranged a rich diet of higher education, first in theology (he was ordained priest in 1848),

---

[2] Bronowski 1973, p. 380      [3] Iltis 1932, p. 48; Orel 1973

**Fig. 1.1.**   Prelate C.F. Napp, Abbot of the Augustinian monastery in Brno from 1824 to 1867 (printed from a negative in the archive of the Mendelianum, Brno).

agriculture, pomiculture, viticulture and natural history in Brno, and later in fundamental aspects of natural science, primarily physics and mathematics along with botany, plant physiology and chemistry, with some palaeontology, at the University of Vienna.[4]

---

[4] Weiling 1967; Czihak 1984; Czihak and Sladek 1991/2

Mendel was away from Brno between September 1849 and the summer of 1853, first teaching in Znojmo (Znain) before spending two years as a student at the Institute of Physics in the University.[5] Napp smoothed his path by making financial provision for him in Vienna, at the same time trusting him to make his own choice of subjects there. In preparing the ground for Mendel's time there, Napp wrote 'I shall not grudge any expense for the furtherance of his training'.[6] Upon Mendel's return to the monastery, Napp invested extra sums of money to his benefit. The year 1854 saw the erection of a substantial greenhouse, which was placed at Mendel's disposal by 1855 for the planned experiments.[7]

The path leading to the experiments was not without pitfalls. While Mendel was away in Vienna, Napp had a tough fight on his hands to keep the educational and scientific aspects of the monastery going. His policy came under direct attack by the Brno Bishop Schaffgotsche who finally, in June 1854, led a two day 'apostolic visitation' to the monastery, on behalf of the Archbishop of Prague, as a result of which he recommended the Holy See to issue a special decree dissolving the monastery. According to his plan, Napp would retain his influential social position in promoting education and science; the fate of the other monks would be decided individually. Where, one wonders, would that have left the impoverished Mendel?

In his letter of defence to Cardinal Schwarzenberg in Prague, Napp expressed his firm conviction that cultivation of science did not contradict the spiritual mission of the monastery. He displayed characteristic political skill in proving the special status of the Brno monastery by reference to a Royal decree of 1802. To the relief of the community, he was successful, the Bishop subdued, the danger overcome and a suitable environment prepared for Mendel to undertake his experiments. Later, in 1870, Abbot Mendel reported 'that cultivation of science in all respects has always been considered as one of the first tasks of the monastery'. But how nearly might this not have been the case, but for the determination and astuteness of Abbot Napp in 1853–54?[8]

Napp had a shrewd sense of business and a strong interest in the useful applications of science. A contemporary wrote that he was 'the supporter of every scientific effort'.[9] By cleverness in economic matters and estate management, he had made the monastery rich, restoring its buildings and maintaining them in excellent condition. Surprisingly for a former Professor of Old Testament and oriental languages, he had acquired a practical knowledge of agriculture and the means of its improvement. Perhaps he viewed the bespectacled Mendel at work in the garden and greenhouse as one who sees an aspect of his own ambition being fulfilled. Napp's interest in breeding was

---

[5] Weiling 1967, 1986    [6] Iltis 1932, p. 75    [7] Orel 1975a
[8] Orel and Fantani 1983; Czihak 1984; Orel and Verbik 1984; Czihak and Sladek 1991/2
[9] Rohrer 1830

long-standing and broadly based.[10] He took particular satisfaction in local progress made in plant breeding which kept Brno abreast of wider European achievements.

There was an earlier phase in Moravian agricultural history that had also caught Napp's attention. We refer to a period at the beginning of his life when Moravia had produced some of the finest sheep in Europe by an advanced system of breeding,[11] which had brought considerable economic advantage and been widely copied. In this context Napp had good reason to call to mind the small but intensively managed estate of Hoštice, in the lush countryside to the north-east of Brno. There, half a century before Mendel was experimenting with peas, Baron Ferdinand Geisslern had bred a newly defined race of sheep, based on imported Spanish Merino stock, superior to any seen before.[12] The Hoštice breed's exceptional qualities and potential for improving less refined flocks, by siring excellent crossbred lambs, was a godsend to Moravian sheep owners, including Napp with his extensive monastic lands to manage. But to Napp and some of his friends, the interest was more than simply economic. Breeding techniques used to create the new race, which were gradually revealed by some of Geisslern's associates, fired their intellectual curiosity. Here we introduce that story; later we shall recount how Napp and others set out to uncover the basis of the new breed's hereditary powers and how difficult they found the task, constrained within the theoretical framework of their time.

Geisslern and Mendel, seignior and monk, separated by time, dealing with heredity in quite different organisms, are linked through Napp, and by events and intellectual developments, which this book will describe. The story relates to commercial breeding in Moravia, particularly in the region of Brno, but also embraces events on a much broader canvas, as will become clear. In the Moravian context it reveals how lessons learned from sheep breeding were applied to the production of new varieties of fruits and vines, field crops and ornamental plants. The last were increasingly in demand from an urban population growing comfortably affluent as a result of a highly successful textile industry, the most important in the Hapsburg monarchy.[13] Napp encouraged all these commercial breeding activities within the monastic environment and also more widely through his influential membership of various committees and societies. There he joined forces with manufacturers and merchants, academic experts and journalists, as well as other landowners and managers. He showed characteristic enthusiasm and single-mindedness, willing even to take a personal hand in some of the breeding experiments when time allowed.[14] More importantly, however, he had the power to command a serious hearing when it came to interpreting the results.

---

[10] Orel 1975b, Orel and Wood 2000b   [11] Nestler 1841   [12] Orel and Wood 1981; Chapter 8
[13] Freudenberger 1977, p. 170   [14] Weiling 1968; Orel 1975b

When Napp arrived in Brno to take up his appointment, he had found a local economy deeply dependent on the clothing industry. The manufacturers and merchants of Brno had particular cause to appreciate the value of high quality wool and to support every means to obtain it. The prosperity of their rapidly growing city, the 'Manchester of the Hapsburg monarchy',[15] was becoming increasingly dependent on locally bred sheep. More and more land was turned over to superior fine-wooled stock. On the monastery lands, wool production was the most profitable activity.[16] From the 1790s onwards, the highly bred Merinos of Southern Moravia had gained a reputation next to none in all the Hapsburg lands, as reliably healthy, robust and capable of producing a heavy fleece of fine yet lustrous wool, predictably uniform in quality. The animals became widely distributed from their original focus in the village of Hoštice, where the best breeding stock was to be found, even still in Mendel's day.[17]

News from abroad—Germany, France and England—told of Merinos degenerating after a few generations away from Spain. It was gratifying to the wool producers of Moravia in the early 1800s that the sheep under the care of Geisslern and his followers were maintaining their quality. For more than 30 years after his death in 1824, it was still so.[18] Not that they found their task easy. There was no room for complacency in such a risky business; selection for carefully defined features had to be constantly maintained. In 1814, leading breeders, with the support of landowners and representatives of the textile industry and business interests, formed themselves into a Society of Friends, Experts and Supporters of Sheep Breeding (hereafter abbreviated to the 'Sheep Breeders' Society') organised as a section of the Brno-based Royal and Imperial Moravian and Silesian Society for the Improvement of Agriculture, Natural Sciences and Knowledge of the Country (hereafter abbreviated to the 'Agricultural Society'), to build upon Geisslern's success.

Before long an important development took place in the Sheep Breeders' Society's discussions. Members began to debate the need for scientific principles to govern breeding, so as to be sure of maintaining or even improving wool quality. The more scientifically minded members attempted to define a theoretical framework to underpin the new methods. The quest for 'sound theory' (*gesunde Theorie*), as formulated by the Vienna-based B. Petri in 1812,[19] or 'pure principles' (*Grundsätze in Reinen*), as expressed by the youthful Moravian expert Rudolph André (1792–1827) in 1816, was pursued.[20] Discussions based on extensive sheep breeding activity were reported in locally edited journals. These were *Patriotisches Tageblatt*, published in Brno from 1800–1805, *Hesperus oder Belehrung und Unterhaltung für die*

---

[15] Freudenberger 1977, pp. 25 and 189      [16] Orel 1996, p. 48
[17] Orel and Wood 1981; Wood and Orel 1982      [18] Settegast 1861      [19] 'Irtep' 1812
[20] R. André 1816

*Bewohner des österr. Staates*, published from 1809 in Brno, from 1813 in Prague and from 1823 in Stuttgart, and, most influential of all, *Oekonomische Neuigkeiten und Verhandlungen* (abbreviated to *ONV*), published in Prague from 1811.

The editor in each case was Christian Carl André (1763–1831), father of Rudolf, former school teacher, author of articles and books on economics, mineralogy and natural history, and influential co-ordinator of scientific activities. His journals had a wide German readership, even outside Moravia and the Hapsburg monarchy. He made sure his readers were kept well informed about the increase in quality wool production in Moravia and, at the same time, the breeding techniques used to achieve the improvements observed, both in quantity and quality. André and his contributors also highlighted areas of scientific controversy: questions about racial differences, racial crosses, inbreeding, the value of pedigree and the concepts of trait recording and progeny testing, all of which demanded to be explained. An ongoing debate took place in the pages of the journals during 1816–19 on the question of how to combine desirable qualities of wool, characterising different races of sheep, within a single breed. On the basis of practical experience and the latest findings in natural science, the writers sought to explain the resemblances and differences between parents and their progeny. The debate, which was conducted under André's editorial control in the pages of *ONV*, culminated in 1818 with a Hungarian landowner and breeder, Count E. Festetics (1769–1847) attempting to define 'genetic laws' (*genetische Gesetze*), based on direct observations of the progeny of crosses between strains showing different traits.[21] These laws, which are analysed later (Chapter 9), have obvious interest in relation to what was to come.

The Sheep Breeders' Society was not the only group to react to a commercial stimulus. A Pomological and Oenological Society was established in Brno in 1816, instigated by André on the pattern of the Pomological Society of Altenberg near Leipzig, which had been in existence since 1803 and of the Horticultural Society of London founded a year later.[22] The achievement of the President of the Horticultural Society of London, Thomas Andrew Knight (1759–1838), in producing new varieties of fruit by controlled fecundation (i.e. pollination) was seized upon as an inspiring example.[23] Napp showed a deep and growing interest in Knight's work as it became translated into German. Knight's experience of breeding cattle and sheep had convinced him of the value of making crosses between stocks of different ancestry and selecting among their progeny. By this means he had originated the grey form of the Hereford cattle breed[24] and a superior flock of fine-wooled sheep of the Ryeland breed. In a paper entitled 'An Account of Some Experiments on the

---

[21] Orel and Wood 1998     [22] Hempel 1818; Orel 1978a     [23] Mylechreest 1988
[24] Heath-Agnew 1983, pp. 48–50

Fecundation of Vegetables', published by the Royal Society in 1799, he recommended a comparable crossing technique for plant breeding. He wrote of 'the close analogy between the animal and vegetable worlds and the sexual system equally pervading both', which induced him 'to suppose, that similar means might be productive of similar effects in each'. As a rapidly maturing plant for studying the effect of crossing, 'none appeared so well calculated to his purpose than the common pea'.[25] A German translation of this influential paper which appeared in *Oekonomische Hefte*,[26] published in Leipzig within the year, was taken by the library of the Agricultural Society in Brno, through which Knight's advice became known to the plant breeders of Moravia as they expanded their interests in wine and fruit production.[27] Pea crossing experiments by two English investigators were published in 1824 and appeared in German translation in 1837 in the Bavarian gardener's journal,[28] reported by Napp in the 1837 meeting of the Pomological Society.

Ever since the founding of the Brno Pomological Society, the *ONV* edited by C.C. André had included details of Knight's activities published in the proceedings of the Horticultural Society of London. The first appeared in 1816.[29] The author, most probably André himself, was fascinated by Knight's claim to have selected entirely new varieties, each uniting various outstanding qualities, among the progeny of inter-varietal hybrids. An article in *ONV* in 1820, on the application of hybridisation in creating new varieties of cereals, included a call to uncover 'the law of hybridisation'. The author, G.C.L. Hempel, Secretary of the Pomological Society of Altenburg, added his thoughts on the personal characteristics that the future discoverer of such a law would have to possess, 'a man of deep botanical learning, of keen powers of observation, and tireless and unflinching perseverance'. It was a pretty good description of Mendel, yet unborn.[30]

André's efforts to promote this idea might have continued except that he was forced to vacate the territory of the Hapsburg Empire in 1821 because of his liberal political views. However, he left behind an able disciple in J.K. Nestler, who in 1823 became Professor of Natural History and Agriculture at the Moravian University of Olomouc, and in 1827 introduced the subject of scientific animal and plant breeding into the University's teaching. Two years later he published his lectures, meanwhile beginning a long and productive association with the newly elected Abbot Napp.

Napp's own involvement with breeding developments can be judged from his responsibilities as abbot. After his election in 1824, when a professional involvement in agricultural matters was thrust upon him in connection with

---

[25] Knight 1797, 1799    [26] Knight 1800, pp. 332–3    [27] Orel 1978a
[28] *Algemeine Deutsche Gartenzeitung*, Orel 1996, p. 77    [29] Anon 1816b
[30] Hempel 1820; Orel 1974a

the monastic estates, there is every reason to believe he welcomed it. Within a year of his appointment, he had been elected a member of the Agricultural Society; two years later he was on its committee (1827); in 1850 he became Deputy Director and in 1864 he became Director of the Society. He held the Presidency of the Pomological and Oenological Society for 22 years, from 1827 to 1849. Because of his wider agricultural responsibility he also developed a close familiarity with published literature dealing with the improvement of wool, and became an active participant at the annual meetings of the Sheep Breeders' Society. His knowledge and experience were recognised by his appointment as a member of the state commission for examining shepherds. The Agricultural Society was responsible for organizing courses for shepherds and for their examination, according to the requirement of the State Government.

Napp and Nestler had much to learn from one another. Napp was among those listening intently to Nestler as he addressed the Sheep Breeders' Society at their annual meeting in 1836. The Chairman, Baron Bartenstein (1769–1838), asked the speakers to define the most important question for wool improvement. Nestler's response was to examine the essence of heredity (*Vererbung*) as a separate issue from the enigmatic process of generation (*Zeugung*). Most German writers did not use the term 'heredity' until later in the century; so Nestler was ahead of his time. The lecture prompted so much interest that further discussion took place in the pages of the journals for almost two years afterwards. Napp became convinced that the problem of heredity would be solved only through physiological research aimed at answering the question 'What is inherited and how?'[31] Motivated by issues in practical breeding, the most critical of which was when to inbreed and when to outcross pure-bred stocks,[21] Napp, Nestler and probably other members of the Sheep Breeders' Society had become interested in heredity as a problem in itself.

Not long after this period of lively debate, interest in fine-wooled sheep declined in Europe. As always, farming practice was dictated by economics. By the 1840s it was becoming cheaper for cloth manufacturers in Brno and other centres to import their raw wool from abroad, mainly from the fast growing Australian market where some of the best flocks had been formed from Moravian bloodlines from the Lichnowsky estates. When in 1840, the Agricultural Society played host to the Fourth Congress of German Agriculturists and Foresters, held that year in Brno, it was natural that sheep breeding should be reviewed. After Napp had introduced the subject, he invited Nestler to speak, who recalled with warmth the innovative work of Geisslern and outlined the impressive progress achieved, which had made Moravian breeding stock so much in demand.[32] Later, as chairman of the

---

[31] Orel 1977    [32] Nestler 1841; Fraas 1852, p. 602

section for fruit tree breeding, Napp confined his remarks to a problem that all breeders faced when attempting to produce new varieties by artificial fertilisation, the significance to be attached to mere chance.[33] Napp was in an ideal position to recognise its potential importance because of his wide and varied practical knowledge. The chance element demanded to be understood and, if possible, controlled. Perhaps even before his protégé Mendel, Napp had recognised 'the wild, the fortuitous, the gambler's throw, invisible in the germ cell', a degree of perception that Eiseley attributes to other naturalists prior to Mendel.[34] Eiseley also asks whether 'the idea lurked hidden in the work of eighteenth century breeders'. His question remains one to ponder in the light of Napp's words uttered in 1840.

Such speculations apart, it is Napp's role of leadership in the search for better quality fruit and vine varieties that can be seen as his greatest contribution to Moravian agriculture and his most obvious scientific connection with Mendel. As English sheep and cattle breeding had inspired Knight, so the triumphs of Moravian sheep breeding, with Geisslern as the 'Moravian Bakewell', were there to inspire Napp. His response to the practical achievement of those around him was to ask a series of theoretical questions about what determines heredity and with what degree of certainty. That much is known from his own words. Equally certain is the fact that nobody then was in a position to answer him.

The focus of the younger man, Mendel, took him in a new direction but gave him reason, nonetheless, to share Napp's awareness of the unanswered questions of the past. Like Napp he became an active member of the Agricultural Society. The elusive law or laws of heredity remained an intriguing mystery, despite much speculation and argument among horse, cattle and sheep breeders in particular. Published books and articles on the nature of generation, fertilisation, racial purity, racial crossing, variation and individual potency in farm animals were appearing thick and fast in the 1850s and 1860s when Mendel was undertaking his experiments. Although Mendel's experimental focus was directed elsewhere, his desire to find a 'generally applicable law governing the formation and development of hybrids' tells us that his mind was open, at least as far as plants were concerned. A wider interest was shown by his attempt to cross different races of bees.

The extent to which Napp, driven to seek answers about heredity posed in an earlier era, encouraged the younger man's experiments on pea plants remains an intriguing question, one which has led us to examine the breeding literature of the period, including the large amount published in Brno. Our analysis, to be explored in the chapters which follow, has convinced us of the significance of Moravian sheep breeding in the progress of ideas about

---

[33]  Nestler 1841, p. 337    [34]  Eiseley 1959, p. 11

heredity. This is not because of any specially close practical or theoretical connection revealed between the sheep breeders and Mendel, or even because of Napp's powerful influence on events, but due to a strong motivation to understand the underlying basis of heredity created by the technical and commercial success of practical breeders, first of sheep, later of cattle, fruit, vines, field crops and ornamental flowers. It drew the search for laws of hybridisation and heredity into the same orbit. Technical success with each of their crops in turn maintained the breeder's demand for a scientific explanation of heredity, in the hope of achieving even greater success.

# 2

# The fleeces of Spain

Spain, as everyone knows, is the true homeland of fine-wooled sheep, from where sheep breeding spread to all cultivated states.

*B. Petri, 1825 (original in German)*

Every lively debate on the nature of heredity taking place in the Brno Sheep Breeders' Society, from its foundation in 1814 and for nearly three decades afterwards, owed its origin to a particular development in international trade. The entry of top quality Spanish Merino breeding stock into Moravia and neighbouring territories after 1775 had provided the essential basis on which production of fine woollen cloth could be greatly expanded there. Flocks of the valuable animals, imported at great trouble and expense, were carefully kept apart and bred selectively for desired qualities. Skilled breeding ensured that, generation by generation, they maintained their ability to produce superfine wool and provided rams for improving the coarse-wooled native sheep through crossing. Although a few Moravian flocks had been taken through early stages of improvement during the previous century by crosses to part-Merino Paduan sheep, transported across the Adriatic Sea from Italy, most were still coarse-wooled or even hairy. The transformation gained pace as Spanish stock became more widespread from the 1790s onwards. It was an exceptionally successful venture both in trade and in breeding, enabling the woollen fabric industry of Brno eventually to become entirely independent of imported Spanish wool, which was replaced by an equally good or better product from locally bred Merino sheep, adapted to local conditions. The great advance in Moravian sheep breeding, and all that followed from it, cannot be imagined without the entry upon the scene of the Merino from Spain. The impetus to gain expertise on the new breed grew rapidly. In search of enlightenment, attention turned to the breed in its native territory.

To hold a sample of Merino wool in the hand was to recognise it to be like no other. Bishop Huet of Auranches was expressing no more than the sober truth when he assured readers of his *Memoirs of the Dutch Trade* in 1706 that 'Spanish wool was ever in high Esteem on account of its Fineness and Excellency'.[1] Aspects of its quality listed by Youatt (1837) lay not only in exceptional fineness, softness and silkiness but also in 'unexpected' felting properties deriving from a degree of curliness and 'in the weight of it yielded by each individual sheep'.[2] It seemed to score highly on every count (pliability and lustre being other characteristics often stressed), which dictated that any country with pretensions to produce woollen cloth of fine quality had to obtain the Spanish raw material one way or another. In this chapter we ask what made the Merino wool and the sheep that produced it so universally in demand and describe how Spain began to lose her monopoly. We shall revisit the question with respect to particular nations and provinces, including Moravia, in Chapters 6 and 7. Before that we shall examine how the impetus to preserve the unique nature of the Merino fleece (while at the same time trying to introduce its qualities into other breeds by crossing) began to provide information on heredity and what influences it. Also to be considered is the experience gained in breeding for non-wool traits.

The Merino emerged into historical prominence under the rule of the Moors, revealed to the world through rich fabrics woven at Cordoba and Seville, at a time when the raw material was hardly known outside Spain. The first appearance of the wool itself on the European market was towards the end of the twelfth century[3] although successive rulers guarded the precious product jealously and controlled its export. By a royal edict of 1303, wool from the vast flocks of Castile (to be joined later by Aragon), became a closed monopoly for almost two centuries. Then, after the expulsion of the Jews (1492) and the consequent loss of their commercial expertise, swift action was required to create an organisation to gather the wool and ship it abroad. The *Consulado*, set up at Burgos in Castile under Ferdinand and Isabella, became what has been justifiably described as 'the first great wool marketing scheme of the modern world'[4]. Neighbouring Italy benefited from some of the best of it although much else found its way northwards to the Low Countries, England and France,[5] where it met tough competition from English wool. Being imperfectly sorted in the home country, and thus of mixed quality, it was relegated largely to village production to be combined with other wools. A century later, however, there was a different tale to tell

---

[1] Smith 1747, ii, p. 97    [2] Youatt 1837, p. 149; see also Low 1845, pp. 134 and 136
[3] Power 1941, p. 13; Ponting 1980    [4] Carter 1969
[5] Youatt 1837, pp. 156–7; Davies 1958; Carter 1969

when English exports began to lose ground to Spanish competition.[6] Commercial documents from 1576 record the transport of 40,000 sacks of Spanish wool to Bruges in Flanders (then under Spanish Hapsburg rule) at 20 gold ducats a sack, and others of a finer kind to Italy, at 50 gold ducats a sack.[7] The Merino product was by then without serious rival, even among the finest of English wool. It had attained what Carter has described as 'its strategic position as a catalyst in the new industrial and agricultural revolution of the succeeding three centuries'.[4] By 1700 England was importing two million pounds' weight of it per annum for weaving superfine cloth.[8]

Attracted by wool of such exceptional quality, foreigners visiting Spain to see its production would remark on discovering that it came from sheep that were far from uniform in appearance. A series of sub-breeds or *cabañas* (stables) differed quite markedly in body shape and size, length of leg, degree of wrinkling of the skin and even in the length and uniformity of the wool.[9] As the Spaniards took stringent precautions to ensure that sheep from different *cabañas* did not mix, the resulting isolation must have encouraged the diversity observed. Some of the *cabañas* ranked more highly than others. *Paular*, *Escurial* (= *Patrimonio*) and *Negretti* (= *Negrete*) were among the top half dozen of more than 20 *prima piles* (first quality wools) that could be listed, each associated with a particular noble family or religious order.[10] Other *cabañas* rated specially highly were *Muro*, *Guadalupe*, *Infantado* and *Montarco*. The massive flocks possessed by the Duke de l'Infantado and some of the other great landowners could exceed 40,000–60,000 head.[11]

Variation apart, Merino sheep had unmistakable features in common (Figures 2.1 and 6.3). The males were characterised by large curved, spiral horns although the females were mostly without them. The breed could seem ugly to foreign eyes,[12] long-legged and hollow-necked in contrast to the straight back and relatively short leg of an English Southdown, Cotswold or New Leicester sheep. The breast and back were narrow and the sides somewhat flat. Scarcely more appealing to those accustomed to admire English sheep was a 'peculiarly coarse and unsightly growth of hair on the forehead and cheeks'. Their greasy coats, easily blackened by dirt, belied the rich whiteness of the densely packed, fine or ultrafine wool beneath. So much dirt was attracted to the pile that when pressed upon, it felt 'hard and unyielding',[13] the fibres being matted and 'stuck together as it were with a varnish, by

[6] Bowden 1962, p. 213    [7] Livingston 1813, p. 41    [8] Ryder 1983, p. 427
[9] Parry 1806, p. 344. The length of wool fibre in a fleece was denoted by its staple; 'long staple' wool consisted of fibres of 12 inches or more in much of the fleece; wool described as 'medium' or 'short' had shorter fibres; long wool was generally not as fine as shorter wool (Bowden 1956, pp. 44–5)
[10] Banks 1808; 'Lincolnshire Grazier' 1833, pp. 232–3; Bonwick 1894, p. 38; Ryder 1983, p. 426; ; Carter 1964, p. 324; Massy 1990, p. 13    [11] Stumpf 1785
[12] Culley 1807, pp. 224–5; Youatt 1837, pp. 148–9
[13] Youatt 1837, p. 148; see also Culley 1807, pp. 225–227; 'Lincolnshire Grazier' 1833, p. 229

**Fig. 2.1.** Rams of two Merino *cabañas*, (A) Negretti and (B) Infantado, both imported from Spain and kept under the same conditions by the Austrian agriculturalist Bernard Petri on his farm at Theresianfeld. They have been drawn to the same scale, as reproduced by Petri 1815.

means of which the interior wool preserves all its cleanness and its white colour, which only acquires somewhat of a yellowish cast from it greasiness'.[14] Folds in the skin about the neck and body of some of the sheep added to their unattractiveness, although such folds were much esteemed by the

---

[14] Schulzenheim 1797, p. 312

Spaniards who believed them to be associated with a heavier fleece.[15] This was to prove an accurate perception for, in the later history of the Merino breed, the skin fold character was selected in controlled breeding programmes that resulted in a great increase in the proportion of body weight contributed by the fleece.[16] Extreme examples were to be found in the USA between 1812 and 1865, particularly due to the activities of Edwin Hammond of Vermont. A parallel development took place in Australia.[17] Imaginative writers likened it to the ridge and furrow system of arable farming, which enhanced productivity by increasing the surface area of fields.

Modern studies reveal that the special fineness of Merino wool rests upon the high density of fibres per unit area, more than any other breed.[18] Packed closely together, more than 95% of them are unusually narrow in cross section, lacking the hollow core (medulla) found in coarser wools, and are rendered exceptionally soft and glossy by a lavish production of 'yolk' or 'suint', a mixture of grease secreted by sebaceous glands in the skin, and sweat.[19] The high density of fibres in the fleece is a constant feature in crosses between modern Merino sub-breeds although other features, such as staple length, may vary. We refer to the fleece as a whole although, as with any other breed, the wool varies in different parts of the body and is generally descibed as short to medium.

The fibres of contemporary Merino fleeces measure on average less than 24 μm (24/1000 mm) in diameter, the finest as little as 18 μm. Measurements made on Merino samples preserved from the seventeenth century reveal mean diameters in the range 21–24 μm, which is typical of present-day Merino wool when unsorted.[20] It may seem surprising that Merino wool has changed little in fineness over such a long period. Probably the breed was approaching its selective peak by the seventeenth century. This corresponds with the time when the first recognisable Merino sheep were figured in Spanish paintings, as in 'El Buen Pastor' by Murillo.[21] The fibres of the finest English wool preserved from the same period (from the Ryeland breed of Herefordshire) measure 26 μm.[20]

## Origin and development of the breed

The uniqueness of the Merino breed has encouraged longstanding curiosity about its origin and diversity. The name itself is obscure, coming into vogue first of all as a description of fine white wool imported into early fourteenth century Genoa from Aragon.[22] By the sixteenth or seventeenth century the name was attached to a particular kind of sheep.[23] Nobody doubts, however,

---

[15] Culley 1807, p. 225      [16] Ryder 1983, p. 596      [17] Massy 1990, p. 15
[18] Ponting 1980, p. 13      [19] Carter 1979, p. 561      [20] Ponting 1980, p. 16
[21] Ryder 1983, p. 797; Massy 1990, p. 11      [22] Massy 1990, p. 10 (quoting Bishko 1982)
[23] Klein 1920, p. 11

that the Merino breed had a more distant beginning. Some early writers considered it to be very ancient, linking it romantically with the fine wool of Colchis on the eastern shore of the Black Sea, made famous by the hero Jason and his argonauts in their search for the Golden Fleece.[24] Samples of fine woollen cloth preserved from a Greek colony in the Crimea, dating from the fifth century BC, examined by Ryder, support the idea of an ancient centre of wool quality in the Black Sea area.[25]

As to Spain, classical literature reveals that fine-wooled sheep were certainly present there before the Romans arrived (i.e. before the Punic Wars, 264–201 BC), when the territory was under Carthaginian rule. The wool was mainly of a colour described as buff, fawn or satin-coloured[26] although some occurred in shades of brown, like earlier Roman wool, or even black.[27] The darker colours may have been evidence of crosses to pre-Roman breeds.[28][29] When much of the peninsula, including what is now Portugal, came under Roman control, the fine-wooled Tarentine sheep from south-east Italy were introduced,[30] to be crossed with the native breeds and probably also with north African stock, freshly imported.[31] The wool was, even then, considered of particularly good quality although afterwards it may have deteriorated since it created no recorded interest outside the country. During the periods of Visigothic and early Arab rule, there was nothing to indicate internationally that Spanish wool was particularly special.

The time when this wool came again into prominence corresponded with the influx of Berber tribes in the twelfth century, introducing fresh breeding stock from north Africa.[32] The manufacturing capacity of Moorish Spain expanded greatly, reaching 16,000 looms in Seville alone in the thirteenth century.[33] In the north of the country, the great reconquests of the thirteenth century by the Kings of Castile led them to grant generous concessions for loyal land-holding nobility to raise sheep in the Moorish manner.[34] The connection with North Africa always seems to have been a factor to consider. Later imports from the same direction made by Pedro IV, King of Aragon (1356–1387), and the Cardinal Archbishop of Toledo, Francisco Ximenes de Cisneros (1436–1517), probably helped to maintain the high quality of the breed.[35] At the age of 73, Ximenes led an army into Africa to capture Oran.

---

[24] Rees 1819 ('Sheep' section); Carter 1969; Ryder 1983, p. 146*ff.*
[25] Ryder 1983, pp. 154–5    [26] Anon 1811d, p. 6; Low 1845, p. 129
[27] Pliny (quoted by Schulzenheim 1797, p. 322)    [28] Anon 1811d
[29] Hawksworth 1920, p. 28; Carter 1979, p. 496 (letter from B. Thompson to J. Banks 20 November 1809)    [30] Anon 1811d
[31] Schulzenheim 1797, p. 322; Youatt 1837, pp. 123 and 145; Russell 1986, pp. 36–8
[32] Klein 1920, p. 11; Power 1941    [33] Youatt 1837, p. 146
[34] Massy 1990, p. 7 (quoting Vives 1969)
[35] Livingston 1810; Cuvier 1817 (English edition, translated by C.H. Smith, 1827, p. 326)

Thus what the Moors (Berbers) had achieved and consolidated, their successors continued and exploited.[36] By transporting the vast flocks appropriated from the Moors northwards on to high pastures for the summer, the reconquering Christians would have exerted selective pressures upon them in ways that can now only be guessed at. Probably they also brought some northern sheep down to be wintered in the fertile valleys of the South,[37] thereby adding further variety to what would become the Merino with its ultrafine, mainly pure white wool.

Meanwhile, for the greater part of the Middle Ages, up to the fifteenth century, it was English wool that dominated the international scene.[38] After that, Spanish wool gradually and progressively took over. Here we refer to wool appearing first of all in Aragon.[39] The sequence of events persuaded the Frenchman Rapin de Thoyras to claim in his *Acta regia*[40] that the Merino breed was derived from English stock. The most persistent story was of a gift of sheep made to King John of Aragon by Edward IV in 1468:

> Upon the occasion of reviewing an ancient alliance between England and Aragon, the King is said to have sent the King of Aragon a present of some ewes and rams, which so multiplied in Spain that it proved very detrimental to the wool trade in England.

R. Bradley, Professor of Botany at the University of Cambridge, wrote (1726), on the authority of Gervaise Markham, 'that it is beyond all doubt that *Spain* had their sheep, which produce the wool which is so valuable, from England.'[41] Romantic stories of even earlier English benefactions (1464, 1437, 1399, 1350) were also circulating.[42] Such gifts between monarchs were not unusual and would sometimes inspire further exchanges. However, the suggestion that a few English sheep could have provided the basis for the vast Merino flocks with uniquely fine wool, was rightly described by the eighteenth century English wool expert John Smith as 'fabulous'.[43]

The nature of the breed must have owed much to the methods of husbandry employed and the selection pressures that resulted. Typical of these sheep is their characteristic habit of crowding together in tightly packed 'mobs', an adaptation to survival when migrating to fresh pastures. Many flock owners in Spain exploited their sheeps' migratory capacity as a management strategy. This was the procedure known as 'transhumance', during which the sheep would often walk great distances between their lowland winter pastures to their upland summer territory where good grazing could be obtained when the lowlands were parched by drought.

---

[36] Youatt 1837, pp. 123–4; Low 1845, p. 134     [37] Anon 1811d     [38] Power 1941, p. 16
[39] Massy 1990, p. 10 (quoting Bishko 1982)
[40] English translation by S. Whatley, 1733, quoted by Smith (1747, i, p. 69) as *Histoire de l'Angleterre*     [41] Bradley 1726, i, p. 161
[42] Youatt 1837, p. 147, quoting John Stow (1631–2) and Sir Richard Baker (1730); Bonwick 1894     [43] Smith 1747, i, pp. 69–70; Low 1845, p. 135

Transhumance had been a traditional feature of the management of many Mediterranean breeds, although not particularly associated with fine wool. In fact, the sheep producing the finest ancient Roman wool, from the Tarentine flocks in south-east Italy,[44] were so delicate that they needed to be kept in sheep cots and were almost always fed there.[45] In Castilian Spain, the practice of transhumance became a national institution for which an administrative structure was set up in the mid-thirteenth century with the granting of two royal charters by Alfonso X ('The Wise'). In 1273 the sheep owners formed themselves into a huge guild for the whole of Castile, established as a powerful tribunal or pastoral court, known as the Honourable Assembly of the *Mesta* (*Honorado Consejo de la Mesta*) to regulate the details of migration.[46] It was divided into four companies (*quadrillos*) with capitals at Segovia, Soria, Cuença and Leon, the first being reckoned the most significant, giving cloth 'the distinguished softness of which the courtier sets so great a value'.[47] With the unification of Spain under Ferdinand and Isabella (1479), the *Mesta*'s influence was extended, although territories outside Castile produced lower quality wools, particularly in Andalusia.[48]

On the spring march to the uplands, the great flocks followed their shepherds for hundreds of kilometres, partly along public highways but largely on specially maintained grassy tracks, like linear fields, 75 m wide (Figure 2.2), or spreading out over the countryside in less populous regions.[49] Four major routes (*cañadas reales*), running roughly North to South, which dated from the sixth century or earlier, were joined by numerous tributary ones,[50] and the streams of sheep which travelled along them were given the highest priority in the agricultural economy of Spain. By the sixteenth century there were at least 50,000 km of such tracks. The planning and execution of the movements took considerable organisation since the individual *cabañas* (races) had to be kept separate.[51] In addition, the larger cabañas were divided for ease of management into flocks of 10,000. Each was under the charge of a manager or *majoral* (from the corrupt Latin *majorinus* or *merinus*—a possible origin for the name 'Merino'[52]—and these great flocks were further divided into smaller units of about 1000 head (*rebaños*), each looked after by a *pastor* with four assistant shepherds and five dogs.[53]

Passing directly through certain towns on the highways, the sheep travelled at 24–29 km per day although the rate could drop to 8–9 km per day over open country. Special areas were designated for resting. The lambs were generally born soon after the flocks had reached the southern pastures, ready to

---

[44] Rees 1819 ('Sheep' article)    [45] Anon 1811d, p. 8; White 1970, p. 312
[46] Martin c. 1849, pp. 55–7; Klein 1920    [47] Schulzenheim 1797, p. 309
[48] Carter 1964, p. 6; Parry 1806, pp. 356 and 377    [49] Davies 1958
[50] Ryder 1983, pp. 428–9
[51] Banks 1808    [52] Banks 1808, p. 549; Rees 1819 ('Sheep' article)
[53] Banks 1808, p. 549; Ryder 1983, p. 428

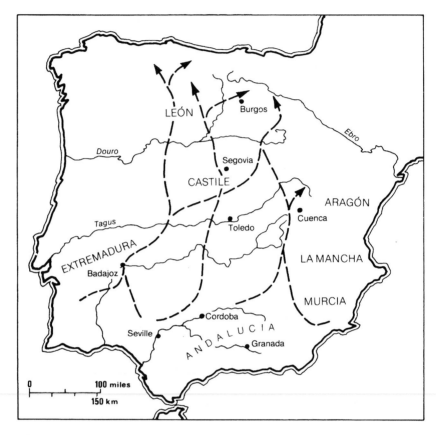

**Fig. 2.2.**   Map of the Iberian Peninsula show *cañadas reales*, the major routes of Merino transhumance (reproduced from Ryder 1983).

be branded the following March. The departure back to the uplands began in mid-April and the sheep were shorn in vast sheds on the way. By late May or early June, the flocks would be back in high pasture for the summer.[54]

Estimates of the number of sheep managed this way reached 3.5 million in the early sixteenth century,[55] rising to 4 million two centuries later.[56] All these sheep were claimed to give the most valuable wool, classified as fine or superfine. By the outset of the nineteenth century, the number had risen to 5 million, and possibly as high as 10 million.[57] Although estimates vary, the number was certainly enormous. As to the splendid fineness of the Merino wool, most experts were in no doubt that transhumance was in some way responsible for inducing it.[58] It must happen, they supposed, because of some

---

[54] Klein 1920, pp. 28–9      [55] Klein 1920; Davies 1958
[56] Bischoff 1842, p. 339; Carter 1964, p. 3      [57] Youatt 1837, p. 151
[58] Rees 1819 ('Sheep' article); Carter 1964, p. 9

aspect of soil or climate, or both. The English encyclopaedist Abraham Rees wrote that 'He who asserted the contrary was regarded by agriculturalists and clothiers as a speculative theorist, only deserving of their pity.'[59] Support for a link between transhumance and wool quality came from the existence of a non-migratory type of sheep in Spain, the *Chunah*, which bore wool considerably longer and coarser,[60] and which commanded only half the market price of Merino wool.[61]

From such evidence, the relevance of transhumance to wool quality might seem to have been proven, although this was not universally accepted. Certainly John Luccock, a wool merchant of Leeds, author of the famous treatise *On the Nature and Properties of Wool* (1805), felt there was good cause to reject the idea. Recognised as a 'careful enquirer',[62] his opinion was influential. The Spanish 'stationary', ie non-migratory, sheep he declared to be simply a different breed, as there are different breeds in any other country. He saw the establishment of Merinos all over Europe or beyond as evidence enough that transhumance under Spanish conditions was not necessary for the breed to show its expected characteristics.

Another reason to doubt that transhumance was essential to the breed was the existence of some flocks of non-migratory (*estante*) sheep in both northern and southern Spain whose wool was claimed by some commentators to be as fine as that shorn from migratory (*transhumante*) sheep. In a published communication to the Board of Agriculture, a Swedish landowner, Schulzenheim,[63] referred to the high quality of such sheep existing in Segovia.[64] Youatt quotes from another communication to the Board,[65] 'that Borgoyne tells us that there are stationary flocks both in Leon and Estremadura which produce wool quite as good as that of the transhumantes'.[66] Foreigners planning to get hold of Merino stock to breed for themselves took heart from such information because they knew that they were bound to dispense with transhumance in their own countries. In earlier years when Merinos were first transported northwards from Spain (initially to Sweden in 1723), such reassuring evidence was not available. To flout the received wisdom that breed and environment were intimately bound up together was then an act of faith. Whether these pioneers are to be regarded as visionary geniuses or lucky optimists, they were willing to trust their judgement and to pay highly for their mistakes. On the other hand, the potential reward for maintaining the breed in a healthy, true-breeding condition in a northern country was enormous.

---

[59] Rees 1819 ('Sheep' article)
[60] Luccock 1805 (quoted by Bonwick 1894, p. 36); Youatt 1837, p. 147
[61] 'Lincolnshire Grazier' 1833, pp. 231–2     [62] Baines 1970 (1858), p. 80
[63] Schulzenheim 1797, p. 316
[64] See also 'Lincolnshire Grazier' 1833, p. 232; Spooner 1844, p. 56; Low 1845, p 136
[65] Vol. VI, p. 2     [66] Youatt 1837, p. 149

At the same time, the influence of tradition continued to leave breeders of static Merino flocks with the fear that their good fortune might not last. Thus the supposed advantage of transhumance remained a point of dispute, even into the nineteenth century.[67] Uneasiness about the matter was encouraged by speculation that breeds were never entirely stable entities, even in their native territory. Thoughts by then were turning to ideas of hereditary change due to evolutionary influence. Considered from a long-term perspective, a progressive adaptational process seemed inevitable, gradual perhaps but pre-destined by Nature. The idea was given substance by the writings of French and English naturalists, such as J.-B. de Lamarck (1744–1829) and E. Darwin (1731–1802). The latter's grandson, Charles, would come to attribute great significance to differences between populations in different environments, and ascribe the mechanism of change to the survival of the fittest individuals under each circumstance.

The selective impact of mortality on a flock of sheep when migrating would almost certainly be more intensive and different from what would happen to the same flock kept in one place. During the long, slow treks between winter and summer pastures, considerable losses were suffered from casualties and disease,[68] and even from the sling shots of shepherds who never, so it was reported, failed to hit a straying sheep with a stone, 'without regard to the size or safety of the animal'.[69] In addition, the shepherds deliberately destroyed no less than half the newborn lambs in order that the remainder should have the advantage of a double number of nurses.[70] In the selection of rams, none of those with tongues or palates spotted with black were tolerated, as they were believed to be predisposed to produce block or spotted lambs.[71] Such levels of mortality, whether by accident or design, indicate a high impact from selection, likely to modify the nature of migratory flocks. The practice of isolating and never crossing the different *cabañas*, let alone allowing them to associate with non-migratory stock, would further encourage diversity and the fixing of types. It is true that outcrosses to imported north African sheep might have added new blood from time to time, but interchange of stock appears to have greatly diminished after the sixteenth century.

Seasonal movements of sheep in modern Spain take place on a smaller scale. As wool has become a less important commodity in Leon and Castile, and cheese and meat have taken over, the need for transhumance is no longer paramount. The result has been that more and more Merino flocks remain in their original province all year round.[72] Two varieties of non-migratory Merinos are recognised, differing in size and wool quality. The distinction

[67] 'Lincolnshire Grazier' 1833, p. 232    [68] Spooner 1844, p. 56; Low 1845, p. 140
[69] Anon 1811d    [70] Anon 1811d; Rees 1819; Youatt 1837, p. 48; Low 1845, p. 140
[71] Parry 1806, pp. 348 and 403    [72] Monserrat and Fillet 1991

from their migratory cousins is not absolute, although the latter still tend to be larger with finer wool.[73] Thus, despite all other selective pressures, migration seems to have retained its influence.

## Export of Merino breeding stock begins

The superiority of Spanish fine-wooled sheep encouraged attempts by the governments of various foreign nations to get hold of breeding stock. In most cases no real success was achieved until the eighteenth century. Since the reign of Alfonso VIII (1169–1214), every King of Castile had prohibited the export of sheep under the strictest rules.[74] For much of history the restriction did not extend to Portugal which, in company with Spain, had a long-standing tradition of excellence in wool production. Even back in Roman times, the wool of Turtedaria (the Roman province of southern Portugal and south-west Spain), had been praised for its 'excellent beauty'.[75] Later this part of Iberia was fortunate to share abundant fine-wooled sheep of a Merino type. Long linked politically with Spain, Portugal was an integral part of that country in the Middle Ages. A particularly significant period for Portuguese wool spanned the years 1580–1640 when the two countries were politically united again after a period of separation. Merino wool was beginning seriously to make its mark internationally at this time. One of the Spanish *cañadas reales* extended into southern Portugal (Estremadura) (Figure 2.2).[76] After spending the winter south of Lisbon, the flocks would be walked to their summer grazing in Spain. With the coming of independence in 1668, the official connection between Spain and Portugal came to an end. Needing then to become self-sufficient, the Portuguese looked to Britain for expertise. A thriving cloth manufacture existed between 1684 and 1713, organised by a group of British clothiers led by an Irishman named Courteen,[77] until it was forced out of business by English competition. To make some of their cloth, these weavers had to incorporate Spanish wool because their own was too short in staple for all types of production. The restricted growth of their wool was attributed to the climate of southern Portugal, where fine-wooled flocks were now grazing throughout the year. It was milder and more humid than much of Spain. Limited scope for transhumance within Portuguese territory was compensated by good lowland grazing available throughout the summer.

Reports of Portuguese wool by English authors suggested that quality was remaining high. Thus in the *British Merchant* (1713), a reporter wrote: 'I am afraid that Portugal as well as Spain has better wool than ever grew in England' and that the 'Portugal wool runs higher than the ordinary sort of Spain'.[78] For a French point of view, we can turn to Savary's *Dictionnaire*

---

[73] Mason 1967, p. 3; Ryder 1983, pp. 428 and 434      [74] Klein 1920
[75] Smith 1747, ii, p. 127 (quoting the *British Merchant* 1713)      [76] Ryder 1983, p. 429
[77] Smith 1747, ii, p. 140*ff.*      [78] Smith 1747, ii, pp. 83 and 137

*Universal de Commerce* published in Geneva in 1742, where three types of wool are listed as superior to that of France, those coming from Spain, Portugal and Great Britain,[79] reported by John Smith.[80] Smith commented that the order in which these three countries' wool is listed in this passage by Savary indicates their 'Rank of Intrinsic Goodness'. Savary stated further that 'The wools of Portugal do not differ much from those of Spain, and they pass commonly for the wool of Segovie. The cloths made of them are very soft in the hand.'[81] It seems that the Portuguese wool retained much of its quality, including softness, in the eighteenth century, without the supposed benefit of transhumance into a fresher summer atmosphere.

It has to be noted, however, that Spanish Merino stock continued to enter Portuguese territory illegally. This is known from reports of Spanish sheep being smuggled through Portugal to third countries. The flow became a flood during the period of economic instability created during the closing stages of the Peninsular War to expel Napoleon, for by the winter of 1809 the great *cabañas* had been reduced to a fraction of their former size. Even today the sheep to be found in southern Portugal are largely of a Merino type.

Outside the Iberian Peninsula, the first direction in which the valuable Spanish sheep were carried in any numbers was eastwards to Italy. For many years and at great cost, the Italians had been purchasing the best of Spanish wool to be woven into the most sumptuous fabrics of the age, as described by Youatt.[82] They included gold and silver cloth, velours and velvets, each type being the specialisation of a different region. Youatt refers to Venice in the period after 1635 as relatively late on the scene compared with central Italy, which became dependent on imported Spanish wool for textile manufacture in villages around Florence as early as the thirteenth century. Carter[83] has noted that the *lana* [wool] *di Garba* from the Sultanate of the Algarve was then the most highly prized wool imported into Florence. Power[84] claims that Garba, Merino, English and Burgundian wools were all vying for influence there, although the Merino had the clear advantage of low transport costs. Later Lombardy was also importing the finest wools as it engaged in high-grade manufacture at various centres.[85]

Italy enjoyed the advantage of a traditional familiarity with the practice of transhumance, which could be traced back to Roman times. Evidence has been adduced by Ryder[86] that Spanish sheep began to reach Italy and thrive there as early as the fifteenth century, being used to upgrade flocks in the south east of the country, giving rise to the *Gentile di Puglia* or 'Apulian Merino' breed (20–24 µm quality). Political constraints against allowing

---

[79] Savary 1742, ii, p. 942     [80] Smith 1747, ii, p. 411
[81] Savary 1742, ii, p. 951 (cited in Smith 1747, ii, p. 417)
[82] Youatt 1837, pp. 142–43 (quoting *Encyclopaedia Londinensis*, 'Wool' article)
[83] Carter 1969     [84] Power 1941, pp. 14–15     [85] Luzzatto 1961, pp. 106 and 142
[86] Ryder 1983, p. 381

Italians access to the valuable stock were eased after 1504 when Alfonso of
Aragon became King of Naples. During that period, vast flocks improved by
Merinos were kept in the southern part of the country.[87] The famous
'Customs organisation for migrant sheep' (*Dogana della mena delle pecore*),
which Alfonso founded in imitation of the age-long practice in Spain,
enriched the crown with new revenue.[88] By the seventeenth century Merino
sheep were reaching northern Italy where they were crossed with the
Bergamo breed,[89] a prolific but relatively coarse-wooled sheep. The result was
to be the highly prized Paduan race, which became renowned for its fine
silky wool, not only in Italy but also in other countries, including territories
of the Austrian monarchy across the Adriatic sea. The first recorded pur-
chase of Paduan breeding stock from that direction was shipped to Trieste
(Figure 7.4) in 1666 to improve the wool of Hungarian flocks,[90] corresponding with the establishment of a weaving mill at Gomba. According to the
Hungarian writer L. Gaál, this importation was organised by Archbishop G.
Szelepcsényi of Eger (now Cheb) in Bavaria, and Prince Eugene of Savoy.[91]

The influence of the Paduan sheep was not slow to spread northwards
through Hungary into Moravia and then into Silesia, the Austrian monarchy's major centre for wool production. From there much of it was exported
westwards through the Baltic ports to the Low Countries and Great Britain.
'The sheep of the Paduan afford a good sort of wool, little inferior to that of
England', wrote the historian Nathaniel Salmon in 1730.[92] As the Paduan
race began to establish its reputation outside Italy from the seventeenth century onwards, the owners of original Merino flocks in Spain maintained tight
control over their breeding stock. An embargo had existed for centuries[93] and
the owners were not going to slacken it without good reason. The *Mesta*
imposed severe penalties on any shepherd found anywhere near the border
with an unlicenced flock. The few animals leaving unofficially, smuggled out
through Portugal or—less commonly—across the shorter and largely moun-
tainous border with France, were generally of inferior quality.

Hostile as was the *Mesta*'s attitude to unofficial exports, the precious stock
could not be entirely contained within the Peninsula. Southern Italy became
a natural exception as a Spanish possession. Another destination for Merino
flocks was on the other side of the Atlantic Ocean, to Spanish colonies of the
New World, particularly Peru as early as 1580 although, having been
expanded to large numbers, they were bred indiscriminately and thus neglected.[94] Later, through Hapsburg family connections, Merino sheep began to
reach the Dutch colony at the Cape of Good Hope, South Africa, possibly as
early as 1689 and certainly by 1724.

[87] Fussell 1972, p. 102     [88] Luzzatto 1961, p. 166     [89] Gaál and Gunst 1977, p. 258
[90] Bökönyi 1974    [91] Gaál and Gunst 1977, p. 258     [92] Smith 1747, ii, p. 218
[93] Carter 1964, p. 7     [94] Bonwick 1894, p. 51; Ryder 1983, p. 583

By the eighteenth century, limited exports were being permitted even to foreign countries outside direct Spanish influence. Confidence was growing among the Spanish owners that the small numbers of their precious stock reaching alien territory would be unable to retain their valuable characteristics when separated from Spanish soil and climate.[93] Probably they were encouraged by the unfortunate experience of those of their own countrymen who had tried to establish Merino flocks in the colonies. Even in Italy, where conditions could be quite similar to the traditional Merino range of habitats, and where transhumance was practised as in Spain, the breed had failed to survive quite in its original form. Spanish confidence seemed justified by previous experience although, in the end, history was to prove otherwise. The export of valuable breeding stock set in progress a course of events that would lose Spain the European fine wool market within a century or less.

## The Merino enters northern Europe

How could an economic outlook be so misjudged? The answer has much to do with the rapid march of technology. Sheep husbandry was being radically transformed in certain countries to the North. A description of what was happening there will occupy the remainder of this chapter and will continue as a thread running through the rest of the book. It is a story of what, in modern terms, would be called applied genetics, arising from necessity: a struggle not simply to keep animals alive, healthy and fertile but to make improvements by selective breeding. Success came from taking what was really a most limited view of animal nature but one which served its purpose well at the time. In the tradition of early breeders of horses and dogs, the new sheep owners saw their animals as self-perpetuating machines that could be modified by selective breeding to become even more efficient.

The first country to reveal the changing situation could be truly described as 'the most unpropitious for the experiment':[95]

> It was in the bleak land of Sweden, the country of Linnaeus, Hasselquist and other great naturalists, that the bold attempt was first made to naturalise the Merino sheep of Spain and improve the native race by judicious intermixtures.[95]

This was in 1723 and the pioneer responsible for this leap into the dark, 'that most zealous improver of the breed of sheep', as Baron Schulz de Schulzenheim[96] described him, was a self-made businessman and former shop assistant, Jonas Torreson (1685–1761), later knighted as Jonas Alströmer of Alingsås (Figure 2.3). It is instructive to consider his unusual

---

[95] Martin c. 1849, ii, pp. 60–61. This quotation refers both to the great Swedish botanist Carl Linnaeus (1707–1778) and to his pupil Fredrik Hasselquist (1722–1752), who died tragically young on an expedition to the Levant, having named 100 new species as he passed through the Holy Land in 1750.                    [96] Schulzenheim 1797, p. 306

**Fig. 2.3.**  Portrait of Jonas Alströmer (1685–1761), businessman and sheep breeder, wearing the Order of Wasa (reproduced from *Svenskt Biografiskt Lexikon*, illustrating an entry by Eli F. Heckscher).

background and how the experiment began, an enormous gamble that not only paid off handsomely for the gambler but inspired an economic revolution. Wool production would never be the same again.

Jonas Torreson (Alströmer), the first large-scale importer of the Merino into northern Europe, showed every sign of anglophilia. As a young man he left his own country to live and work in London. Born in Alingsås, 40 km north-east of the port of Göteborg, he had a poor childhood but finally gained sufficient education to advance from being a shop assistant to an office job copying documents. In 1707, at the age of 22, he managed to get to London and obtain a position as a book keeper. One supposes that he must have been an excellent linguist, for within only three years he had founded a successful trading company and become a British citizen. Dealing in cloth, he could not fail to appreciate the wonderful quality of English and Spanish wools in contrast to the poor hairy product grown in his native country. Determined to produce cloth of English quality in Sweden, he managed in the late winter of 1715 to smuggle a number of 'fine-wooled English sheep' across the North Sea.[96] The venture was not, it seems, without problems, however, because he was back in England by 1719–1720 making enquiries about breeding techniques. As Swedish Consul in London from 1722, he was able to make commissions on trade to the State of Sweden as well as to private Swedish companies. His position gave him the opportunity to visit France, Holland and Germany, thereby adding to his knowledge of textile manufacture and sheep breeding. Grown rich by this time through his business activities, he finally returned home in 1724 to found the Swedish textile industry in Alingsås. It was a propitious moment to do so. In the great northern wars with Russia (1700–1718), Sweden had lost most of her Baltic colonies and needed to find new sources of revenue within the Swedish state. Soon Alströmer had smuggled textile machinery from Holland and France, and recruited foreign experts to work for him.

Already one year earlier had come an action that would make him famous, when he secretly transported the first Merino sheep to Sweden directly from Spain by sea; he had landed this small consignment at Göteborg and moved them to Berga Sateri. These sheep left Spain officially with the approval of the Spanish King, Phillip V.[93] [97] [98] Evidently the Spanish authorities had every confidence that their sheep would not thrive or even survive in Sweden. It seemed no more than common sense, but this was to ignore the genius of a very unusual individual. With the benefit of past experience of foreign sheep, and drawing upon advice gained during his travels, Alströmer was able to breed this little stock of Spanish sheep into a healthy and productive flock. Opinion was united in acclaiming this to be a startling achievement in such a hostile climate. 'It was a hazardous—it appeared to be a presumptuous, an almost insane attempt; and this spirited individual had to struggle with almost insuperable obstacles', wrote Youatt.[99] Too late the Spanish authorities put a ban on further exports to Sweden, but Alströmer's agents found a way to circumvent it. Smuggled shipments of sheep from Spain followed at

[97] Ryder 1983, p. 427
[98] K.-H. Suneson, personal communications 4 January 1990 and 10 March 2000. By 1746 Linnaeus was proposing that all university students should be compelled to study natural history, including the care of Spanish sheep and silkworms. Starting in 1747, university professorships were established 'in practical economics based on natural science' to impart such information (Koerner 1999, pp. 107–8).          [99] Youatt 1837, p. 137

intervals of two to three years until the breed had become established in Sweden (Chapter 6). Other foreign nationals were also supplying valuable breeding stock illegally to Sweden. The practice was encouraged by a Swedish royal decree on 25 May 1736 that a premium should be awarded to any breeder prepared to import sheep from Spain, England, Ireland or south-western France (Bayonne). The decree was kept confidential to avoid disrupting foreign relations. In 1736 the King established a special school for sheep experts at Hojentorp. During the following years, 55 people completed their training there, one of whom was placed in every Swedish *lan* (county) and three in Finland.[98]

In 1739, the King appointed Alströmer to the position of *kommerseråd* (commercial councillor) and director of sheep breeding and wool manufacture for the whole country. In the same year, with crown permission but at his own expense, the new director of sheep breeding established, near Alingsås, a school for shepherds under his direction, for which he provided 'a salary for a master shepherd brought from abroad'.[100] His enterprise paralleled the activity of the Swedish naturalist, Carl von Linné (Linnaeus) (1707–1778) in his drive to establish exotic plants in Sweden. Together with three others, they joined in founding the Swedish Academy of Sciences in the spring of 1739, with Linnaeus as first President.[101] This small group of likeminded patriots initially called itself the Economic Society of Science because of their common concern to apply natural knowledge for the good of the State. In 1751 the new King, Adolfus Frederic, raised Alströmer to the nobility as Knight of the Order of Vasa,[102] when he first assumed his new name. The ceremony took place in connection with the coronation, giving Alströmer a prominence that underlined his value to the nation. The growing industrial town of Alingsås had by then become a national institution, benefiting from state subsidies and trade barriers. Linnaeus, who visited there in 1746, wrote that 'Everything and everybody seemed to revolve around the industry . . . there are no beggars here; when they do come, they are immediately put into the factories.'[103] (figure 2.4)

When Alströmer was showing off his sheep to Linnaeus, he had different flocks in various stages of improvement. He was proud to be building up mixed flocks based on crosses with Swedish sheep which in their native state produced very poor quality wool (Chapter 6). In 1758, the School for Sheep Experts moved from Hojentorp to Alingsås. After the move, Alingsås became the dominent centre for wool production. The school was closed in 1765–6 when its job was considered to have been completed.[98]

---

[100] Schulzenheim 1797, p. 306    [101] Uggla 1957    [102] Schulzenheim 1797, p. 306
[103] Gullers and Strandell 1977. Being strongly influenced by Cameralist ideas on economics, Linnaeus was much in favour of 'taming' exotic species in Sweden. He was convinced that his friend Alströmer was manufacturing cloth from home-grown Merino wool as good as 'ever other nation abroad' (Koerner 1999, p 101), an obvious exaggeration but one that fitted the euphoric optimism surrounding the Swedish enterprise (Koerner 1999, pp. 2, 6,10 and 127).

**Fig. 2.4**  Carl Linnaeus, on his Västgöta Journey, July 6th 1746, accosted by a beggar as he approached the prosperous town of Alingsås, dominated by the smoking chimneys of its weaving mills. The foreground is occupied by exotic sheep imported from Spain by his friend Jonas Alströmer (based on a coloured wood engraving by Carl Larsson, published in 1882).

The crossing scheme that Alströmer and his successors employed was a copy of English practice, according to a French economic textbook of 1755, *Nouvelliste oeconomique*.[104] The author claimed that Alströmer took this course of action 'having observed that the English had made hundreds of millions [of francs] by refining their sheep with Spanish breeding stock introduced into the country'. An earlier French publication (1730), quoted by Bourde, had reported how Spanish sheep had entered England in the fifteenth century and how regulations were established to preserve their quality.[105] If true, the most likely route of mediaeval introductions would have been through the influence of Cistercian livestock improvers, as remarked upon by Trow-Smith.[106] The Cistercians were famous for wool production, and the Order expanded rapidly in England, growing to 50 abbeys by 1152.[107]

English writers ready to acknowledge a Merino influence do not produce much in the way of evidence. In the case of the Exmoor breed (Dunface), Culley[108] claimed a connection with the sheep of Andalusia to be 'almost out of dispute' on the basis of physical similarity alone, but his explanation that the original sheep (*Ovejas marinas*) had swum ashore from ships of the Spanish Armada wrecked along the English coast seems fantastical. There is better evidence of import of Merino stock in the sixteenth century to improve Gloucestershire wool, concerning which Bonwick writes of Henry VIII having procured no less than 3000 through his friendship with Charles V of Spain.[109] In the event, however, they seem to have created very little lasting impact, for 'within a few generations the merino staple had lost its quality', according to Trow-Smith.[110] More than occasional appearances of Spanish sheep in the English countryside are difficult to substantiate and, being contrary to the rule of the *Mesta*, must be viewed with caution, if not dismissed as the stuff of legend. Evidence of a major influence of Spanish blood is not to be found in English records, either official or commercial.[111]

Alströmer, however, preferred to believe otherwise. He himself wrote, in a treatise on sheep breeding published in 1733, about English sheep having been 'enobled' by Spanish crosses. It is interesting that his action in importing Merinos may thus have been partly inspired by a myth. Perhaps he had personal experience of observing or hearing about Spanish sheep in the countryside around London. Charles, Earl of Tankerville is known as one landowner with a small flock of them, kept on his estate near London, listed as 'one ram, two ewes, three lambs—Spanish kind' in the inventory of his estate in 1722.[112] However, Alströmer would certainly have realised that

---

[104] Published in 22 volumes 1754–8; see Bourde 1953, p. 232
[105] N. Chomel 1731 in *Dictionnaire oeconomique* Vol. 1, p. 436*ff.*, quoted by Bourde 1953, p. 136        [106] Trow-Smith 1957, p. 112; see also Darby 1936, pp. 242–3
[107] Darby 1936, p. 185        [108] Culley 1794, pp. 160–61        [109] Bonwick 1894, p. 55
[110] Trow-Smith 1959, p. 39        [111] Smith 1747, i, p. 71
[112] S.J.G. Hall, personal communication

the term was used loosely in England, even to include the four-horned 'Jacob' breed. 'Spanish' was not necessarily to be equated with 'Merino'; neither was all Spanish wool equally superior.

Of greater and more pressing importance to Alströmer was what he could learn about the principles and techniques of breeding currently practised in England, which he could apply in Sweden. The profitable art was more advanced in his adopted land than anywhere else at that time. Living evidence was to be seen all around of the transforming power of selective breeding, not only in sheep but also in cattle and, above all, in horses, witnessed by the early and dramatic success with the 'thoroughbred'. At the races held at Newmarket or Ascot, great sums of money were won or lost on the breeders' skill with Arab, Berber or Turkish bloodstock. The horse breeders had developed the technique of 'grading up' by crossing males of a superior exotic breed to local females for several successive generations (Chapter 3). It was of obvious significance to Alströmer, contemplating the importation of an exotic breed himself, to take note of this. England gave him an invaluable schooling and he proved himself a diligent pupil. Lessons learned about selective breeding in his adopted country gave him confidence to import sheep from Spain into Sweden and to find a way to keep them healthy and thriving. His actions would make an important contribution to expanding understanding of how to maintain the foreign breed, to the considerable benefit of Merino breeders in other countries who took courage from him. Later we shall describe the course of events in these various territories, including Moravia. But first we turn to England where we consider in more detail how the guiding principles of breeding were defined at the time when Alströmer began his epic task.

---

[113] Smith 1743, ii, pp. 412–13 (quoting an article in French, dated 1719

# 3

## Sheep breeding sets new standards

Not one man in a thousand has accuracy of eye and judgement sufficient to become an eminent breeder

*Darwin, 1859*

For the greater part of the Middle Ages, England drew pride and economic strength from a reputation as 'the largest and most influential source of fine wool'.[1] Because of the perceived uniqueness of the English product, royal ordinances forbade it to be mixed with any foreign wool, even the finest from Spain. Eager consumers demanded more and more of it until, by the third quarter of the sixteenth century, shortages began to develop. The domestic textile industry had grown sufficiently to absorb all home production. It was a situation that demanded a change of attitude to foreign wool. Exports of English textiles abroad continued to be buoyant only because the wool deficit was made up by imports from Spain and from central European countries such as Silesia, Poland and Hungary via Danzig and other east Baltic ports.[2] Then, as those countries began to produce their own cloth at competitive prices, the demand for English textiles began to fall, causing a loss of revenue that was much discussed and agonised over. 'The Decay in Trade is in every Bodies [*sic*] Mouth from the Sheepshearer to the Merchant', wrote the author of a document presented to Parliament in 1641.[3] There were calls from all sides for urgent action to redeem the situation. The policy that was finally adopted, and the theories of selective breeding that emerged from it, form the subject of this chapter.

[1] Power 1941, p. 16; see also Bischoff 1828, pp. 46–51 and Youatt 1837, p. iii
[2] Ramsay 1982, pp. 19–20; Swain 1986, p. 142
[3] Smith 1747, i, p. 179 (quoting Henry Robinson)

First it was necessary to identify the cause of the decline, which stimulated speculative writing on a wide scale before the truth began to emerge. The Reverend John Smith of Lincoln, a Church of England cleric with unique expertise, was the first to state it convincingly. Graduating from Trinity Hall, Cambridge in 1725, he devoted more than 20 years of his life to discovering everything he could about wool. Having made a comprehensive collection of relevant books, tracts, pamphlets and documents, he reproduced them, in whole or part, with extensive notes and comments in two heavily packed volumes under the title *Chronicon Rusticum Commerciale* or *Memoirs of Wool* (1747). A Victorian writer, I. Culloch, author of *Literature of Political Economy* (1845), judged this work to be 'most carefully compiled and valuable',[4] and few would now disagree with that analysis. Smith concluded that the root of the problem was that English fine wool had lost its competitive edge in an increasingly open market to which it was ill adapted. A change of fashion had greatly increased demand for finer types of longer fibred wool, which could only be satisfied by importing it from abroad, particularly from Spain, at considerable extra cost. In written sources published over two centuries, Smith found abundant indications of the competitive quality of Spanish wool, confirmed by statistical evidence of relative wool prices.[5]

With carefully marshalled evidence, Smith demonstrated how access to the best of all foreign wools had allowed more genuine comparisons to be made with the home product, in terms both of quality and price. He piled example upon example to convince his readers that English fine wool had become overrated, in comparison with that from Spain. Even when reaching its highest quality in Elizabethan times, English wool had not been able to dominate the market. Even then, the English were 'far from being the Clothiers of the World', as many writers 'in very serious sort but most foolishly, imagined or knavishly represented'.[6] Smith was convinced that believers in the dominance of English fine wool are simply ignorant of the facts, and he was not slow to suggest why—it was the fault of English education for ignoring the march of technology:

> Our youth (of liberal education) never reading anything of Manufacturing etc. in Homer or Virgil, or their College Notes; and being from thence carried to other studies, our Men of Learning are generally silent in this matter, or do speak of it with contempt. . .[7]

Thus the accusation levelled at those in authority is that they have allowed England's traditional eminence as a producer of fine wool to slip from their

---

[4] Clarke 1897; see also Carter, 1979 for a positive reaction to Smith's book by Sir Joseph Banks, elected President of the Royal Society in 1778: Carter 1979 p. 528, Banks 1787 *My pamphlet*; pp. 523–524, Banks (undated) *Anecdotes of the Revnd. John Smith L.L.B.*; see also Mann 1987 (1971), pp. 268–9          [5] See also Bowden 1956; Ramsay 1982
[6] Smith 1747, i, p. 127 According to Bischoff (1828, pp 51–52), the deterioration of English wool was first recorded officially in 1622.          [7] Smith 1747, i, p. 342

grasp though ignorance and complacency. Smith quotes compelling docu-
mentary evidence that for almost a century, from about 1660 onwards, the
finer type of Spanish wool had always commanded a substantially higher
price than the finest produced in the British Isles. The price of English wool
rose and fell with the price of Spanish wool but was always substantially less.[8]
For Smith this is the central issue accounting for the pattern of trade.
Although he recognises a number of other impacts—social, political and
fiscal—upon the decline of England's greatest economic asset,[9] it is the
ascendancy of Merino wool that provides the yardstick by which to judge
their combined effect. That much was to becoming increasingly evident to
wool experts of all nationalities, including the astute Swedish businessman
Jonas Alströmer as he assessed the situation in early eighteenth century
London and made his plans for the future. The rise in significance of Spanish
wool could no longer be ignored, even in the British capital.

The import of Spanish wool into England had started in the twelfth century although always in small quantities, 'without detriment',[10] until the felt
hat came into fashion in the reign of Henry VIII (1509–1547). Then, by
virtue of its superior felting qualities, the Spanish product gained a larger
place in the British economy.[11] By the sixteenth century, clothiers as well as
hatters were using it on an increasing scale.[12] As well as its felting properties,
it was distinctly softer than the best Britain could then offer, the reverse of
the position two to three centuries earlier, in both respects.[13] Softness was a
quality growing in favour from the 1560s onwards as traditional types of
cloth gave way to lighter fabrics—the 'new draperies'.[14] As the scarcity of
British fine wool became chronic, the weavers of Wiltshire, Somerset and
Gloucestershire had no option but to make increasing use of wool purchased
in Spain at a high price. By 1628, England was already exporting annually
3000 'Spanish cloths', and other material woven partly from Spanish wool,
and by 1640 the number had risen to over 14,000.[15] Produced in a variety of
colours, this material was greatly in demand for the luxury trade. Other
manufacturing nations, notably the Netherlands, France and Italy, were
competing for the precious Spanish wool, so elevating the price.

---

[8] Anonymous trade tract 1677, quoted by Smith 1747, i, p. 312
[9] For a modern assessment of which, see Ramsay 1982    [10] Klein 1920, p. 34
[11] Bonwick 1894, p. 176; Bowden 1962, p. 47    [12] Bowden 1962, p. 213
[13] Power 1941, p. 14
[14] Youatt 1837, p. 74; Wilson 1965; Perkins 1977; Ramsay 1982, pp. 14–15
[15] Wilson 1965, p. 78. Even the designation 'Spanish cloth' is a confusing term. It did not
necessarily imply the use of Spanish (i.e. Merino) wool except in the more expensive kinds. The
major characteristic of English manufactured 'Spanish cloth' was its lightness, achieved by using
a weave requiring only half the wool needed previously. It was greatly improved after 1650, and
the King's decision to wear it in 1675 gave rise to great sales (Mann 1987 (1971), pp. xiv–xv, 11,
14 and 15).

## British sheep breeders react to the changing situation

By the time Alströmer decided to import Merino sheep into Sweden, beginning in 1723, the quality of Spanish wool available in England was rumoured to be in decline. To use the words of two British clothiers around 1700, consulted by Lisle,[16] it was 'not so fine and not so good as of late years'. Even so, the situation offered no encouragement to British farmers to increase fine wool production because the price of home-produced wool was still consistently lower. They found greater profit from keeping sheep in richer pastures to encourage growth, despite the unsuitability of such stock for the production of fine wool. Even the Ryeland breed of Herefordshire, bearing England's finest wool, 'Lemster [Leominster] ore', was being brought down from the hills on to better land.[17] Any sheep failing to adapt to richer soils would be outcrossed to a breed that possessed that capacity.

This break with long tradition came from a desire to take control over breeding, which could not be done with unfenced flocks on common land or upon the open hillside. The change was made possible by the development of new forage crops (cabbages, carrots, rape, turnips and clover) and the intelligent use of crop rotation and irrigation.[18] More and more land was being enclosed. By keeping sheep conveniently close to farm buildings, the owner aimed at getting them to fatten more quickly under secure conditions. He could then draw more profit from the carcass, as well as selling the wool. While the change in practice made economic sense, it brought inevitable consequences for wool quality.[19] The product from sheep kept on low ground was longer and coarser than that from hill sheep: 'hill country hay ... will bring finer wool ... than vale hay', wrote Edward Lisle (1666–1722). Mr Anthony Methwin, 'a great clothier', advised him that 'Fallows always produced better wool than the very same ground when laid to grass.' It was recognised that 'The longer a ground [was] laid to grass ... the ranker [was the] food and ... the ... coarser [the wool].'[20] 'Clover grasses' were said to raise a particularly coarse wool. Later in the century the turnip would also be accused of a coarsening influence.[21] Although everybody agreed on the effect, nobody could stem the tide of change. English wool had already lost its medieval fineness by the 1600s,[22] after which changes in husbandry had accelerated an ongoing trend.

[16] Lisle 1757, ii, p. 357
[17] Trow Smith 1959, pp. 39 and 132. 'Lemster ore' was the 'golden' product grown on the sheep of Herefordshire which, in mediaeval times, had rivalled all others at the looms of Flanders and Florence.                    [18] Kerridge 1967, p. 323
[19] Bowden 1956
[20] Lisle 1757, ii, pp. 358; see also Trow Smith 1959, p. 39, Bowdon 1956
[21] Rudge 1807, p. 396; Sinclair (quoted by Mitchison 1962, p. 114); Carter 1979, p. 158: letter from T. Meech to J. Maitland (Chairman of the Committee of the (London) Wool Trade) 28 July 1788                    [22] Power 1941, p. 22

British sheep farmers were forced to reassess their options. Rather than meeting the foreign challenge head on, they chose to build on their strengths and devise alternative strategies. The declining status of locally produced fine wool focused attention on the unique characteristics and value of another type of English wool, the exceptionally long, lustrous combing variety. Ideal for weaving hard-wearing worsted cloth, English long wool was much admired abroad where it found few equals. Its value was further enhanced by scarcity, as a result of prohibitions placed on its export by the government. In order to bring about further improvement in this type of wool, English farmers made their first obvious big advance in selective breeding, to produce larger sheep bearing heavier fleeces.[23] Consequently 'What was regarded initially as a disaster, the lengthening and coarsening of British fleeces, proved finally to be a blessing.'[24]

An appreciation of the advantages of larger size and rapid growth rates in sheep is often thought of as an early eighteenth century phenomenon. In fact the practice went back a long way in Britain. Thomas Tusser (1524–1580) wrote about it in his *Five Hundred Pointes of Good Husbandry*:

> For gaining a trifle, sell never and store
> What joy to acquaintance, what pleasureth more?
> The larger the body, the better the breed,
> More forward of growing, the better they speed.[25]

The more intensive diet, which enabled sheep to be kept more and more in the lowland pastures and rich marsh grazings, produced larger carcasses of more tender mutton and longer, coarser wool, especially suitable for hard-wearing worsted manufactures.[26] As some sheep breeders began consciously to select their breeding stock for larger size, in order to maximise fleece weight and depth of pile, Gervase Markham, writing in 1631, surveyed the variety of English sheep and revealed that those of the Midland plain were already larger than average, with the longest known wool.[27] Markham remarked: 'In the choice, therefore, of your sheep, choose the bigger boned, with the best wool. . . .'[28] Large English sheep made their appearance always in lowland areas, in association with changes in land use and cropping. The absence of skeletal remains of similar animals uncovered at medieval sites indicates when the trend towards larger size began.

Investment in mutton became especially popular in the Midland counties of Leicestershire, Nottinghamshire and Warwickshire. There seemed no limit to the market for it in the expanding communities of the industrial heartland

---

[23] Fussell and Goodman 1930; Mantoux 1961, p. 161 (quoting Eden 1797); Russell 1986 pp. 176–88    [24] Hartwell, 1973, p. 330
[25] Tusser 1573 (quoted by Johnstone 1846, p. 152)    [26] Russell 1986, p. 156
[27] Russell 1986, p. 158    [28] Johnstone 1846 p. 153 (quoting Markham 1648, 7th edition)

and the great city of London. The profit to be gained from the appetites of miners and mill workers with rising expectations and money in their pockets could not be neglected.

## The matter of heredity

It was thus that sheep breeding in the British Isles changed with the times. The turning point came in the sixteenth century and gathered pace in the seventeenth, while the eighteenth saw new vistas opening in all directions. Change was inevitable among farmers who had learned to adopt meaningful breeding plans and to practise their husbandry pragmatically rather than on the basis of tradition. Furthermore, as the more intelligent among them recognised theoretical implications in their actions, aspects of heredity were gradually clarified by trial and error, through shared experience or practical experiments.

The terms 'heredity' (*hérédité*) and 'inheritance' had come into use in a biological sense as early as the sixteenth century, in relation to human monstrosities, dispositions and maladies, transmitted from one individual to his descendants.[29] Heredity was a concept readily extendable to domestic animals, stimulating the breeder to ask how best he could exploit its potential. He was offered contrasting views of the living world at the time. The idea of the animal body as an integrated material mechanism, governed by 'matter and motion',[30] was one concept. An alternative view sought to relate individual nature to powerful external forces expressed in terms of local climatic conditions, altitude, soil and foodstuffs. It is possible to find both philosophies reflected in the attitudes and actions of breeders. No individual would stay long in business by ignoring the potential influence upon his stock of natural forces in the outside world. At the same time, a view of animals as machines allowed for the possibility of repairing them from such influences. Edward Lisle alluded to the idea of 'mending' sheep or cattle in his report of discussions with farmers around 1700.[31] He was referring to the repair of a *breed* during successive generations, not simply the restoring of health to an individual animal. Later in the century Robert Bakewell would make reference to his highly bred New Leicester sheep as 'machines for turning herbage . . . into money' (Chapter 4). The British, as the first nation to be infected by the heady excitement of the industrial revolution, embraced the mechanical aspect of nature with enthusiasm, wherever they thought it could be profitably applied. By the time of Bakewell, if not earlier, the mechanical concept had become a truly dynamic one, Newtonian rather than Cartesian,[32] as selective breeding brought about inherited changes far beyond anything seen

---

[29] Gayon 1996, pp. 61–5     [30] Roger 1997, p. 117     [31] Lisle 1757, ii, p. 154
[32] For definitions in this context, see Farley 1982, pp. 23–25; see also Schiebinger 1993, p. 108

before. A growing awareness of the capacity of animals to change their nature under Man's guiding hand challenged the Cartesian notion of animals as senseless automata.

Changes of attitude to heredity could not come about without opposition or even struggles of personal conscience. Breeders who were tempted to ally themselves with the machine age had to deal with long-held beliefs concerning the divine plan for the world. A major aspect of this natural theology, most powerfully expressed in British Protestantism, was 'the perfection in the adaptation of all structures and organic interactions as evidence for design'.[33] Selective breeding had somehow to reach an accord with a theology which saw all of nature as 'the finished and unimprovable product of divine wisdom, omnipotence and benevolence'. Acknowledgement of the divine order would dictate that selective breeding could be justified only on the grounds that the perfectly adapted type had degenerated and needed to brought back to its lost perfection. It was in such terms that the English natural philosopher John Ray (1628–1705), author of the *Wisdom of God in Creation*, allowed for the idea of 'plastick Nature' which would relieve God from directly controlling every small deviation from the perfect type although still retaining ultimate responsibility for such deviations.[34] A more radical approach was required to come to terms with the demand for traditional breeds to be replaced or even surpassed. Could it be that the domestic situation, being man-made and largely cushioned from natural forces, was not subject to the same laws as the natural world outside? In more and more cases it was observed that domestication allowed an animal's nature to be changed by design, appropriate to almost any given set of conditions, so the answer appeared to be yes, the domestic situation did encourage variation. Even greater variation might result from crossing the new varieties together and selecting among their progeny.

Naturalists of the period were greatly taken up with the enigmatic process of generation and how it was determined. The manner in which a new individual assumes its existence in the general form of its own species or race remained a mystery. The ovists and animalculists (spermists) developed their separate theories of preformation.[35] However, neither idea was particularly attractive to breeders, whose experience told them that both sexes must be involved. The only body of theory of interest to the average breeder was that which supported his own actions and was delineated in terms of the breeding success he achieved in co-operation with his friends and relatives. Collectively these pioneers have to be regarded as novice practitioners of an evolving art that was destined to generate its own theoretical base. Their activities derived little from outside example, most from practical necessity.

---

[33] Mayr 1982, p. 372     [34] Willey 1940, pp. 35–6
[35] Farley 1982, pp. 22–3; Pinto-Correia 1997, pp. 16–104

As sons of an enlightened age they discovered that the best results in breeding come from actions taken on the basis of close observation and carefully controlled experiments. Intrigued by the logic of their experimental results, reinforced by the promise of approbation by their fellows, they behaved, in a way, like scientists although, unlike scientists they were reluctant to reveal their methods, preferring to be judged by their achievements. In their hands we see foundations being laid for the science of genetics but are tantalised by incomplete information about the rules of breeding they discovered. Being professional secrets, such rules are not to be revealed to rival breeders until full commercial advantage has been gained. The scientific explanation might come later but to the farmer it is of minimal importance compared with the achievement itself, except as a guide to future action. As Lerner has stated with reference to agricultural progress in general in 1660–1780, 'The arrow of influence appears to have led from agriculture to scientific progress.'[36] Accepted practice in breeding was far from being the expression of some fully formed theory. Rather it created its own theory and demonstrated it to be true.

Advances in breeding were sufficiently dramatic to attract a host of commentators and propagandists. Persistent in their efforts to gain information, they made every effort to build up the total picture, bit by bit, in the face of breeders' reticence. Interpreting the secret world of breeding was not a task for the faint hearted. Hence the modern writer seeking a reliable view of breeding practice may find himself quoting from sources written decades after the events under consideration. The rest of this chapter relates these early events, leading to the body of principles that finally emerged in the first half of the eighteenth century. An important stimulus to enlightenment is provided by the rising significance of Merino wool, the demand for which outstripped supply, inspiring attempts to breed Spanish sheep outside their home territory and to cross them with native breeds, in different European states. At the same time there were adventurous sheep owners in the British Isles who were concentrating their efforts towards increasing body bulk and fleece size, and improving meat production in terms of growth rate. Success in either direction demanded high skills of husbandry with respect to diet, weather protection and choice of breeding stock. As we shall see, progress was patchy and gradual, although—in some soon-to-be-famous hands—dramatic.

## Like begets like

An observant traveller around the seventeenth or eighteenth century European landscape would have noticed sharp contrasts between sheep native to particular territories or districts: a great diversity of form, size, colour, horn

---

[36] Lerner 1992, p. 12

characteristics and wool quality. Such local differences were seen at their sharpest in the British Isles, which had been untroubled by foreign invasion for many centuries. It was a phenomenon that the agricultural expert Arthur Young (1741–1820) would later designate as 'the circumscribed locality of every distinct Breed of Livestock'.[37] A frequently quoted example is illustrated in a letter written in 1773 by Gilbert White of Selbourne in Hampshire, in which he reflects upon the two quite different breeds of sheep to be found on the opposite banks of a narrow river: white fleeced, white legged and horned to the west, and black faced, speckled legged and hornless to the east. The situation was viewed by local people as fixed for ever:[38]

> If you talk with the shepherds on this subject, they tell you that the case had been from time immemorial and smile at your simplicity if you ask whether the situation of these two breeds might be reversed.

In the face of such diversity, yet consistency within any one area, it was natural to conclude that 'like begets like', an adage often quoted in these or similar words.[39] Any animal could be seen to have the characteristics of those around it—its breed. What could be more natural than to assume that the production of like by like was intimately dependent upon the district in which the animals were raised, the climate and pasture they shared in common? This would be even more so if distinctive breeds had a long ancestry in the same place. 'Every soil has its own stock', wrote William Pearce in his *General Review of the Agriculture of Berkshire*,[40] which William Marshall, who assessed Pearce's report a few years later, considered to be accurate and judicious.[41] It was an axiom that could be traced back even to Roman literature[42] and, as a metaphor of nationhood, it had inspired England's greatest poet and playwright:[43]

> And you good Yeomen,
> Whose limbs were made in England, show us here
> The mettle of your pasture, let us swear
> That you are worth your breeding.

In the case of sheep, it was wool quality and body size that seemed most dependent on pasture. A self-proclaimed authority on the matter at the University of Cambridge, Professor R. Bradley, author of *A General Treatise of Husbandry and Gardening*,[44] drew an analogy between the growth of wool in the skin and that of plants in the soil. On this basis he found it natural to conclude that wool quality must depend entirely on nourishment.[45] But could this be true? It represented a challenge to the breeder's art.

---

[37] Young 1811, p. 6    [38] White 1977 (1789) pp. 80 and 82    [39] Russell 1986
[40] Pearce 1794, p. 46; see also Carter 1979, p 129: letter from C. Chaplin to J. Banks 25 October 1787    [41] Marshall 1818, Vol. 5, p. 58
[42] White 1970 p. 312 (quoting Varro)    [43] *Henry V*, Act III, Scene i, lines 25–8
[44] Fussell 1947, pp. 106–7    [45] Bradley 1726, i, pp. 159–61

One line of evidence on the extent of environmental influence arose from trade in bloodstock, when animals were transported from one district to another. How did their subsequent performance and the survival of their type in the new situation fit with the locality-based concept of race? It was a matter researched by the already mentioned Edward Lisle, who sought opinions from some of his more successful farming contemporaries on the basis of their practical experience, and whose notes reveal a dilemma concerning the best approach to ensuring the quality of stock. While farmers could sometimes make a short-term improvement to flocks or herds (either to 'improve' or to 'mend' them) by introducing rams or bulls from another district,[46] they commonly found that, in succeeding generations, the descendants of such animals would revert back to the local type.[47] A case in point was the transportation of long-horned cattle 200 km from Lancashire to the more equable climate of Leicestershire. 'In the third descent [i.e. third filial generation] they had their Leicestershire breed again', wrote Lisle.[48] Examples such as this seemed to support the traditional locality-based view of race [although since the imported animals were frequently crossed into the local strain, the loss of their characteristics was hardly surprising]. It had to be admitted that evidence for local adaptation was strong and seemingly based on common sense. What could be more logical than that the traditional breed of a district would be more in harmony with the local environment and better adjusted to its potential hazards than alien stock introduced artificially? Local adaptation seemed like a perfect expression of divine design. Such was commonly believed among farmers, as explained by Russell,[49] who has reviewed the evidence in the writings of Gervase Markham among others. The idea of a close and *directed* relationship between an animal's environment and its outward form was widely accepted. In Chapter 2, we considered it in relation to the Merino. Russell has traced it back to its classical roots, as well as to natural theologians led by John Ray in England in the seventeenth century.[50]

However, there was also evidence supporting the idea of breeds being changed by forces other than the natural environment. In the words of Marshall:[51]

> We have, at present, through time and the industry of our ancestors, various breeds; some of them adapted, though not perfectly, yet in very considerable degree, to the soil they are upon, and the purpose for which they are wanted. . .

Two causes of variation are thus identified, one natural (time), the other artificial (the industry of our ancestors). Shortly afterwards, James Hutton

---

[46] Lisle 1757, ii, pp. 154 and 159    [47] Lisle 1757, ii, p. 99–101
[48] Lisle 1757, ii, p. 100    [49] Russell 1986, pp. 193–4    [50] Ray 1691;Russell 1986, pp. 16, 105–110
[51] Marshall 1790, i, 463; the idea was expressed in French by Maupertuis (1745) who wrote: 'Nature holds the source of all these varieties: but chance or art sets them going'.

(1726–1747), a Scottish farmer and geologist,[52] would write, with reference to dogs, 'The form best adapted . . . will be the one most certainly continued.' Although Hutton does not distinguish human action from a natural trend as separate causes of variation, the context of his words allows for both influences, as in the case of Marshall.

Evidence of Man's capacity to breed animals selectively provided the context of a conversation between Lisle and Benjamin Clerk who farmed at Dishley in Leicestershire before the Bakewell family took over the lease (Chapter 4). In his talk with Clerk, which took place around the year 1700, Lisle learned that reversion of superior stock to an inferior breed could be avoided by careful breeding and management, if it was economic to do so. Regarding the example of long-horned cattle from Lancashire, the problem in Leicestershire, so Clerk informed him, was that farmers were not so particular about breeding and management as those in Lancashire. The Lancashire men chose their bulls more carefully, used them only when young and weaned the calves on a richer diet. The reason that Leicestershire farmers failed to follow suit was, according to Clerk, entirely economic: 'They kept but small dairies, and therefore it would not be worth their while, when they milked but a few cows, to go to such a price for a bull.'[53] So the Leicestershire farmers quickly lost the quality features of the breed. We may note the implication that without selective breeding and good husbandry, new blood would have no lasting influence, but that when animals were carefully bred and looked after, the outcome might be more favourable; the Lancashire breed *could* be preserved in Leicestershire if desired. Here was a hint of something not yet generally much written about, that heredity might be controlled by human action to resist the effect of environmental pressure from the outside world, at least for a few generations. To ascribe a dominant role to the natural environment was beginning to seem overfatalistic, contrary to the spirit of improvement. The voice of experience that had led farmers to conclude that an imported breed would inevitably revert to the local type was being challenged, as stock was moved more frequently.

## The wider movement of stock

There was a clear need to establish the limits of natural environmental influence on heredity, a matter to be resolved by accumulated experience. Every occasion of moving stock between districts, whether with the intention of crossing them with the local breeds or of keeping them as a pure strain, provided further evidence. The value of protecting an animal against extremes of natural variation had long been understood, but changes that had been brought about by the breeders' own actions were beginning to indicate an inherent link between parents and offspring even when conditions of nutrition or climate

---

[52] J. Hutton c.1796 (unpublished), ii, p. 739    [53] Lisle 1757, ii, p. 100

were varied. As the distances over which animals were transported grew longer, the influence of parentage was brought into even sharper focus. Breeders with no concern except practicality and profit could see for themselves, in the manner of scientists conducting experiments, what really did happen. Their experience allowed them not only to appreciate the strength of family resemblances, even when conditions of nutrition and climate were dramatically altered, but also to define more clearly the true limits of external influence. Some breeds did indeed suffer when moved, or even failed to survive, but not all of them. The conditions in which an animal was kept had certainly to be considered. That was a lesson well learned, and it led to various actions on the part of the breeder to mitigate the impact of a change in environment. Starting with efforts to cushion the animals against the worst excesses of environmental pressure, his efforts culminated finally in selective breeding to maintain the type in the face of degeneration, or even to bring about improvements. Certain breeds were to become changed almost beyond recognition.

The first major test of the true limits of environmental influence on the characteristics of a breed came not from farm animals but with the transportation of equestrian stock from the southern Mediterranean countries into northern Europe, which took place a century before the time we are considering.[54] Lord William Cavendish, Duke of Newcastle (1592–1676), 'perhaps the most famous and widely respected of all authors on equestrian matters in the England language in the seventeenth century',[55] believed temperature and humidity to be the most important factors in raising a horse of true quality. In his book entitled *A New Method and Extraordinary Invention to Dress Horses. . .* (1667),[56] he offered the following advice about foals of Turkish, Neapolitan and Spanish origin imported into northern countries:

> It matters not what kind of pasture they feed in providing it be but dry . . .
> The secret then in bringing horses rightly up in cold countries is nothing else
> but keeping them warm in the winter, and feeding them with a dry kind of
> food, and in turning them out in the summer to dry pastures.

Cavendish, 'whose interest in physiology or the theory of inheritance was negligible' according to Russell, selected both stallions and mares for breeding in terms of their temperament as well as their physical form, including their 'beauty', meanwhile noting that 'good' and 'bad' horses came in all colours and shapes.[57] Nevertheless he was convinced that by taking these precautions he could retain the valuable qualities of these hot-blooded animals in Britain, which indeed proved to be the case. Equestrian writers after the time of Cavendish were inclined to place increasing stress on an animal's pedigree beyond all other considerations.

---

[54] Wilson 1912, pp. 2–8     [55] Russell 1986, p. 77
[56] Russell 1986, pp. 86–7 (quoting Cavendish 1667)
[57] Russell 1986, pp. 77 and 84 (quoting Cavendish 1667)

For the early importers of Merino sheep from Spain into northern Europe it was vital to know whether these sheep would respond as favourably as exotic horses had done. The omens for sheep were not good. If climate and diet truly produced degenerative changes in breeds transported between districts, how much greater would the effects be when they were transported across the length of Europe? In the face of such uncertainty, Alströmer took an enormous and potentially expensive gamble when he imported Merinos into Sweden in 1723. Undoubtedly he gained heart from the horse breeders he had heard about or met in England. He may have picked up practical tips from them, for he adopted precisely their policy of warm, dry housing (Chapters 2 and 6). But how many times must he have been told by fellow farmers, with a knowing shake of the head, that his Spanish sheep were bound to deteriorate in the alien environment of Sweden? Many years afterwards, there were still misinformed people using this argument to discourage import of Merinos into England. This is well illustrated by a letter from an English merchant in Madrid, Henry Hinkley, to a London-based colleague, Mr Collier, written in 1787:[58]

> The expense of purchasing them in New Castille and bringing them over the Biscayan mountains to Bilboa would be considerable and no good end would afterwards be obtained. These animals immediately degenerate, or at least their wool does, on their being prevented from going through the two long journeys they annually undertake from Castille to Estramadura . . . and the King of Spain is so perfectly satisfied of this that he has permitted flocks of these sheep to be sent to Berlin and Paris whose wool in a couple of years has turned precisely the same as that of other sheep of the same country.

Mr Hinkley was, of course, being wildly pessimistic if not entirely incorrect. Fortunately for the Kingdom of Sweden, Alströmer possessed the confidence to trust his own judgement. His experience revealed that degeneration was not a problem in the sheep he introduced, and did not usually occur in their immediate progeny when bred to one another. When problems did arise in subsequent generations, Alströmer found ways of dealing with them. To avoid degeneration it was his practice to import fresh breeding stock from Spain regularly, at first every two or three years. This policy may have been influenced by a taboo against close inbreeding. Some experts believed that the very act of inbreeding accelerated the process of degeneration. So the practical question for Alströmer was how quickly the degenerative changes took place, and therefore how often it was necessary to reimport stock from Spain in order to compensate for the damage done. This led to a second question: could such changes be delayed in order to reduce the frequency of reimportation? Eventually Alströmer would find a way to do it, just as the horse

---

[58] Carter 1979, p 116: letter from H. Hinkley (merchant) to Collier (merchant) 15 July 1787

importers were doing. And ultimately, well before the end of the century, it became unnecessary to import any further Spanish stock at all.

The measure of Alströmer's success may be judged from the reports of foreigners, the most influential of whom was the French expert C.P. Lasteyrie who visited a successor to Alströmer, Baron Schulz de Schultzenheim, who farmed at Gronsoe in the Province of Upland. Schulzenheim had a Merino flock based on stock introduced in 1745, which was in excellent condition when Lasteyrie had the chance to inspect it in 1800. He wrote that he 'did not find it inferior in either beauty or fineness' and that 'Mr Shultzenheim preserved the descendants of sheep which he imported from Spain to the fifth generation, and a comparison of fleeces proved that they had not in the least degenerated.'[59] Lasteyrie's book (1802) had a major influence on opinion internationally, being translated into several languages.

## Generation, degeneration and selective breeding

The threat of degeneration, where breed and climate or soil were seen to be competing for influence, presented a challenge to the best skills of breeders. Out of their progressive successes grew the belief that controlled husbandry and selective breeding were complementary techniques. Applied effectively, they could be used either to ensure stability between generations or to create inherited changes. The evidence pointed to a link between parents and offspring independent of climatic or other natural influences, a powerful connection between generations. This was the essence of a breed (race), its 'blood',[60] which would resist a change of climate or soil if correctly nurtured. The idea that an animal's own inner nature ('blood') had a chance of becoming independent of the outer world of nature reflected the logic behind selective breeding. Otherwise to breed selectively would have been pointless.

The pinnacle of understanding about breeding reached in the eighteenth century was expertly summarised at its close by the Scottish farmer and economist James Anderson. He confidently stated that variations 'accidentally produced' are inherited, as well as the 'general characteristics of the parent breed'.[61] Anderson is indicating that breeding should be directed not simply towards maintaining the traditional quality of a breed but at improving it. As he stressed, such selection can be accomplished even in a breed normally stable in its characteristics, one that 'appears not to be liable to be changed by climate or other extraneous causes'. He stressed that his view arose not from the exercise of logic but from experience. That such variations existed and could be selected had been proved, though Anderson could not explain their

---

[59] Translation of Lasteyrie's French text (quoted by Rees 1819, in his section headed 'Sheep')
[60] Marshall 1790, i, p. 299    [61] Anderson 1796, p. 23

origin, writing 'they have arisen from circumstances that have utterly eluded our observation'. He was no wiser on this point three years later,[62] still unwilling to commit himself to any conclusion that was not strictly based on experience.

To view continuity between generations in terms of 'blood' was nothing new. The idea was one that even Aristotle had accepted as traditional. It stated that semen is a transformation of the blood or, in the words of Diogenes of Appolonia, 'a foam of the blood arising from disturbance produced by heat'.[63] However, the concept of blood being transformed into semen begged many questions for the farmer, arising out of uncertainty about what affected it:

First there was the relative significance of male and female to consider. Was the female weaker in her influence, because she lacked male semen, or were her own genital secretions seminal in nature? A conviction that both sexes produce a form of 'seed' can be linked to the writings of Lucretius, who is famous for using the pleasure argument, hinting at a connection between shared pleasure and shared ejaculation.[64] Leonardo da Vinci is credited with reaching the same conclusion after observing the results of sexual intercourse between a white woman and an African male, from which he inferred (c.1490) that 'The seed of the female was as potent as the male in generation.'[65] The matter was to be taken up by eighteenth century naturalists, including Linnaeus in Sweden and Buffon (George Louis Leclerc, Comte de Buffon (1707–1788) in Paris, the latter an ardent admirer of Lucretius' writings.[66] Each became convinced that offspring must proceed from both male and female seminal fluids or substances,[67] the product of which then becomes subject to various influences during development. A major extra influence to be considered was the nutriment provided in the womb directly from the mother's blood, reinforced later by her milk.

Concerning the environment of the womb, factors that might affect it at the moment of conception included shocks received or even sights seen as 'maternal impressions'. Many other questions arose about this mysterious organ, with regard both to conception and foetal development. Was the intensity of sexual activity, including the degree of heat produced, important? Was the state of the womb influenced by previous conceptions by the same or another male? What of possible influences of disease in the mother?

---

[62] Anderson 1799, p. 87
[63] Brown 1987, p. 184. This was an idea that persisted for many centuries. In Rome it took a form attributed to Galen, of the 'ebullition [boiling, effervescence] of the blood' during coitus. The same idea was widely circulated in a printed pamphlet for sale in London in the 1730s, said to be based on the writings of the thirteenth century alchemist Michael Scott (Pinto-Correia 1997, pp. 85–6).                                    [64] Brown 1987, p. 320
[65] Grant 1956     [66] Glacken 1967, p. 139     [67] Farley 1982, pp. 23–5

Finally, there was the environment in which breeding stock was kept, including climate and altitude, the crops on which the farmer fed his stock and the natural herbage of the hedgerows, which he could not control.

To what extent was the blood of the new generation in danger of being influenced by these different factors? Could conditions at conception, fetal development or life after birth accumulate their effects over generations? For the intelligent breeder these issues could not be ignored. Experiments were called for in all directions. With so many matters for him to consider, he had to be ever watchful, to note anything at all which seemed significant. The central issue was the extent to which nature could be moderated by nurture. What power was in the hands of Man? Which procedures really worked? The good husbandman controlled what he could, trying to avoid extreme conditions and attempting to impose some order upon nature by breeding as selectively as he could afford.

Nutrition and housing were re-evaluated as enlightened breeders came to realise that the only chance they had of picking out the best individuals from the rest was to keep all of them in the same high-grade environment. The best parents were those that would give progeny as good as or better than themselves, *under the same conditions*. As Clerk of Dishley had assured Lisle, breeding and management were both important for a successful outcome.[68] Once farmers like Clerk had recognised selective breeding as a powerful agent for preserving the qualities of a herd with its traditional characteristics, the next step was to apply the same technique towards improving it. Lyle's published notes reveal that experiments were already proceeding around the year 1700 on characteristics that could be changed in the cause of improvement. The burst of activity associated with the name of Bakewell, to be considered in the next chapter, identified a whole range of further inherited characteristics. Variation was available due to many earlier rounds of selection by a host of unknown breeders during the course of the 11,000 years or so (about 4000 generations) since the sheep was first domesticated.[69] Breeders were now beginning to exaggerate this variation in new directions.

## Cross-breeding as an aid to selection

It had long been obvious that the descendants of the wild progenitors of a domesticated species varied much less among themselves than their domesticated cousins. The Greeks had explained the contrast in terms of a conserving force in the natural environment that eliminates all deviations from the perfection of the average type.[70] The act of domestication seemed to 'release' variability from this constraint, the consequence being the creation of strong regional and local differences between breeds, as a kind of 'second nature'

---

[68] Lisle 1757, ii, p. 100    [69] Bökönyi 1974, 1976; Ryder 1987
[70] Zirkle 1941; Mayr 1982, pp. 448–9

created within nature.[71] The more that breeding was regulated, the greater seemed the variation.[72] As the Victorian geologist Charles Lyell wrote later:[73]

> The best authenticated examples of the extent to which species can be made to vary may be looked for in the mystery of domesticated animals and cultivated plants.

The increase in variation thus created provided the potential for breeds to be modified, if considered expedient. The most potent way to generate variability rapidly was by crossing the different varieties and selecting among their progeny in the second and later generations. As in other aspects of breeding, the pioneering example was to be found in the racing stable. The expense of importing valuable exotic running horses into Europe in the seventeenth century meant that crossing was inevitable. Whatever the risk of losing quality, the experiments had to be attempted, to maximise the use of the expensive stallions. Leading breeders like Sir John Fenwick began at once to introduce characteristics of the north African 'Barb' or 'Arab' into English stock by successive generations of crosses, choosing a stallion as close in conformation to the English or Irish mares as possible.[74] The reason for making a close match was fear of producing a useless mongrel, with a disproportionate mixture of traits, only too easy when crossing breeds. The best crossbred dams, carefully selected, were matched to pure Barb or Arab stallions. The procedure was repeated for as many generations as necessary to produce the desired strain. This was the process of 'grading up', a sequence of backcrosses designed to increase the proportion of 'noble' blood. It was the accepted pathway to success, a breeding method by which the English race horse was brought into being in the seventeenth century. The first cross produced progeny with half the stallion's blood. Mating him to his daughters introduced a quarter more of his blood into his grandchildren. After four generations of breeding to the same male or a first-degree relative of his, the proportion of desired blood would almost reach 95%. This figure was recognised and recorded by Cavendish in 1667[75] although almost certainly there were breeders with this knowledge before that.

Any venture into crossing was fraught with potential pitfalls. Wisdom dictated that even when a crossbreed appeared outwardly normal or even superior, breeding from it could introduce problems simply because it represented a deviation from the natural order of things. For a noble animal to consort sexually with a common one was like an aristocrat uniting with a peasant, subject to all the risks of misalliance, so it was believed. At the same time, favourable experiences with some hybrid horses led to optimism. The impact of early success in cross-breeding, hints of which were emerging even from

[71] Glacken 1967, p. 138    [72] Lawrence 1819, p. 253    [73] Lyall 1834, ii, p. 354
[74] Cavendish 1667, pp. 62–3 (quoted by Russell 1986, pp. 85 and 96)
[75] quoted by Russell 1986, p. 86

the secretive world of the racing stable, was tremendously significant. Progressive breeders of farm stock took good note of it. Parallel advances were being made in the breeding of dogs. The key to success proved to be selection of carefully defined features of both sexes in each generation, using stock strictly matched for size and compatibility. Such crosses, followed by reselection, produced results that reinforced the reality of heredity most strongly. Alströmer in Sweden was an early contributor to this demonstration with his Merino sheep.

Increasingly there were breeders ready to surpass Nature in her measured pace, to reconstruct their flocks. By the time Marshall (1790) in England and Anderson (1796) in Scotland came to report this great story, they were both clearly aware of the power of heredity reflecting the qualities of both parents, to provide a double source of variation upon which changes could be based. Coupled with artificial aids to survival in the form of shelter and supplemented diets, selective breeding provided the means for an enterprising farmer to 'design' a breed of animals for his own purpose. Thus was 'Nature' confronted by 'nurture'. In the usage of time, 'nurture' included all actions on the part of the farmer that deviated from the natural order of things, including the careful matching of male and female in controlled matings.

## Breeding larger sheep

In the drive to increase the size of sheep as a response to changing economic priorities in the British Isles, the rates of increase varied in different parts of the country. This may be explained partly on the basis of differences in pastures, although selective breeding was certainly involved in some cases. In the Lincolnshire area, farmers were ready to claim that they had been at great expense and trouble in improving their sheep. That was before 1750, although in other areas of the country the increase in size probably occurred later than this.[76]

The ease with which a character is selectively enhanced depends on its heritability, defined as the proportion of the total variance in that character that can be transmitted to the next generation when rearing conditions remain the same. Mature body weight has a substantial heritability in sheep, as in most domestic species. In unselected seventeenth or eighteenth century stock its heritability would have been particularly high. Fleece yield also has a substantial heritability which might seem a desirable feature in itself except for an association sometimes found between large size and coarseness of wool. Many believed this to apply to Lincolnshire sheep although it was strongly disputed in a letter written by Henry Butler Pacey FRS, Member of Parliament for Boston in Lincolnshire, to the President of the Royal Society,

---

[76] O'Connor 1995

Sir Joseph Banks, in 1782.[77] Pacey was convinced that after enclosing sheep on improved pastures, conditions in which they grow larger, 'The staple of the wool is deeper and finer than formerly, which renders it of more value than when short and coarse.' Opinions clearly differed on this matter, probably reflecting different outcomes in different regions.

During the same period an effort was being made to increase the lustre of the wool of the Lincoln sheep and to remedy the bareness of their legs and bellies.[78] Each of these characters responded to the selection imposed. As the seventeenth century progressed into the eighteenth, the heavy, long-legged and long-wooled animals become even more abundant in their expanding pastures. Russell has evaluated the achievement in measured terms:[79]

> The drive for size as a response to better markets and improved nutrition was undoubtedly naïve compared with the sophisticated analysis of the major components in animal productivity provided by Bakewell in the eighteenth century, but against the background of the small, bony and primitive sheep of the medieval period, the large lowland and marsh breeds were more efficient animals.

The increase in weight must have appeared rapid in the later stages because Sir John Sinclair, writing in 1791, was of the opinion that none of the large breeds went back more than half a century.[80] This fits in with published records[81] and osteological evidence[82] that size increased gradually and on a piecemeal basis and was mainly an eighteenth century phenomenon. By the time detailed descriptions of the different English lowland pasture and marsh breeds (Lincoln, Leicester, Romney Marsh, etc.) are given by William Marshall in 1790, they seem to be almost interchangeable: massive, flat-sided animals, hornless in both sexes, often with a topknot and with very white, fairly coarse, long wool. The long combing wool, provided by the great animals, was in constant demand and its exportation to foreign countries strictly controlled. Nevertheless, no small quantity was secretly smuggled out across the English Channel.[83] The number of sheep kept in England and Wales rose, according to the best estimates (Fussell and Goodman 1930) from more than 16 million in 1741 to more than 26 million in 1809.

## Breeding for other qualities

Edward Lisle wrote in one of the notes published by his son[84] that hornless sheep would often be preferred on grounds of nutritional efficiency 'because so much of the nourishment doth not go into the horns'. Another, more

---

[77] Carter 1979, pp. 72–3: letter from H.B. Pacey MP FRS to Sir J. Banks 28 March 1782
[78] Ernle 1936, p. 179    [79] Russell 1986, p. 157
[80] *Proceedings of the Society of British Wool*, 31 January 1791, quoted by Housman 1894
[81] Fussell and Goodman 1930    [82] O'Connor 1995
[83] Hill 1966, pp. 117–18; Mann 1987, pp. 269–70    [84] Lisle 1751, i, p. 155

significant, trait for a breed's survival was fecundity. In Tudor times, if not earlier, the shepherd would note the ewes that gave birth to twins and save her lambs to breed from, as advised by Tusser:[85]

> Ewes yearly by twinning, rich masters do make
> The lambs of such twinners for breeders do take.

Early dropped lambs were also favoured for breeding.[86] If something extra was required—either improvement of a quality already valued or the achievement of an imagined ideal—the animals selected for breeding would have to be exceptional. Success rested ultimately on the breeder's vision and his skill and judgement in realising that vision. The principle of breeding at this level of genuine improvement was expounded in an unpublished manuscript by the progressive Scottish farmer and geologist James Hutton, summarised in the following brief extract:[87]

> The first or great principle of the breeding art is this, that animals as well as plants, propagate their like. But in the *refined* practice of that art, the principle on which we are to proceed is this, that in the propagation of species there are seminal varieties – varieties that may be useful in our economy, and that may be directed in the propagation by the skilful applications of the art. Hence a nice distinction is required in the artist, first to perceive those useful qualities that may occasionally appear, in the extensive breeding of a species; secondly to see the value of a *nascent quality*, which otherwise might be over-looked or neglected; and lastly to understand the ultimate perfection of that quality or the extent to which it may be carried in successive propagation. . .

The seminal varieties to which Hutton refers are individual animals of a deviant form, i.e. with an unusual quality or qualities: 'particular varieties among the individuals'.[88] How difficult it must have been, in practice, to see the value of a 'nascent' quality or understand the true limits of selection in animals deviating from the accepted norm. Most breeders were cautious for fear of losing what they already possessed. Hutton had in mind any quality that made an animal (or plant) superior in a commercial sense, different qualities for different species. Examples included radical changes in growth rate or wool quality, which some breeders were attempting by the eighteenth century.

## The question of incest

Although the successful breeder paid attention to the qualities of both sexes, he gave special attention to choosing the sire because his progeny might be many while that of the dam would inevitably be few. Using the grading technique developed for horse breeding, the inherited qualities of a male could be

---

[85] Johnstone 1846, p. 152     [86] Marshall 1796b, i, pp. 265–6
[87] Hutton c.1796, ii, pp. 744–5, his emphasis     [88] Hutton c.1796, ii, p. 733

evaluated quickly. A high-quality male could bring improvement to an inferior flock within a single season. Then by continuing to breed from the same male, or from males of equal quality, improvement had a chance of becoming progressive.

Still there were many farmers with a serious objection to using the same male over several seasons. To do so would be to mate father with daughters or granddaughters, and brothers with sisters or mothers, a practice that conflicted with a strong and widely held objection to 'incest' in sheep and cattle, as in humans.[89] The matter had earlier caused difficulties among the horse breeders but had been dismissed by the much respected Lord Cavendish (1667) in the following words:[90]

> But you cannot breed better than to breed to your own mares that you have bred; and let their fathers cover them; for there is no inceste [*sic*] in horses: and thus they are nearer, by a degree to the purity, since a fine horse got them, and the same fine horse covers them again.

Adherence to the contrary and traditional view that incest certainly did apply to animals as well as to humans meant that a fine male could be neglected in favour of those of lesser quality, merely to avoid repeated matings. It needed the strong mindedness of a Cavendish to flaunt convention and oppose this trend. Yet experience was to show that once the decision had been made to continue to breed from the same male, a skilful breeder had the chance of accelerating the changes he wanted, and stabilising the characters selected. His success in concentrating good qualities in his stock then depended on how far he was prepared to recognise the possibility that he might also be concentrating *unfavourable* ones, and to find ways of avoiding or minimising such undesirable consequences by selective breeding. The potential for injurious effects from consanguinity was known about and feared with good reason. At the same time it was impossible for a serious selective breeder to avoid such matings. Inbreeding and selection were two sides of the same coin and had to proceed together. The most rigorous culling was required to weed out individuals exhibiting undesirable traits. Lord Cavendish, who was one of the first writers to make a big issue of selection, realised that there must be sufficient stock to allow for the inevitable wastage, 'something no previous author had made explicit', according to Russell.[91]

Few sheep breeders were willing to take the chance with such radical and expensive practice, although by the mid-eighteenth century the pace of activity increased. Many mistakes must have been made before the more skilled or fortunate individuals were able to achieve satisfactory results, beyond previous experience. Selection at this level, to change a breed towards a preconceived

---

[89] Culley 1786    [90] Cavendish 1667, p. 93 (quoted by Russell 1986, p. 85)
[91] Russell 1986, p. 86

ideal, was an extremely complex matter requiring the qualities of an artist, as well as profound technical expertise, to make the correct judgement.[92] As Charles Darwin (1809–1882) was later to write:[93]

> If selection consisted merely of separating some very distinct variety, and breeding from it, the principle would be so obvious as hardly to be worth notice; but its importance consists in the great effect produced by the accumulation in one direction, during successive generations, of differences absolutely inappreciable to the uneducated eye.

The intensive way of breeding that so impressed Darwin will be considered in the next chapter. First, it is instructive to summarise what Alströmer could have learned about sheep breeding when he was actively seeking information during his years in England, prior to 1726. Then we shall see more clearly the advances made thereafter, advances that would be copied by a new generation of Merino breeders, mainly outside Britain, who would carry Alströmer's achievement to new heights by the nineteenth century.
The key principles were as follows:

(1) The quality of an animal depends both on nature and nurture, so the ambitious breeder must start with the best possible stock and look after them, being especially careful to cushion them from unfavourable climatic or nutritional influences.

(2) Animals imported from a different district or even a far-off country can usually be kept successfully and without deteriorating greatly during their own lifetime. In the case of sheep, special attention must be given to their diet and protection from extreme climatic conditions, both of which can influence wool quality.

(3) The descendants of imported animals will tend to degenerate. A way of overcoming this is to reimport fresh breeding animals every two to three years. A radical alternative is to try to delay or stop degeneration by carefully selecting all breeding animals for desired features. The latter was an idea gaining ground but not yet generally accepted.

(4) The characteristics of an imported breed may be introduced into a local breed by crossing, to produce a mixed breed with stable characteristics. The process of grading (or grading up) to introduce maximum blood from a foreign breed into local stock had been successfully demonstrated in horses. Use of an imported breed in this way depended on the extent to which it was believed that its characteristics could be retained. In practice, success in doing so with horses had depended on skill in selecting and matching parents for corresponding or complementary traits.

---

[92] Hutton c.1796    [93] Darwin 1872, p. 23

(5) The characteristics of a breed, pure or mixed, may be changed, according to the desire of the breeder, by careful selection of breeding stock and mating like to like. As selection became more systematic it led to the exploitation of unusual individuals ('accidental varieties') for breeding. Intensive selection could imply consanguineous mating, a practice that was not to be openly supported by farmers until Bakewell's day.

These five principles, recognised by enlightened farmers in the early eighteenth century, were among the first generalisations about breeding based on empirical observations. They shed new light on the nature of heredity among those who held the principles to be true. Based on experience, they have stood the test of time, in contrast to various vaguely defined beliefs based on traditional ideas (such as telegony [the supposed influence of a previous sire on the progeny of a previous sire from the same dam], maternal impressions or separate effects of male and female) which, one by one, were rejected. As Vorzimmer has written:[94]

> In terms of practical necessity, it was essential for breeders to formulate empirical generalisations from which predictions could be made on the results of certain matings. These guiding maxims accumulated gradually as a sort of lore, and the closing decades of the eighteenth century saw the advent of a new group of naturalist experts, breeding specialists.

The greatest advances in breeding were made under the stimulus of economic pressure, particularly during the early stages of the industrial revolution, which occurred first in England. There was also the stimulus of sport, the ruling passion of the English country gentleman. It has been said that for a century or more after the English Civil War (1642–49), during a period of relative peace for the British Isles, the breeding of horses, dogs, cattle, sheep and pigs took the place of war.[95] At no stage was progress much influenced by scientific institutions. The 'arrow of influence' continued to fly in the reverse direction throughout that century and even the one that followed. Neither Lamarck nor Darwin seriously influenced the actions of breeders, although both took note of breeders' achievements when reaching separate explanations for the origin of new life forms over geological time.[96] The pioneers of breeding were endeavouring to substitute procedures based on observation and deduction for a changeless tradition, as each of them accumulated knowledge for his own particular purpose. They represented what Mantoux[97] has written of Jethro Tull: 'if not the scientific spirit proper, at least something akin to it—the enlightened empiricism which has so often led men to discoveries'.

---

[94] Vorzimmer 1972, p. 23     [95] Paget 1987 (1945), p. 9
[96] Wood 1973; Ruse 1975     [97] Mantoux 1961, p. 158

The enlightened empiricist most often mentioned in the context of selective breeding is the Englishman Robert Bakewell. By considering his life and achievement, we can span the gap between the level of understanding open to Jonas Alströmer at the beginning of the century and the enlightened statements written by James Hutton and James Anderson at the end. In one sense Bakewell was an unlikely model, for he depreciated the significance of wool in favour of mutton, which risked alienating him from the dominant international trend. But his outstanding originality stimulated not only his own countrymen but breeders of talent in many foreign nations who knew him only as a legend, and who applied his ideas to improving the very product he neglected, namely wool.

# 4

## *Bakewell's new system*

By harnessing generation in the female animal's body, man has achieved power over the functions of his domestic animals, and only by this means did the well known Bakewell produce deliberate changes in the shape of his animals.

*Albrecht Thaer, 1825, p.25 (original in German)*

Successful farmers are usually too busy to publicise their achievements in print. This is left to those who claim agricultural reporting as their profession. One such writer, the value of whose considerable body of work is judged to have been exceptional, both in accuracy and in soundness of opinion, was William Marshall (1745–1818).[1] This shrewd and perceptive Yorkshireman, who set out to survey English farming from North to South, reached the Midland counties in 1789, where he found much that was original to capture his interest. No district seemed to be without progressive farmers with 'many advantages over lower orders of husbandman', including, he noted with approval, their readiness to travel and to mix with 'men of science and fortune'.[2] But while recognising a wealth of genuine talent, he still found himself commenting at length upon one particular individual. An especially influential figure had emerged from this central area of England, one farmer who was challenging and changing accepted opinion on a controversial and economically significant subject at a faster pace than his contemporaries. Marshall felt bound to acknowledge Robert Bakewell of Dishley, near Loughborough in Leicestershire, as 'the principal promoter of the art of breeding' and 'the grand luminary of the art'.[3]

---

[1] Cooper 1893; Fussell 1950, p. 114    [2] Marshall 1790, i, p. 116
[3] Marshall 1790, i, pp. 294–5

**Fig. 4.1.**   Robert Bakewell (1725–95), pioneer animal breeder. Detail of an equestrian portrait (oil on canvas by John Boultbee (1745–1812)) belonging to the Leicestershire Museums, interpreted in an original drawing by Richard A. Neave.

In this chapter we introduce this exceptional farmer (Figure 4.1) and analyse his methods as he applied them to sheep. With the aim of relating practice to theory, we ask in what particular respects he advanced sheep breeding methodology and by what means his achievement led to a more reliable perception of heredity. Caution is necessary because it is clear that some of the developments generously attributed to him by his admirers, what one of them referred to as 'The Bakewellian System',[4] had their roots before his birth[5] and were extended by other talented breeders.[6] If the image that emerges seems over-colourful and excessively heroic by modern standards, so be it. For the contemporary view is what we seek—how he was seen in his own lifetime and in the three or four decades that followed, when his achievement still seemed revolutionary and he was credited with founding a 'new school of breeders'.[7]

---

[4] Anon. ('J.L.') 1800, p. 203. [This anonymous author is John Lawrence (1753–1839), writer on horses.]          [5] Russell 1986, pp. 146–8, 169 and 196–205
[6] Chambers and Mingay 1966, pp. 66–7          [7] Parkinson 1810, title page

## Robert Bakewell and the management of his farm at Dishley

Entering the world in 1725, the only one of three brothers to survive child-hood, Robert Bakewell lived a full span of 70 years on the family farm known as Dishley Grange. Covering about 450 acres (182 ha),[8] this fertile stretch of land had been held on lifetime lease by two previous generations of Robert Bakewells, considered by their landlord to be careful and profitable tenants.[9] Describing themselves as 'graziers', they made a speciality of fatten-ing sheep and cattle for the big city meat markets, and their trade benefited from Dishley's convenient position astride a main road to London, 109 miles to the South (Figure 4.2).[10] Extensive, well kept buildings, lush meadows, neat hedges, perfectly maintained fences and gates in willow, and intensively cultivated fields, almost weed-free it was claimed,[11] combined to present an attractive sight to those who passed along the road.

This was the background against which the last of the Robert Bakewells of Dishley, powerful in physique and confident in speech,[12] began experiment-ing with selective breeding. Starting to breed sheep even as a youth, he extended his skills to cattle and horses, and eventually to pigs. He was still a young man when the farm came totally under his control in all aspects, many years before his father died in 1760.[13] Eventually he was to give up the family's grazing business altogether and concentrate exclusively on selective breeding, the only farmer then to do so.[14] It was a brave decision, a strategy to be chanced only by the toughest, most professional and competitive of farmers. His achievements were placed under a continual spotlight as rivals sought to challenge his competence. This was no ordinary farmer but a true 'Prince of Breeders',[15] a title conferred upon him by no less an authority than Arthur Young (1741–1820), Secretary to the Board of Agriculture in London.

In support of his breeding programmes, Bakewell prided himself on being self-sufficient in animal food production. A major improvement to the Dishley land that made this possible was an extensive irrigation system. Started by his father, it was an asset that Bakewell exploited to the full as he greatly extended and elaborated it.[16] To increase his water supply, he con-structed a canal to divert the course of a mill stream that ran along the southern border of the farm. The mill was used to regulate its flow, for which Bakewell paid a hire fee to the owner[17] until finally, to give himself sole

---

8 *Leicester Journal* 3 February 1781; Young 1786a, p. 490
9 Pitt 1809, p. 31; de Lisle 1975, 1993   10 Owen 1764, p. 100
11 Pitt 1809, pp. 32–3
12 Sir Richard Phillips (quoted by Pawson 1957, pp. 28–9); Anon (J.L.) 1800, pp. 208–9
13 Anon 1795; Anon. ('J.L.') 1800, p. 200; Young 1811   14 Marshall 1790, i, p. 303
15 Young 1791b; Young's *Tours* (1932), p. 302
16 Young 1771a, p. 124; Young 1771b, pp. 109–10; Young 1786a, 1811; Marshall 1790, i, pp. 284–6. The floating of water meadows became a growing practice in England before the 1670s (Kamen 1984, p. 150; Watts 1984, pp. 157–8) when seeds of nitrogen-fixing crops such as clover were becoming available at lower prices.   17 Young 1786a, p. 491

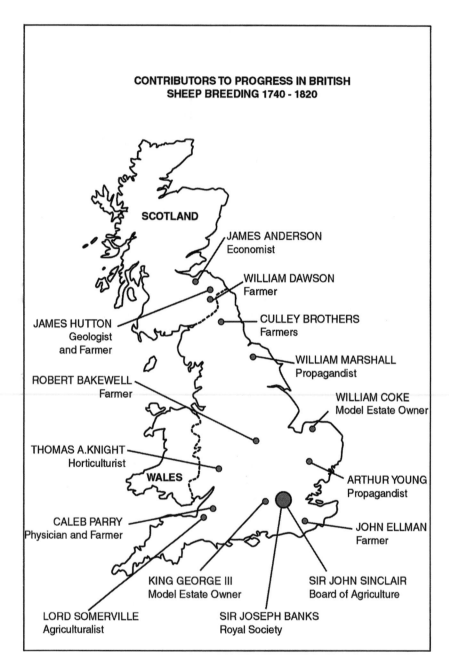

**Fig. 4.2.**    Map of Great Britain showing the location of individuals making a particular impact on sheep breeding during the period 1750–1820.

command of the water, he gained the tenancy of the mill,[18] together with some land that went with it.[19] By digging a series of channels, controlled by sluices,[20] he was able to flood much of his holding whenever he wished, thereby increasing crop yields, particularly of grass.[21] When his meadows were flooded in the autumn, the first spring grass appeared a month early, in March rather than April; the extra fodder produced in this way gave him greater flexibility in overwintering his stock. Determined to exploit water management to the full as an aid to intensive stock breeding and feeding experiments, 'He kept a skilful irrigator as a foreman, for many years, at high wages.'[22]

Much of the extra benefit Bakewell obtained by controlling the mill depended on the artificial canal he constructed, 15 feet (4–6 m) wide and eight or nine feet (2.4–2.7 m) in depth, which grew to more than a mile (1.6 km) in length.[23] By 1795, the year of his death, extensions to the system had allowed him to bring an extra 200 acres (81 ha) under irrigation, in addition to the 80 acres (32 ha) improved by channels cut directly from the stream.[24] The canal, 'cut' or 'carrier trench', as it was variously called, provided Bakewell with a uniquely efficient transport system to and from the farm buildings, by which to move manure in one direction and the produce of his fields in the other. Turnips were thrown directly into the canal to be carried to the barn for storage[25] 'quite clean and fit for use', as Bakewell delightedly reported to his close friend, former pupil and fellow breeder, George Culley.[26] Other crops were loaded into a pair of 5 ton (5.1 t) boats purchased specially for the purpose,[27] also used to transport liquid manure, 'the drainings from the yard', to his meadows, allowing him to harvest four crops of hay per season. 'He has the mowers constantly at work, and says he keeps cutting till Christmas', reported two visiting farmers from Sussex in 1793.[28]

Irrigation on such a scale provided the essential basis for a system of farm management that allowed Bakewell to undertake precisely controlled selective breeding. It was vital for the animals to be maintained in a uniformly high nutritional regime. Even in summer he was opposed to folding sheep in large flocks, believing this to be a barbarous practice because the weaker individuals would suffer.[29] Difficulties inherent in caring for and evaluating specially selected animals led him to keep them singly or in limited groups. His answer was to divide his farm into 10 acre (4 ha) fields, separated by neat

---

[18] Pitt 1809, p. 203    [19] Wykes c.1995, unpublished manuscript
[20] Young 1771a, p. 126; Young 1771b, p. 110
[21] Marshall 1790, i, pp. 285–6; Pawson 1957, p. 113: letter from Bakewell to Culley, 30 June 1787    [22] Young 1811, p. 15
[23] Pitt 1809, pp. 32 and 201
[24] Young 1771a, pp. 124–7; Young 1786a; Young 1811, pp. 15–16; Pitt 1809, pp. 203–6
[25] Marshall 1790, i, p. 286; Anon (J.L.) 1800, p. 204
[26] Pawson 1957, p. 106: letter from Bakewell to Culley, 8 December 1786
[27] Pawson 1957, p. 152: letter from Bakewell to Culley, 17 February 1790
[28] Redhead and Laing 1793, p. 21    [29] Young 1786a, p. 481; Lawrence 1805, pp. 318–19

hawthorn hedges,[30] within some of which he was to erect 'different small folds in different parts of the farm, for animals of different ages, kinds and strengths'.[31] Near the house were parcels of pasture land of less than 1 acre (0.4 ha), well fenced round 'for the convenience of keeping by themselves small lots of sheep or cattle' for comparison.[32] Some of his pasture was divided into 'pieces so small' that Young 'could not but enquire into his motive'.[33] Bakewell's answer was that it made 'the food go so much further'. He used other small plots, which he called his 'proof-pieces or panes, for the admission and comparative exclusion of water',[34] to demonstrate the favourable effect of irrigation on grass production.[35] The proof pieces, each of 1 perch (16.2 m²) in area, were compared in blocks of 20, bounded by carefully levelled channels to spread water across the land, some being raised 'on little wooden aqueducts above streams running at a lower level'. The intricacies of the system were much appreciated by a visitor from France, Count François de la Rochefoucauld, who recorded in his diary that there were more than 10 or 12 miles (16.1–19.3 km) of channels of one size or another.[36]

Bakewell was lucky to have enlightened landlords, the March Phillips family of Garendon, who allowed him to mix grazing and breeding with arable farming,[37] a privilege denied by most Leicestershire landlords of the period.[38] Bakewell's success must have owed much to the freedom of action allowed to him in this respect, to grow all the animal fodder he needed on his rented land. One particular 'wonder' to be seen at Dishley was Bakewell's 'breed' of very large cabbages on which he fed his sheep, and the seed of which he supplied to other farmers.[39] Determined to preserve the integrity of this variety, the first record of which appears in 1783, Bakewell took great pains to protect the plants from cross-pollination, 'being careful to suffer no other variety of the brassica tribe to *blow* near his seed cabbages', by which means they were kept 'true to their kind'.[40] A favourite way of isolating cabbages was to plant them in the middle of a field of wheat.

## Stock improvement strategy

The much irrigated Dishley farm was managed with a well defined strategy in mind, one which Bakewell stated explicitly in 1780 in a petition for financial backing (see Chapter 5). It was 'a plan for improving the breeds of horses

---

[30] Pitt 1809, p. 32; Scarfe 1995, p. 34
[31] 'Lincolnshire Grazier' 1833, paraphrasing Young 1786a, p. 481    [32] Pitt 1809, p. 32
[33] Young 1786a, pp. 495–6
[34] Young 1811, p. 16; see also Culley 1786, pp. 203–5; Young 1786a, p. 491
[35] Pawson 1957, p. 113: letter from Bakewell to Culley, 30 June 1787
[36] Scarfe 1995, p. 34. Further information on the irrigation system is provided by Pawson 1957 and Stanley 1995.    [37] Young 1771a, i, pp. 123–4
[38] Ernlé 1936, p. 201
[39] Marshall 1790, i, pp. 260–61; Pawson 1957, p. 133: letter from Bakewell to Culley, 22 October 1788    [40] Marshall 1790, i, p. 261

for cavalry, harness and draft, and also of meat cattle [the English long horned breed] and sheep'.[41] His large black horses were built for strength and stamina. In the case of sheep and cattle, and later pigs, Bakewell envisaged and brought into existence breeds specifically designed to satisfy the increasing public demand for cheap meat. He needed animals that would grow quickly on the minimum of food, giving tender flesh with sufficient fat to meet popular demand. As Young wrote in 1771, 'The only object of real importance is the proportion of grass to value.'[42]

It was with his sheep (Figure 4.3) that Bakewell gained the greatest market advantage. Because of the rapidity with which his wethers (castrated rams) put on weight, they were killed to best profit at two years old and a few months, after their second shearing.[43] Their superiority in this respect represented a significant commercial advantage compared with 'unimproved' sheep which took three or even four years to reach the same condition.[44] As the value of Dishley breeding stock rose progressively, the evidence of

*New Leicester or Dishley Sheep.*

**Fig. 4.3.**   A Dishley ram drawn by Samuel Howitt and published in *Memoirs of British Quadrupeds* by the Rev. W. Bingley in 1809.

---

[41] Pawson 1957, pp. 181–3: Bakewell's bankruptcy petition
[42] Young 1771a, i, p. 111;Young 1771b, p. 104       [43] Pitt 1809, pp. 258–4
[44] Pawson 1957, p. 62; Kerridge 1967, p. 324; Macdonald 1974, p. 237: letter from Culley to Sinclair, July 1792

Bakewell's skill stimulated his supporters, led by Arthur Young, to trumpet every increase with enthusiastic delight.

In the cause of more efficient meat production, Bakewell bred selectively for certain defined physical traits, which may be judged by the final appearance of his breeds (Figures 4.3 and 5.1). Both his sheep and his cattle were immediately recognisable by their broad, barrel-shaped bodies, straight short flat backs, small heads and short small-boned legs: 'the value lies in the barrel, not in the legs', wrote Young.[45] In extreme cases of sheep selection, individuals would even assume a 'tortoise like shape',[46] which was described as 'much wider that it is deep; and almost as broad as it is long'.[47] Selected changes of an extreme kind could be reversed if necessary on the basis of experience. The fine-boned legs, which were at first bred to be very short, were later lengthened[48] although the tail remained small. Bakewell's sheep were white-wooled and hornless in both sexes, with clean heads (i.e. they had no topknot) and small thin ears. Hornlessness was said to have an economic rationale, as the energy necessary for the growth of horns was saved. Several detrimental characters emerged, including an over-thinning of the pelt because of too vigorous selection for thinness in the ears.

Drawn to Dishley by Bakewell's reputation, visitors were intrigued to discover than he was testing the worth of his innovations by secret experiments. His manner of keeping sheep or cattle in small groups or singly in stalls, wintering all his stock 'tied up in houses and sheds',[49] up to '170 beasts of all sorts',[50] kept 'as clean as race horses',[51] set him quite apart from traditional practice. Although unattached to any academic body or official institution, he seemed to be conducting himself as a man of science, gaining knowledge through critical and systematic experiments, trying to discover principles of breeding on which he could rely in the future. 'This was the grand source of his powers', wrote the Victorian cattle expert William Housman.[52] With his animals kept side by side in stalls or in neighbouring small plots, with their food uptake closely regulated, Bakewell could compare examples of his own stock with other kinds.[53] Young, who inspected some of the trials, believed Bakewell had 'made experiments on most of the breeds of sheep in the Kingdom'.[54] There were also some animals from abroad, among the more exotic of which were Merinos (or Merino crosses) from the Cape of Good Hope.[55] Late in his life, Bakewell extended his interest to what would now be called $F_1$ hybrids. By breeding into further generations he could sometimes produce

---

[45] Young 1771a, i, p. 112    [46] Lawrence 1805, p. 391    [47] Marshall 1790, i, p. 389
[48] Marshall 1818, iv, p. 302    [49] Young 1786a, p. 487
[50] Young 1771a, i, pp. 120–21    [51] Throsby 1791    [52] Housman 1894, p. 9
[53] Pawson 1957, p. 106: letter from Bakewell to Culley, 8 December 1787; Young 1771a, i, p. 119; Young 1786, p. 48; Redhead and Laing 1793, p. 22    [54] Young 1786a, p. 477
[55] Young 1786a, pp. 482–3; Pawson 1957, pp. 176–7: letter from Bakewell to Young, 23 December 1791

a 'mixed breed sort', as he did in the case of his pigs.[56] Although he made every effort to keep the results secret from his rivals, both Young and Marshall managed to gain a degree of information.[57]

A favourite experiment consisted in keeping a series of individual animals on exactly the same food and noting the differences in growth rate among them. Marshall describes such a trial which came to his notice in 1785, made by a farmer he discretely refers to as 'a Midland grazier' who—judging by evidence from an independent visitor—was almost certainly Bakewell himself.[58] Having purchased a number of ram lambs in their first year at the same price, although from different sources, he had kept them on uniform feed for nearly 12 months. Marshall reports that 'There is now a difference in their respective values, merely from a difference of constitution, of at least ten shillings [50 p] a head, being one of many instances that prove the great propriety of attending closely to the inherent properties of livestock.'[59]

When writing about such trials, Marshall made free use of terms like 'breed', 'blood', 'constitution' and 'inherent properties'. He was ready to reveal clearly the extent to which he and, by implication, Bakewell were prepared to acknowledge the influence of heredity in creating the differences observed. He seemed to be under no doubt that an animal's nature enjoys a substantial degree of independence from the external world in which it lives. He was unequivocal in asserting 'that the Midland breeders rest everything on BREED; under a conviction that the characteristics *beauty* and *utility of form*, the quality of the *flesh*, and it propensity to *fatness* are, in the offspring, the natural consequences of similar qualities in the parents'.[60] The two farmers from Sussex already mentioned, who visited Bakewell in 1793, stated the case even more strongly when they wrote: 'Keep the breeds pure and Mr Bakewell is satisfied that no material change will be affected by either climate or pasture. This not being generally believed, or understood, deserves to be further investigated and attended to.'[61]

For the traditional farmer the idea of dismissing the influence of climate and pasture was heresy indeed. Practical books on sheep breeding had invariably stressed the advantage of choosing the local breed, adapted to local conditions, as indeed they still do. It is good advice, although Bakewell had the confidence to set limits upon it. He was encouraged to do so by his observation that different kinds of sheep would remain different when kept side by side under the same conditions.[62] It was not that he failed to recognise changes in flesh or wool created by climate or pasture, e.g. 'that the same kind

[56] Marshall 1790, i, pp. 452–3    [57] Young 1786a, pp. 482–3; Marshall 1790, ii, p. 65
[58] Scarfe 1995, p. 34, on Rochefoucauld's report of 1785; Pawson 1957, p. 171: letter from Bakewell to Young, 26 February 1785 (incorrectly dated 1783 by Pawson)
[59] Marshall 1790, ii, p. 65    [60] Marshall 1790, i, p. 299 [Marshall's emphasis]
[61] Redhead and Laing 1793, pp. 23–4    [62] cf. Boys and Ellman 1793, p. 135

of sheep would not produce such fine wool on rich as in poor pasture',[63] but that such changes seemed minor to him and of a temporary nature, not 'material', compared with the power of selective breeding. The truth of this generalisation has since been firmly established, even with respect to wool.[64]

## The origin of the Dishley sheep

The way in which Bakewell could create a new breed of animal more economically successful than the one it replaced, within only a few generations, created astonishment among those encountering the situation for the first time. He was not dealing simply with trivial characters, such as might distinguish varieties of pet dogs or fancy pigeons, but with traits central to the animal's very constitution, affecting multiple aspects of appearance and performance. His Dishley (New Leicester) sheep, with their extraordinary capacity to mature and fatten quickly, differed dramatically in appearance from any type of sheep known before in Leicestershire, or indeed anywhere else. The new breed remained as he wished it to be, generation by generation, even when kept side by side with traditional breeds, under the same conditions and given the same food. Bakewell's neighbours and visitors had no option but to conclude that he had in some way harnessed and concentrated the power of heredity. But how had he done it? If by selective breeding (as he claimed), from where had he drawn the most unusual combination of features in the new breed? Had the radical transformation been achieved exclusively on the basis of the traditional Leicester breed? Or had he made crosses with other breeds and selected among their descendants? If crosses had been involved, were they made between 'kindred' breeds, resembling the traditional Leicester type, or had he introduced stock from 'alien' breeds from distant places?

The crossing of alien breeds was a topical subject because of what the racehorse breeders had achieved with Arab stock, and what was happening on the European continent with the cross-breeding of Merinos. Dog breeders and pigeon fanciers were also sometimes crossing established breeds to produce new types. The capacity of Bakewell's sheep to breed consistently for their special form and qualities was all the more remarkable if they had a mixed origin. To the disappointment of all enquirers, Bakewell steadfastly refused to reveal in full detail the primary makeup of his Dishley stock. This was one secret, above all others, that he was determined to keep from his rivals. What little they were able to discover we shall now consider, with the added advantage of being able to draw information from all possible sources and to check one opinion against another.

---

[63] Redhead and Laing 1793, p. 24; Pawson 1957, p. 51
[64] Carter 1958 (a summary of experimental work published earlier by himself and co-workers)

The Leicester longwool breed of Bakewell's youth (the 'Old Leicester') was a type of sheep well adapted to his gently sloping fields and low lying meadows. Its large size and flat-sided body and the deep staple of its very heavy fleece were matters of frequent comment, being the result of selective breeding for generations on rich pastures during the previous century (chapter 3). The pamphleteer and novelist Daniel Defoe, who could claim special expertise as the son of a butcher, noted that Leicestershire sheep were 'without comparison of the largest and bear . . . the greatest fleeces of wool on their backs of any sheep of England'.[65] He wrote these words one year before Bakewell's birth.

In a bid to create something new from this well known breed, Bakewell made his first move in the late 1740s.[66] He seems to have begun in a conventional way by picking up the best sheep he could find as he travelled round his own and neighbouring Midland counties, 'a selection of all the best kinds of long or combing woollen sheep, wherever to be met with', as Marshall describes them.[67] For rams he extended his search outside the Midland counties, in particular to the uplands (wolds) of Lincolnshire, where he picked out animals at the smaller end of the size range.[68] In favouring smaller Lincolnshire rams, Bakewell was following a practice recorded by Edward Lisle in notes he had taken around 1700 of a conversation with a previous tenant of Dishley farm, Benjamin Clerk, who rented the land before Bakewell's grandfather.[69] It was a tradition sufficiently well established for Bakewell to find himself in competition with other local farmers for the same type of ram. He got his pick of the best by bribing the 'jobbers' who imported them.[70] He ended up with a breed substantially smaller than the traditional Old Leicester type, which was precisely his intention so it seems, although it represented a practice contrary to the mainstream of English longwool breeders whose sheep seemed to get larger every season.

From a letter Bakewell wrote to the Lincolnshire breeder Charles Chaplin,[71] we learn that his habit of importing rams from that county had ended by about 1768. What he did not reveal, a matter which is still unclear today, is whether those forays into Lincolnshire represented the limit of his outside purchases. This was a matter upon which he declined to comment, although the possibility exists that he made 'wider crosses' to introduce some of the features observed in the Dishley breed that finally emerged. So greatly did his sheep differ from others in Leicestershire, or indeed those of Lincolnshire,

---

[65] Defoe 1724, ii, p. 89; see also Bischoff 1828, pp 41, 53    [66] Young 1811, p. 4
[67] Marshall 1790, p. 383; see also *Farmer's Magazine* 1804, p. 301; one breeder from whom Bakewell obtained ewes was Webster of Canley in Warwickshire. The evidence comes from Webster's shepherd (Carter 1979, pp. 179 and 574).
[68] Young 1771a, i, p. 117; Young, 1811, p. 9, quoting Culley    [69] Lisle 1751, ii, p. 159
[70] Pitt 1809, p. 249; 'Lincolnshire Grazier', 1833, p. 217
[71] *Annals of Agriculture* Vol. 10 (1788), p. 571: letter from Bakewell to Chaplin, 15 November 1788 [Charles Chaplin (1759–1821) was a landowner and stockbreeder of Tathwell, Lincolnshire]

small or large (Figure 5.1), that some commentators were unable to believe that the changes he brought about could possibly be based solely on traditional practice. Could the power of selective breeding be so great? A Dorsetshire farmer and stockbreeder, John West Parsons, wrote to Sir Joseph Banks in 1799:[72]

> In justice to the memory of Mr Bakewell I must say that I believe he improved his sheep upon the right principle by crossing the filthy, foggy Leicesters with a Mountain Sheep.

Opponents of this view included Bakewell's pupil George Culley, who knew the situation as well as any one could, having spent some months in residence at Dishley, and was convinced that selective breeding from long-wooled stock was the whole story, as he reported:[72a]

> The gradual process of amelioration is founded in nature who rarely deviates out of path, so that with care and proper choice of individuals, it becomes no difficult task to approximate towards that point to which the improvement of breeds is to commence.

Marshall reached the same conclusion. After acknowledging that improvements might sometimes be attributed to crossing, he could not believe this to be so in Bakewell's case. Bakewell's achievement did not arise from 'mixing alien breeds but by uniting the superior branches of the same breed'[73] or 'by selecting individuals of *kindred* breeds, from the several breeds or varieties of long-wooled sheep, with which Mr B was surrounded'.[74] If pressed to speculate on which alien breed Bakewell *might* have employed for crossing, Marshall and other reporters could not agree on what it would have been. All we have from Bakewell himself is a statement, made in 1788, that the rams he then had for hire 'were without a cross to the Durham or any other kind'.[75]

Marshall expressed the hope that Bakewell might communicate the complete facts before his death 'for the government of future improvers'. To general disappointment, no further evidence was forthcoming, which convinced Sir John Sinclair (1754–1835), first President to the Board of Agriculture, to add his weight to the conclusion that selection, rather than crossing to 'alien breeds', must have been largely responsible for the changes.[76] Bakewell, he concluded, had sufficient variability to work with, 'having spared no pains or expense in obtaining the choicest individuals from all the best kinds of long and combing wooled sheep wherever they were to be met with'. In reaching

---

[72] Carter 1979, p. 314: letter to Sir J. Banks from J.W. Parsons, farmer and stockbreeder of Dorset, 6 July 1799                         [72a] Culley 1807, p261

[73] Marshall 1790, i, p. 301         [74] Marshall 1790 i, p. 383

[75] *Annals of Agriculture* Vol. 10 (1788) (quoted by Stanley 1995, p. 36): letter from Bakewell to Chaplin, 12 September 1788

[76] Sinclair 1832 (1817), p. 95 and note 121 on his Chapter 2

this opinion, Sinclair acknowledged a letter he had received from Sir John Saunders Sebright, politician and agriculturalist (1767–1846), in which he had written:[77]

> The alterations that can be made to any breed of animals by selection, can hardly be conceived by those who have not paid some attention to this subject; they attribute every improvement to a cross, when it is merely the effect of judicious selection.

Whatever might be the precise truth in Bakewell's case, he himself welcomed the publicity when journalists began to describe his breed as the 'New Leicester', linking it with the true old Leicestershire breed. As he wrote to Culley in 1786, 'We are obliged for this epithet to a paragraph that appeared in the papers.'[78] To leave this idea in the public mind was expedient at a time when a man's pride in his own county and its products was very strong.

Part at least of Bakewell's reluctance to speak in detail of his breed's origin probably resulted from his impatience with the concept of pedigree (so dear to horse breeders). He gave every evidence of judging a sheep for breeding purposes more by its qualities, and those of its immediate relatives, than its ancestry. This is captured in a remark attributed to him by the *Leicester Journal*.[79] In a reported dispute with another breeder, he is said to have retorted that 'It is of less importance where the breed came from than where they are now.'[80] His concern was with the desirable traits that an animal shares with other members of the flock.[81] Most important was its proven capacity to transmit these traits to its existing progeny, to the maximum degree. Bakewell was looking for particular features in both ram and ewe which, when matched together, would result in all their lambs putting on weight quickly with the minimum ingestion of food. To quote Young, 'He was the most careful feeder of stock that ever I met with, and who made his food go the furthest.'[22]

The business of choosing and combining his basic sheep stock took Bakewell up to 15 years,[82] which suggests that he made some early mistakes and false starts. However, by putting a certain combination of bodily characters together, he was progressively creating a new breed, 'an ideal perfection which he endeavoured to realise' from an image in his mind.[83] Lord Somerville wrote that 'It would seem as if [Bakewell] had chalked out on a wall a form perfect in itself and then had given it existence.[84] Others saw him producing his breed as an engineer invents and builds a new machine.

---

[77] Sebright 1809, p. 26
[78] Pawson 1957, p. 121: letter from Bakewell to Culley, 20 November 1787
[79] *Leicester Journal*, 10 October 1789        [80] Wykes c.1995, unpublished manuscript
[81] Marshall 1790, i, p. 418        [82] Culley 1786, pp. ix–x
[83] Ferryman (quoted by Pitt 1809, p. 249)
[84] Somerville 1806 (quoted by Bischoff 1842, i, p. 380)

Charles Darwin was later to write that 'breeders habitually speak of an animal's organisation as something plastic, which they model as they please',[85] an attitude which Bakewell would surely have understood as he determined to get animals to gain weight rapidly in parts of the body where it mattered most.[86] Of course it did not always work out exactly as he wanted, leading him on one occasion to write to Culley that 'The sheep I sent is not without some places I would wish better but know not how to alter them.'[87] But although Bakewell recognised his limitations, most observers credited him with supreme confidence in what appeared to be his single-minded progress towards an ideal.

## Breeding in-and-in

> The foundations being thus laid, the means of carrying up the superstructure are evidently those of breeding in and in, and selecting with judgement, the superior individuals produced; having ever in view the idea of perfection.

These words of Marshall's[88] place in context what is acknowledged to be a central pillar of Bakewell's approach to animal improvement, the practice of close inbreeding. Once Bakewell had secured what he felt to be the best breeding stock, with the necessary potential, he was not afraid to see his flocks becoming progressively more inbred with each generation, as he transformed them, step by step, towards the ideal he had in mind. The degree of inbreeding was by all accounts extreme, including matings between first-degree relatives for several generations. As Young put it,[89]

> he entirely sets at naught the old ideas of the necessity of variation from crosses: on the contrary, the sons cover their dams and the sires their daughters, and their progeny equally good, with no attention whatever to vary the race.

Marshall[90] explains that the practice had become "so long established as to have acquired a technical phrase to express it: 'BREEDING IN AND IN'".

George Culley, a hard-headed farmer who, with his brother Matthew, pioneered stock breeding in Northumberland, wrote a practical book about breeding that set out Bakewell's system in some detail. This was *Observations on Livestock, containing Hints for Chusing [sic] and Improving the best Breeds of the most useful Kinds of Domestic Animals*, the first edition of which appeared in 1786. Culley did not really feel competent to write the book, as he reveals in his introduction. He had wanted to find a much more suitable author

> ...who I believed to be well able to perform it; and in particular one whom it is not necessary here to name, for whose abilities I have the highest respect,

---

[85] Darwin 1872 (1859), pp. 22–3        [86] Young 1771a, i, pp. 110; Pawson 1957, p. 50
[87] Pawson 1957, p. 122: letter from Bakewell to Culley, 28 October 1787
[88] Marshall 1790, i, p. 465        [89] Young 1786a, p. 488
[90] Marshall 1790, i, pp. 300–301 [Marhall's emphasis]

whose whole life has been employed in breeding and improving stock, and has carried it to a very great perfection, from the experience and close application of his whole life, spent in the pursuit of breeding useful stock.

But having been unsuccessful in persuading Bakewell to do it, he reluctantly takes the task upon himself and produces a popular work with far reaching influence, of which Bakewell thoroughly approved.[91]
Three major principles of breeding emerged from Culley's book:

(1) Animals imported from one district into another or 'from any part of the world' can be employed to obtain or preserve the best breeds.

(2) To bring about improvement, the best animals must be selected and bred only among themselves, even if it means close inbreeding.

(3) The quality of an improved breed does not depend on its being 'refreshed' from outside.

In support of principles (2) and (3), Culley informed his readers (in 1786) that Bakewell had introduced no fresh blood into his flock for 20 years or more.[92] Evidently by the early or mid 1760s, Bakewell considered his foundation stock as good as he could make it. The intervening years of experimentation had given him plenty of opportunity to obtain the best possible animals from which to weed out undesirable traits as he commenced and intensified his inbreeding programme.

Culley was unsympathetic to breeders who were reluctant to engage in close inbreeding. He was convinced that unreasonable opposition was holding up progress, some farmers having 'imbibed the prejudice so far as to think it irreligious. . .'[93] But for Marshall the pros and cons of inbreeding were not so clearly defined.[94] He wrote of how many breeders expected bad consequences to flow from it, sure that it must 'weaken the breed', although he joined Culley in emphasising how, in Bakewell's hands, it did no harm. To think otherwise was a 'vulgar error'. So what was the explanation? Did Bakewell possess some secret that allowed him to use a technique that others, with strongly held reasons, feared to use? Marshall found the answer in Bakewell's exceptional skills in selective breeding. Close inbreeding concentrates deleterious traits as well as favourable ones. And just as the latter are to be preserved, the former are to be rigorously excluded. Bakewell's own words on the subject were emphatic: 'Nothing but first rate loins,[95] thighs and scrags[96] can support in-and-in breeding.' To this he might add the oracular pronouncement of an old farmer of his acquaintance, 'strong loins, strong

[91] Pawson 1957, p. 102: letter from Bakewell to Culley, 11 April 1786
[92] Culley 1786, pp. ix–x; since 1768 according to another source (see note 71)
[93] Culley 1786, p. viii *ff.*      [94] Marshall 1790, i, p. 321
[95] 'loins': either side of spine, between ribs and hips      [96] 'scrags': necks

constitution'.[97] Only animals in robust good health were suitable for inbreeding at this level.

Bakewell told Young that the confidence he placed in close inbreeding arose from the earlier experience of pigeon fanciers and cock breeders.[98] Culley noted that the technique was also frequently adopted by those breeding the best gun dogs[99] while Marshall claimed that breeding in-and-in originated in Newmarket, by which he meant that the racehorse breeders ('gentlemen of the turf') had established the practice.[100] The latter does not have the support of Russell (1986) who has made a careful examination of the horse breeding literature, and states:[101]

> Inbreeding of the type practised by Bakewell . . . was certainly not applied in the origin of the Thoroughbred. A restricted group of foundation animals were employed, but they were genetically unrelated and apart from occasional close matings, the ancestry of many early Throughbreds showed a careful avoidance of incestuous breeding.

Bakewell is on record as criticising breeders of horses and hounds for their slowness in adopting the practice.[101] So bird breeders emerge as the most likely influence upon him, as he himself claimed. The truth is that by the 1760s a number of breeders had become convinced, from personal experience, that inbreeding was essential to making selected improvements. They were transforming not only sheep but also cattle, horses, pigs, greyhounds, foxhounds, pigeons, rabbits and a whole variety of other species. By bringing rare recessive traits to light, the technique extended the breeder's options for selection, either to preserve them or cull them from the stock, as desired.

A number of good breeders benefited from Bakewell's example, although by no means every one who wished. For those with only a shadow of his judgement, inbreeding could bring drastic penalties. Potential problems included loss of fertility and various constitutional weaknesses and disorders, not easily eliminated. As more and more breeders jumped on to the inbreeding bandwagon, serious warnings were issued by experts like Sebright and Sinclair.[102] Sebright was convinced that rarely did there exist a single sheep untainted by 'some defect in constitution, in form, or in some other essential quality'.[103] Sinclair contrasted the risks attached to inbreeding with what his friend Sir Thomas Andrew Knight (1759–1838) had been discovering about the favourable influence of outbreeding, in terms of increased vigour in the first-generation hybrids of both animals and plants.[104] Meanwhile, Rev. Henry Berry tried to strike a balance in a prize winning essay in

---

[97] Dixon 1868 (quoting from Bakewell's 'Memoirs', a collection of written material in the hands of the proprietor of Dishley Farm at the time of Dixon's visit)
[98] Young 1786a, p. 488      [99] Culley 1786, pp. x–xi      [100] Marshall 1790, i, p. 300
[101] Russell 1986, p. 104
[102] Sebright 1809; Sinclair 1832 (1817), pp. 93–5; Blacklock 1838, pp. 106–7
[103] Sebright 1809, p. 11      [104] Knight 1799

which he suggested that, although close inbreeding may 'increase and confirm valuable properties', it becomes potentially more risky as selection is intensified, when it 'impairs the constitution, and effects the procreative powers'.[105]

A defect ready to reveal itself more strongly with every generation of inbreeding required strong action to correct it. Selection 'towards an improved state' could then depend on the existence of another family of the same breed into which to make a cross. Sebright, supported by Sinclair, claimed this to have been precisely Bakewell's own practice, as Marshall had also implied.[106] Sinclair wrote about keeping two or three 'streams of blood' quite distinct.[107] Berry wrote about making an occasional 'strong cross'.[105] Youatt, who reviewed the sheep breeding literature 'from the earliest period to the present day' in 1837,[108] defined the kind of outside breeding animal to be used in such circumstances as one which

> should be as near as possible of the same sort; coming from a similar pasturage and climate, but possessing no relationship—or at most a very distant one—to the stock to which he is introduced. He should bring with him every good point which the breeder has laboured hard to produce in his stock, and, if possible, some improvement. . .

Bakewell's solution was evidently close to that proposed by Sinclair. He took every opportunity to exchange rams with other farmers who favoured the Dishley breed, including the Culley brothers, whose stock almost rivalled his own but was kept at their farms in Northumberland in a markedly different environment from Leicestershire. Such matings within a stock already strongly selected, brought substantial benefit with a manageable degree of risk, and it set the pattern for other breeders to follow. A degree of inbreeding was still vital to continuing improvement. No successful breeder could afford to dispense with the procedure for, as Rice (1926) pointed out, 'inbreeding has been extensively used in the formation of most of the breeds of livestock.'[109] Certain plant breeders of Bakewell's day also valued the technique. One of these left a published record. This was Joseph Cooper of Gloucester County, New Jersey, who reported in 1799 that he had bred squash, peas, asparagus and potatoes on the principle that 'a change of seed was not necessary to prevent degeneration'. He points out, however, that the procedure was only effective when he restricted himself to seed from the best plants he had already selected. He claimed this to be a practice his family had been following since 1746.[110] The Cooper family's experience places this particular aspect of the breeding doctrine associated with Bakewell (inbreeding

---

[105] Berry 1826 (quoted in footnote to Youatt 1834, p. 526); see also Berry 1829
[106] Marshall 1790, i, pp. 383–4; Sebright 1809; Sinclair 1832 (1817), pp. 94–5
[107] Sinclair 1832 (1817), p. 93, note 109    [108] Youatt 1834, p. 525; Clarke 1900
[109] Rice 1926, pp. 184 and 198
[110] Cooper 1799; Priestly 1797; Bajema 1982, pp. 37–40

is safe only if accompanied by selection for general characteristics of fitness) in an interestingly wider context.

## Trait selection

Finding no alternative to breeding in-and-in as an essential aspect of selection, Bakewell was bound to choose every breeding animal, both male and female, with the greatest care. Apart from the particular physical characteristics that he favoured for meat production, he had also to ensure the good health of his sheep (absence of 'disorders'), their hardiness, including resistance to bad weather and above all, their procreative ability and mothering qualities. The time scale indicates that the business of combining the various desirable qualities within a single stock was long and drawn out. As soon as he saw some particular trait he wanted, he had to be sure to make the best use of it. 'In short', wrote Marshall, referring on this occasion to Bakewell's long-horned cattle, he was 'soliticously seizing the superior accidental varieties produced'.[111] It seemed to Marshall that no breeder expressed his skill with greater eagerness and concern for the outcome.

At its base Bakewell's approach to breeding had much in common with the grazier's way of picking out lean beasts most likely to fatten for the market. Features that would be indicative of fattening quality ('fatting') had been written about in English farming books as early as the 1530s.[112] Marshall remarked on how this grazing skill was particularly well developed in the Midland counties where Bakewell and his collaborators mainly operated.[113] Bakewell's own father had made a successful living out of it. The very qualities in an animal that the grazier looked for were those Bakewell wished his breeding stock to inherit and transmit.[114] A grazier's skill in distinguishing intrinsically fast and slow fattening animals opened the possibility of producing fast fattening *breeds*. Only a committed experimentalist like Bakewell was prepared to take this route, which was unacceptable to the majority of farmers because of loss of income in the short-term.

Identifying superior young breeding animals, accurately and reliably, depended on how well they were looked after and whether they were treated equally. It was Bakewell's particular skill to combine a uniformly high feeding regime with selective breeding, as Sir John Sinclair relates:[115]

> Robert Bakewell. . ., having acquired great skill in grazing by which he was enabled to preserve his breeding stock in the highest possible condition, and having called into his aid all the skill and experience that the butcher had acquired, was thus enabled to ascertain the principles, not only of breeding domestic animals so as to answer the common expectation of the farmers

---

[111] Marshall 1790, i, pp. 383–4    [112] Russell 1986, pp. 122–3
[113] Marshall 1790, i, p. 393    [114] Marshall 1790, i, pp. 372–4; Culley 1794, p. 230
[115] Sinclair 1832 (1817), p. 92

but also of bringing them to a degree of perfection, of which, before his time, they were scarcely supposed capable. . .

Bakewell considered his hands to be more useful than his eyes as he picked out young animals for breeding purposes.[116] He and Culley agreed that they would rather judge an animal with their hands in the dark than only by sight in the daylight.[117] Sheep with a tendency to fatten were said to 'handle well', showing a 'kindliness' to the touch, with a skin that felt 'mellow', 'soft, yet firm'.[118] Marshall noted how sheep were judged by different criteria from cattle.[119] He attempted to describe the points to be looked for under the conviction that 'the ground work of this art, like every other, is reducible to science.'[120] It must be so, he argued, because of the undeniable changes brought about by selective breeding on the basis of such points. As the breeder picks the characters he favours in the parents, the same features appears more frequently, and in a more extreme form, in the lambs or calves.

The first priority for Bakewell was to breed from compact, fine-made, small-boned individuals. He had learned from experience, building probably on information from his grazier father, that such sheep not only carried an extra proportion of edible tissue, weight for weight, but were also quick to mature. As he asserted to Young, 'the smaller the bones, the truer will be the make of the beast—the quicker she will fat.'[121] In his quest 'to substitute profitable flesh for useless bone',[122] limbs and neck were the major target for reduction. The stress placed upon smallness and fineness of limbs was in marked contrast to the traditional stock-breeding axiom of Leicestershire, recalled by Young, that 'a great bone was a great merit'.[123] Another target for Bakewell was the head, which he believed should also be made smaller. At the same time the ribcage was to be greatly expanded. Arguments in favour of such skeletal proportions were summarised later by Sinclair.[124] A wide chest was essential for creating the barrel shape which Bakewell favoured for all his meat-producing stock. One advantage was an increase in lung capacity, as pointed out by the eminent London surgeon Henry Cline.[125] Sebright pointed to the narrow chest of the Merino as an explanation of why this breed could not readily be made fat.[126]

Judging the quality of breeding stock was a continuing preoccupation for the successful breeder. Of the 20–40 lambs Bakewell saved each year for breeding, only the best would become 'ram getters'.[127] From July or August,

---

[116] Culley 1786, pp. 13–15; Marshall 1790, i, pp. 365–6 and 372
[117] Young 1786a, p. 469, see also Pawson 1957, p. 158: letter from Bakewell to Culley, 6 September 1791
[118] Sebright 1809, pp. 21–2; 'Lincolnshire Grazier' 1833, p. 34; Sinclair 1832 (1817), p. 91
[119] Marshall 1790, i, pp. 393–4    [120] Marshall 1790, i, pp. 365–6 and 393
[121] Young 1771, i, p. 111; see also Culley 1786, p. 178; Young 1786a, p. 478; Young 1811, p. 5
[122] Anon. (J.L.) 1800, p. 205    [123] Young 1811, p. 10
[124] Sinclair 1832 (1817), pp. 86–7    [125] Cline 1805    [126] Sebright 1809, p. 22
[127] Marshall 1790, i, pp. 418 and 422–3

when the first selection was made, to shearing time during the first week of June in the following year, the growing lambs enjoyed a rich diet of clover, turnips, hay and corn. To give himself an advantage, Bakewell made detailed measurements and kept accurate records of growth rates.[128] Such of his shearlings that failed to show the necessary form and fatness were sent for slaughter, and only the best retained as breeding stock. Marshall outlined what Bakewell and fellow breeders were hoping that the favoured lambs would reveal as they matured:[129]

(1) *beauty of form* ranked highly, particularly at the outset of improvement. Although beauty may seem a non-economic criterion, it was of considerable importance in attracting customers. The ideal was to link beauty with utility;

(2) *proportion of parts* (*utility of form*), i.e. the proportion of 'valuable parts' to offal (non-edible tissue, including bone). This was often spoken of as the most fundamental selective criterion. Culley linked it with beauty when referring to the 'true symmetry of the parts';[130]

(3) *flesh*, i.e. texture of the muscles, was important for cooking quality and taste. Marshall was convinced that it was 'clearly understood that the grain of the meat depends wholly on the BREED, not as has been heretofore considered, on the SIZE of the animals' [Marshall's emphasis]; and

(4) *fatting* [sic] *quality*, i.e. the ability to reach a state of fatness at an early age.

Each of these four qualities is 'found to be hereditary', wrote Marshall, 'depending in some considerable degree at least, on BREED, or what is technically termed BLOOD, namely on the specific quality of the parents' [Marshall's emphasis].

We may note that Marshall wrote nothing about grandparents or more remote ancestors, i.e. he gave no consideration to the matter of pedigree. Ancestry was much more to be considered when exploiting an established race than when creating a new one. The statement of Rev. Berry in his prize-winning essay that the 'excellences' of the male are 'the accumulated acquisitions of many ancestors'[131] summarised the belief of breeders who came after Bakewell, although it was not a view attributed to Bakewell himself. Using Marshall's words, it is possible to summarise the Bakewellian attitude to heredity (= breed = blood) as follows:[132]

---

[128] Young 1786a, p. 483; Redhead and Laing 1793, p. 22
[129] Marshall 1790, i, pp. 297–9 and 419      [130] Young 1811, p. 5
[131] Berry 1826, p. 34 (this is a different version of the Berry essay from that quoted by Youatt 1834, having been 'kindly rewritten' for the *British Farmer's Magazine*)
[132] Marshall 1790 i, p. 299

Thus it appears, that the Midland breeders rest everything on breed; under a conviction that the *beauty* and *utility of form*, the quality of the *flesh* and its propensity to *fatness*, are, in the offspring, the natural consequences of similar qualities in the parents. And, what is extremely interesting, it is evident from observation, that these four qualities are compatible; being frequently found united, in a remarkable manner in the same individuals.

The fact of all four qualities being 'compatible', i.e. inherited together, may be seen as one effect of inbreeding, now recognised by geneticists as having the effect of sorting heterogeneous genetic material into various gene complexes, open to selection. This being the case, ancestry further back than the parents was undoubtedly a significant factor in Bakewell's achievement. Evidently, however, Marshall thought it of insufficient consideration to be mentioned, being a natural extension of the idea of breeding in-and-in. Moreover, he would have recognised that for a restless originator like Bakewell, the priority was to look always ahead, to bring about progressive change rather than preserve one particular stage of selection. This required him to be constantly on the watch for 'accidental varieties' (i.e. new combinations of genes) that he could select, to make the improvements he had in mind. In the case of Dishley sheep, preservation of the race became a priority only after the master's death.

Among the more radical features that Bakewell claimed to be inherited, his 'criterion of excellence',[133] was the ability of his sheep to thrive on a comparatively small amount of food. Young quotes him in the confident claim 'that the same disposition to be fat will . . . prevail, let the food be what it may.'[134] Marshall was cautious about accepting such a claim without proof, although he recognised that a lower food intake by faster growing sheep 'seems probable'.[135] Banks was waiting for hard evidence before he would give the claim credence, although he accepted it as 'a matter of material importance', as he wrote to Young in 1791.[136] Another of Young's correspondents, using the pseudonym 'Agricola', stated it to be 'an absurdity not worth taking notice of'.[136] The first independent evidence made public was in a trial carried out by James Crowe of Norwich in 1792–3, who had come to selective breeding of sheep through earlier success with greyhounds. Crowe appeared to confirm the greater fattening capacity of the improved Dishley sheep on a lower intake of food. He compared 10 individuals of the Dishley breed with 10 of the improved Southdown and nine of the unimproved Norfork breed, all prime examples of their kind. Each group was put on 1 acre of wheat stubble and fed turnips, topped and tailed. Weight gains were recorded for 18 weeks and 2 days, during the winter of 1792–3. The results

---

[133] Redhead and Laing 1793, p. 23    [134] Young 1786a, p. 479
[135] Marshall 1790, i, pp. 397–8
[136] Carter 1979, p. 226: letter from Banks to Young, 9 December 1791; 'Agricola' 1793, pp. 536–40

showed the improved Dishley and Southdown breeds to be equally efficient in converting fodder into flesh, much more so than the traditional Norfolk sheep.[137] Improvements to Dishley and Southdown breeds had been carried out independently, the latter by John Ellman (1753–1832), who acknowledged his debt to Bakewell's example.[138]

Other experiments followed, and the accumulating evidence was enough to convince William Pitt that 'the Dishley breed will live where many other breeds would starve',[139] a remarkable achievement of selective breeding if true. The hard-headed Marshall remained cautious, believing that only the breed's 'cooler advocates' accepted it, although there was 'some show of reason on their side'.[140] Sinclair quoted what he claimed to be Bakewell's own down-to-earth words on the matter of livestock: that they were 'machines for converting herbage, and other food for animals, into money'.[141] If they not only grew faster but also consumed less food in so doing, what efficient machines they must be!

Successful breeders knew of bodily characteristics (sometimes referred to as 'marks' or 'nicks') believed to be associated with 'the essentials' they wanted to enhance by selective breeding. Young was quick to stress that such nicks had little or no connection with 'fancy points', such as colour or horn length, that distinguished traditional local breeds, 'objects of the smallest importance' to which traditional breeders have been 'so attentive in the past'.[142] Culley was more cautious claiming that characters called 'non essentials, the legs, ears, horns, tail etc., . . . are not to be quite disregarded, for although they are not properly essential, yet they are often strong marks or indications of good or bad.'[143] Nicks proved to be one of the most contentious of issues,[144] causing Young to write ruefully 'in this business of breeding, *assertions* are endless and you look in vain for experiment.'[145]

Renowned for the sensitivity of his hands in picking out sheep for breeding purposes,[146] Bakewell enjoyed the reputation of having brought 'nicking' to a fine art. By drawing upon a wealth of knowledge of associations between nicks and economic traits, he enhanced his capacity to reduce the proportion of non-edible tissues. However, just to increase the proportion of flesh to bone was not sufficiently ambitious for him. From the very beginning his aim had been to produce animals that would make the greatest weight in parts of the carcass which were most valuable. The principal, he told Young, 'is to gain a beast, whether sheep or cow, that will weigh most in the most valuable joints'.[86] These were the 'roasting pieces' located in 'the backward

---

[137] Crowe 1793; Young 1801    [138] Ellman 1834, p. 569    [139] Pitt 1809, p. 251
[140] Marshall 1796a, p. 36    [141] Sinclair 1832 (1817), p. 83
[142] Young 1811, p. 11    [143] Culley 1786, p. 186
[144] Marshall 1790, i, p. 298; Carter 1979, pp. 225–6: letter from Banks to Young, 9 December 1791; Young 1811, p. 11; nicking in horses was discussed by Wall 1758, whose attitude to breeding is reviewed by Russell 1986, pp. 106–107
[145] Young 1791a, pp. 290–91    [146] Young 1786a, p. 44

upper quarters' rather than the 'lower boiling pieces'.[147] With this in mind, he never relaxed from selecting breeding stock with traits which caused the carcass to be, as he put it, 'well made' and to approach ever closer to the ideal he had in mind. Young was quick to point out the originality in Bakewell's approach.[148] 'This doctrine is new, and of very great importance to graziers', he wrote. To summarise what Bakewell achieved we may quote a Frenchman, Léonce de Lavergne:[149] 'At the end of a certain number of generations, following always the same principle, the parts selected in all the reproducers, both male and female, become permanent and thus the breed is established'.

Bakewell's capacity for originality was happily acknowledged by all who viewed him as a hero. But there were others, more conservative in outlook, who were deeply disturbed by the radical intensity of his approach to selective breeding, as a threat to the natural order: 'Changing things to extremes as some breeders have attempted, is setting themselves in opposition to their creator by endeavouring to destroy his works', wrote Richard Parkinson, steward to Sir Joseph Banks in Lincolnshire.[150]

Parkinson was not alone in predicting disaster as a result of what Bakewell was doing and, for a time, the pessimism of these critics seemed to be justified. Shortening of the legs, which Bakewell brought about in the first stages of selection, raised problems with the suckling of lambs, and even with walking. At this juncture, Bakewell changed the direction of selection. As Marshall remarks, 'the legs of the improved breed have been considerably lengthened since their first stage of improvement, and with good effect: they now are better nurses, and better able to travel to market.'[151] Rumour had it that the correction was achieved by crossing to the Wiltshire breed,[152] which Marshall accepted as 'not improbable', though he made no proposal that Wiltshire sheep had anything to do with the *origin* of the Dishley breed.[153]

An unforeseen consequence of Bakewell's selection of a quick fattening breed was a reduction in internal hard fat ('tallow loaf'), a valuable material for making candles and some kinds of soap. This made the early slaughtered Dishley sheep unpopular with butchers. The new breed had only half the usual weight of tallow.[154] As Marshall wrote, 'Tallow is a kind of boon which, if not forthcoming, incurs a disappointment the butcher cannot brook.'[155] Also to be considered was the quality of Dishley wool. Several writers pointed out that, within a few years of Bakewell's death, the wool of his breed had changed its nature. Too weak to comb, in contrast to traditional Leicestershire wool, it was designated as 'feathery'.[156] Some degree of

---

[147] Young 1771a, i, pp. 110–11; Young 1786a, pp. 468–9; Redhead and Laing 793, p. 20

[148] Young 1786a, p. 469    [149] de Lavergne 1855, pp. 19–20; see also de Lavergne 1854

[150] Parkinson 1810, p. 267    [151] Marshall 1790, i, p. 409

[152] Wedge 1794, p. 32    [153] Marshall 1818, iv, p. 302

[154] *Annals of Agriculture* Vol. 19 (1793), p. 537    [155] Marshall 1790, i, p. 401

[156] Parkinson 1810, p. 261

change seemed inevitable to Bakewell and his followers, and not necessarily to be regretted.[157] They justified it in terms of a competition for sustenance between wool and flesh, in which flesh had to be given the priority.[158] It was not as though Leicestershire wool was particularly valuable, even at its best. Employed in the production of coarse worsteds, stockings, bays, blankets and carpets,[159] it suffered a depressed price in Bakewell's time. Farmers found difficulty in making a reasonable living from it. As Bakewell was prone to remark, 'a pound of mutton is worth more than a pound of wool.'[160] He even pointed to the potential advantage of producing sheep with no wool at all. It was a proposal from the 'great breeder' that met with a strong reaction from Banks, who feared it might become a destructive trend.[161] John Ellman, proud of the wool of his own Southdown breed, wrote in a letter to Lord Somerville, 'I once remember B-kw-ll say that he wishes to breed a sort of sheep that produces no wool at all, a sort of doctrine I could not understand.'[162]

The outrageousness of the idea of naked sheep made it widely reported, although Marshall recognised it as nonsense in relation to Bakewell's actual practice. From a comparison between improved and unimproved flocks, it was clear to Marshall that even the rumours of Dishley wool having deteriorated at all were suspect. 'The *fact* is,' he wrote,[163] 'this breed of sheep, when *seen* and *examined*, are not *greatly deficient* in wool' [Marshall's emphasis]. He noted that some commentators even claimed that Dishley sheep 'not only produce more mutton but more wool, by the acre, than any other breed of sheep'. It seems likely that wild talk of naked sheep was encouraged by 'the breed's own advocates', as Marshall referred to them, in order to imply an extra investment in alternative and more valuable qualities.

Support for Marshall's defence of Dishley wool appeared in the *Farmer's Magazine* in 1803, in the form of an indignant letter to the editor ('conductor') from an anonymous Northumberland breeder (not one of the Culley brothers), who claimed that 'the wool produced by the new Leicester makes more money *per head or acre* . . . than the finest wool grown by any breed of sheep on this island, because although the fine wools are sold much higher by the pound, the other more than makes up by weight.' In agreement, the editor commented that 'this claim may almost be received as an axiom'.[164] A more measured opinion came from Pitt, who surveyed Dishley farm in 1809 under post-Bakewell management. While he readily acknowledged the

[157] Stone 1794, p. 59     [158] Marshall 1790, i, pp. 402–4
[159] Marshall 1790, i, p. 412     [160] Parkinson 1810, p. 261
[161] Carter 1979, p. 196: letter from Banks to Young, 1791
[162] *CBA* 1800, p. 549: letter from Ellman to Somerville, 25 April 1799
[163] Marshall 1790, i, p. 403–4
[164] Letter to Conductor of *Farmer's Magazine* Vol. 4 (1803), pp. 164–8

superior productivity of Dishley wool per acre,[165] he was not prepared to agree that the sheep made more money per head for their wool, noting that individual pelts were light and the 'wool fine of its kind'.[166]

The truth of the matter seemed to be much as Pitt was reporting. The high productivity of Dishley sheep per acre could not be allowed to conceal the fact that, when considered fleece by fleece, the wool was in decline.[167] On evidence supplied by a farmer who had recorded wool weights for 20 years, Parkinson reported that Leicestershire fleeces had fallen in weight by one quarter between 1798 and 1808. Although the dates lie outside Bakewell's own lifetime, there seems little doubt that it was he who had set the trend in motion.

Concerning the health of Dishley sheep, Marshall was equivocal, although Sebright, writing two decades later, had no hesitation in designating them as 'tender'.[168] Thinning skins, brought about by an unfortunate consequence of Bakewell's preference for thin ears, opened them to torment from flies[169] and sensitivity to extremes of temperature, even to the extent that the pelt would crack in the sun. Critics also spoke of a premature loss of vigour in both sexes, due to their fatness, which made them susceptible to a variety of diseases. In some flocks Parkinson reported undersized, sick-looking sheep sometimes with bulging eyes ('frog eyed'), what he calls 'dunks',[170] although Marshall[171] comments that he could find no other reports of such extraordinarily degenerate animals. Young and Ellman add to the mystery when revealing that it was Bakewell's intention to accentuate the prominence of the eyes of his sheep, which Leicestershire breeders 'considered to be an indicator of good breeding', i.e. a useful nick.[172] Parkinson had no qualms about associating the production of such sheep with Bakewell's practice of breeding in-and-in, which he considered to be 'not only an improper, but an idle way of breeding useful animals'.[173] But he admitted, in the same report,[174] that he also saw very good examples of the New Leicester 'species', as he called it.

Breeding capacity was another matter for divided opinion. When Pitt appraised the breed in 1809, he wrote that 'New Leicester sheep were sure lamb getters', able to serve one hundred ewes in a season.[175] On the other hand, he also observed that they were less vigorous and lived less long than the Old Leicester type, being generally 'worn out' at four years old, while the old breed retained its vigour to more than twice that age.[176] It may be accepted that the young rams must have had a good breeding performance to

---

[165] Pitt 1809, p. 253    [166] Pitt 1809, pp. 250    [167] Parkinson 1808, p. 130
[168] Marshall 1790, i, p. 420; Sebright 1809, p. 14
[169] Crutchley 1794, p. 15; Parkinson 1808, p. 130
[170] Parkinson 1808, 1810, pp. 314–5, 1811, pp. 243
[171] Marshall 1818, iv, p. 439    [172] Young 1813, p. 404; Ellman 1834, p. 570
[173] Parkinson 1811, p. 259    [174] Parkinson 1811, p. 243    [175] Pitt 1809, p. 267
[176] Pitt 1809, p. 267

have been so much in demand and gained an international prominence. Concerning the performance of the ewes in 'lamb getting', opinion was more divided. Marshall stated that the 'modern breed' gave birth with less difficulty than other long-wooled sheep because the lambs' heads were 'finer' (smaller).[177] By contrast, several other experts claimed low fecundity, and also that the ewes made poor nurses, being deficient in milk.[178] Marshall admitted that New Leicester ewes had a shorter lactation time, although he detected little detriment to the lambs.[179] Both Lawrence and Pitt saw the milk deficiency as more serious. Critical comments about milk, wool, lambing and various physical features increased after Bakewell's death (1795) when the Dishley flocks came under the management of his relatives. But long before that, even Bakewell's closest admirers had to admit that his sheep presented a challenge to breed as a pure blood stock.

## Progeny testing

The Dishley breed of sheep, which was unique in appearance and fattening quality, had reached its first and major stage of evolution by the 1760s. As time passed, subflocks were established by Bakewell's friends and acquaintances. The first was probably that of the Culley brothers who introduced the breed directly from Dishley on to their farm at Fenton in Northumberland in 1766.[180] The spread of Dishley flocks around the country gave Bakewell the opportunity to hire out his rams, to be mated to high quality ewes outside his own flock. Later he would claim from the hirer such lambs as might be suitable for own flock. The practice enabled him to evaluate his rams against a much wider variety of ewes than he could possibly have kept on his own farm.[181] Letting out rams for a season and then reclaiming them was, according to Young, a practice 'first heard of in Lincolnshire', although Sinclair makes clear that it 'had never been carried out to any great extent till adopted by Bakewell.'[182] Marshall had no doubt of Bakewell's reliance upon the technique: 'the great improvement which has been made in the stock of this [Midland] district (horses and cattle as well as sheep) ... may be accounted for in this practice ... A superior male ... became, through this practice, a treasure to the whole district ... Such of his sons as prove of superior quality are let out in a similar way; consequently the *blood*, in a short time circulates through every part' [Marshall's emphasis].[183] Note that the sons have to prove their superiority: it does not arise automatically because of their blood.

---

[177] Marshall 1790, i, p. 436
[178] Redhead *et al.* 1792 (quoted by Wykes c.1995, unpublished manuscript); Lawrence 1809, p. 273; Pitt 1809, p. 273                [179] Marshall 1790, i, pp. 436–7
[180] Bailey and Culley 1805, p. 150        [181] Pitt 1809, p. 256*ff.*
[182] Young 1811, p. 5; Sinclair 1832 (1817), p. 935 (paraphrasing Marshall 1790, i, pp. 303–5 and 417)                [183] Marshall 1790, i, p. 305

Bakewell kept the rental charges for his rams as high as possible. The reason was not simply to recover his costs but to ensure that his rams were sufficiently valued to be mated only to the best ewes.[184] Culley explained the advantage of the policy in a letter to his brother Matthew, late in 1784: 'The very high prices Mr B has, and may, let at, staggers [sic] all his neighbours and what it will come to I know not, but the higher he lets the better for those that have of the same family, and the higher the sheep will stand in credit.'[185] In stressing the importance of the ewe, Bakewell is underlining the bisexual nature of heredity. Marshall realistically adds a warning at this point, that since the principal breeders were in fierce competition, it was important for the one in the lead not to rent his rams to a rival whose ewes were equal, or almost equal, to his own.[186]

Like other principal breeders of his region, Bakewell held a show of rams at his farm, keeping open house for several days, 'commencing by common consent on the eighth of June'.[187] The season for private shows ran from April to August. Before Bakewell's own show, members of the Dishley Society (see below) had a preview on 4–7 June.[188] The private exhibitions, at which all the important rentals were arranged, closed with a public show at Leicester on 10 October when the leftovers were dealt with, some to be sold, although most to be rented for a relative low price.[189] The rams privately rented were already in the hands of their hirers by October. Starting in the middle of September, they were transported in two-wheeled sprung carriages, hung in slings, up to four animals per carriage,[190] sometimes carried to flocks two or three hundred miles away, travelling 20 or 30 miles per day, to be returned at the beginning of December. When Marshall made his survey of the Midland counties in 1789, Bakewell was the only farmer who confined his business solely to breeding and letting; so transportation was a major concern and expense for him.[191]

The practice was to let out as many shearlings ('sharhogs'), i.e. male sheep of 15–18 months old, as possible. These were intensively prepared in order that they should be seen to fatten at an early age. After weaning in July or August of the previous year, they had been fed first on clover (the 'most forcing' food for sheep), then turnips, cabbages, colewort, hay and corn. This was the 'Art of Making Up',[192] which Bakewell took from traditional practice when preparing rams of all ages for sale: 'Fat, like charity, covers a multitude of faults.' wrote Marshall.[192] Any animals not rented, sold or kept for breeding

[184] Monk 1794, p. 29; Pitt 1809, pp. 259–61 and 263; Young 1811
[185] Wykes c.1995, unpublished manuscript: letter from George Culley to Matthew Culley, 18 December 1784                    [186] Marshall 1790, i, pp. 423–4
[187] Marshall 1790, i, pp. 302 and 406
[188] Pawson 1957, p. 153: letter from Bakewell to Culley, 9 July 1790
[189] Marshall 1790, i, pp. 302–3 and 334–5        [190] Marshall 1790, i, p. 431; Lawrence 1805, p 391
[191] Marshall 1790, i, p. 303        [192] Marshall 1790, i, p. 419

were fattened still further for the butcher, to be slaughtered at 27 months, i.e. after their second shearing.

The procedure for letting and exchanging rams became modified and regularised when a select group of breeders of Dishley sheep formed themselves into a kind of breed association known as the tup club or Dishley Society.[193] Its rules were defined in their final form at meetings held between 1787[194] and 1789 when Bakewell became the Society's first President with his nephew Robert Honeyborn as Treasurer.[195] The rules of the club on letting and exchanging rams were still in operation when Pitt wrote his County Report 20 years later. Through this exclusive organisation, which allowed members favourable access to the best Dishley sheep, Bakewell's breeding stock was given a high profile, to every participant's benefit. It was a rule of the Society that members had the first choice of one another's rams.[196] Bakewell's own stock represented the apex of a pyramid of quality. By restricting himself to a few rams 'of particular excellence' and leasing only these to fellow members,[197] Bakewell added extra precision to his breeding activity.[198] This early version of the progeny test, designed to assess a male's breeding value, was acknowledged by later breeders as a highly significant step in technique. The American breeder Eugene Davenport was later to remark that 'individuals of the same ancestry differ marvellously in their breeding powers . . . The line of descent runs only through the few that can produce *breeders of breeders*, not simply performers.'[199] His own experience told him that 'it is necessary to have a large herd in order that the worth of the male may be tested.' Bakewell had privileged access to all the flocks of the Dishley members on which to test his rams. From the point of view of progeny testing, the Dishley agreement was equivalent to his possessing an enormous herd. A most important rule of the Dishley Society, remarkably acute for its genetic implications, was that 'No member shall give his rams, at any season of the year, any other food than green vegetables, hay and straw.'[200] Bakewell and his friends clearly appreciated that extravagant feeding on some farms, but not on others, would confuse the genetic picture.[201] Their willingness to co-operate represented a radical departure from the traditional practice of 'making up'.

The practice of letting rams at 15–18 months old allowed the farmer to evaluate their breeding potential before they got too fat for profitable slaughter.

---

[193] Pawson 1957, p. 73
[194] Pawson 1957, p. 123: letter from Bakewell to Culley, 20 November 1787
[195] Pawson 1957, pp. 148–9: letter from Bakewell to Culley, 13 November 1789
[196] Pitt 1809, p. 255     [197] Marshall 1790, i, pp. 384–8
[198] Marshall 1790, i, pp. 302–5     [199] Davenport 1907, p. 575
[200] Youatt 1837, p. 317
[201] Pawson 1957, pp. 78–9 and 92; a simple plan of feeding was adopted and laid down for all members of the Dishley Society. M.M. Cooper, who contributed a chapter to Pawson's book, comments on this plan devised by Bakewell as 'a commonsense step on the part of a far-seeing man'

For business conducted within the Society, it was especially important that ewes as well as rams were of the highest possible quality. Written rules, agreed by members, ensured that this was so and that the rams were properly looked after when away. In case the rules might be flouted, the high prices charged were a strong guarantee of co-operation. Bakewell certainly felt no need for undercover payments to shepherds. As Marshall notes, 'Mr Bakewell, at present, has the name, at least, of being parsimonious, even to the shepherds of the flocks on which his rams are employed.'[202] The fact that his parsimony in no way inhibited his success says everything about the quality of his stock. He is quoted as saying 'If they will not speak for themselves, nothing that can be said for them will do it.'[203]

It has sometimes been suggested that Bakewell was the very first breeder to evaluate a 'ram-getter' in this manner, but this cannot have been the case. Neither is it possible to imagine that Bakewell was the first breeder to control matings. Any sensible breeder with knowledge of the parentage of his lambs would be bound to take such evidence into account when evaluating them for further use or disposal. But for Bakewell it was particularly critical to do so because he was inbreeding his sheep so strongly. The originality of Bakewell's approach was to use the Dishley Society to make the procedure specially efficient. Because of the trust built up between members around the country, he was able to test rams against a variety of ewes kept under diverse conditions. A good ram, such as his favourite 'Twopounder', was capable of serving 100 ewes in a season.[204] The highest number recorded was 140 when they were brought to the ram singly and mated only once,[205] which was the method Bakewell employed. Bakewell's success had a nationwide impact. The value to the farming community of access to good male stock for rental could not be overestimated. As one early nineteenth century breeder remarked:[206]

> Look to the good arising in our breed of horses from this system, which in fact public stallions exemplify. Look how the districts in England where the renting of bulls and rams by the season exists, have far outstripped the rest of the island in the excellence of the stock which they possess.

## Dishley crosses

Marshall described what Bakewell and his members were doing as 'uniting the superior branches of the same breed', which he considered had brought about as much improvement as breeding in-and-in'.[207] Others around the country were using Dishley rams to improve their own local breeds. The County Reports commissioned by the Board of Agriculture are full of

---

[202] Marshall 1790, i, p. 428    [203] Pawson 1957, p. 12    [204] Young 1791b, p. 588
[205] Marshall 1790, i, p. 432    [206] Boswell 1829, p. 38
[207] Marshall 1790, i, p. 301

examples. Even at extreme distances from Leicestershire, farmers would eventually gain access to the stock. Just a few years after Bakewell's death, the value of the new breed was placed in context by a respected outsider, in terms meeting with warm approval from Marshall.[208] The judgement came from T.A. Knight, who was quoted in the Report of the County of Hereford by Duncombe (1805) as follows:[209]

> If the sheep be to remain what I apprehend Nature has made it, a mountain animal. . .; and if wool still be considered an object of national importance . . ., I have no hesitation in asserting that . . . the Ryeland or the Southdown deserve preference to that of Mr Bakewell.

> If on the contrary, we are . . . to convert our richest and most productive tillage into pasture. . ., we cannot fail to pronounce Mr Bakewell's sheep the best on the island. Its merits in fattening easily and abundantly, and being ready for the market at an early age, and in considerable weight, cannot be denied.

Knight's statement reveals his concern that the demand for quick fattening sheep would deprive the nation of its valuable arable land. In the event, the distribution of Dishley stock was not as geographically confined as predicted because these sheep were principally used for crossing. Breeders everywhere, in upland as well as lowland areas, were experimenting with introducing Dishley blood into their flocks. Although it could be a 'dangerous instrument' to deal with, requiring the determined attention of a skilled flockmaster, it brought great rewards to those who used it cautiously.[210] Its favourable qualities were inherited particularly strongly in first-generation hybrids with other long-wooled and middle-wooled breeds. Most farmers, however, were content with less than half Dishley blood, since they did not require 'the very highest bred sheep', a point underlined in a published letter to Young in 1792.[211] The writer, 'Agricola', a farmer himself, pointed out that the small quantity of tallow in the most highly bred stock made them uneconomic to less well off farmers, the majority in the country, who had no ambition whatsoever to compete with Bakewell and his friends. A quarter or less of Dishley blood was enough for ordinary farmers. Top breeders of that period recognised two kinds of ram, 'ram-getters' and 'wedder-getters" (a 'wedder' is a wether, i.e. a castrated ram). The former, the purest breeding stock, would be 'everywhere cleaner and finer',[212] the latter (no more than half-bred) would be used to produce sheep suitable for farmers like 'Agricola'.

An area of particular activity in the use of Dishley rams was in the Scottish border region. Farmers there used them to cross with local breeds, of which

---

[208]  Marshall 1818, ii, pp. 350–51      [209]  Duncombe 1805, p. 126
[210]  Parkinson 1810, p. 283; Boswell 1829, pp. 31–2; Trow-Smith 1959, pp. 66–9 and 269–74; Walton 1983                                    [211]  'Agricola' 1793, pp. 536–41
[212]  Marshall 1790, i, pp. 422–3

there were many different varieties, suited to various degrees of exposure to rough weather. Even sheep for the high hills could benefit from just a little Dishley influence, being ideally 'in the third or fourth generation from Bakewell's full blood'.[213] Dishley breeding stock was obtained either from Bakewell himself or, more conveniently in terms of distance, from the Culley brothers in Northumberland. One of the pioneers in making the crosses was William Dawson (1734–1815) of Frogden near Kelso in Roxboroughshire, who was well known to the Culleys and at Dishley due to a trick he had once played on Bakewell. One year in the 1750s[214] he had hired himself to Bakewell *incognito*, 'in the guise of a Scottish ploughman', to learn his method of growing turnips, which he turned to considerable profit on his return to Scotland.[215]

In the first outcross of the Dishley stock, improvements were claimed both in size and in temperament.[216] Evidence about growth rates in relation to food consumption came from comparative trials like one carried out by the Gloucestershire farmer John Billingsley, who reported his results to the Bath and West of England Society. Six New Leicester sheep did comparatively poorly over one year compared with an equal number of $F_1$ hybrids of New Leicester $\times$ Cotswold (Gloucestershire), which seemed 'a hardy, useful sheep'.[217] Eventually some completely new breeds were created by taking the $F_1$ through to further generations, although it took great persistence with selective breeding. Most farmers were interested only in what was produced 'in the first instance'. A typical example of what happened when an $F_1$ was inbred is quoted by Young's son, Rev. Arthur Young:[218]

> Above 30 years ago, Lord Sheffield gave 50 guineas . . . to Mr Bakewell for the use of one of his rams; the ewes were of the Southdown breed, and the cross appeared *at first* to have answered well; but he soon found he had sheep of no character . . . and the wool was very indifferent.

In this case it proved impossible to pluck a new breed from the highly variable ('sheep of no character') $F_2$ or from subsequent generations.[219] In a minority of other cases the result might be more favourable, although to achieve success it would be essential to select both sexes most carefully.[220] As we have seen there were experts ready to declare that this was precisely the way Bakewell had created the New Leicester breed in the first place.

The creation of new breeds was a controversial issue in relation to divisions observable between races/varieties of sheep on a geographical basis. In this context, the terms race and variety seem to be interchangeable. Henry Home (Lord Kames), Scottish lawyer and author of *The Gentleman Farmer* (1776),

---

[213] Robson 1987 (quoting Wight 1778, ii, pp. 403–4)
[214] before 1759 (Hardly 1953)    [215] Anon. (W.T.T.) 1875    [216] Lawrence 1809, p. 27
[217] Billingsley 1795    [218] Rev. A. Young 1813, p. 477
[219] Marshall 1818, iv, p. 504    [220] Lawrence 1809, pp. 25–8

saw the more differentiated of them as natural divisions that had existed separately for all time. It was a view that encouraged the conclusion that they
would remain different.[221] The counter argument, which Kames accepted in
relation to Man, as being compatible with scripture, was that one perfect
species originally created could degenerate into various forms giving rise to
modern races.[222] The issue was reviewed by Marshall (1790) who was forced
to admit that

> whether, in the Animal Kingdom VARIETIES are altogether *accidental* or
> *artificial*, or whether they are not, or have been originally, *natural subdivi-
> sions* of SPECIES would, with respect to DOMESTICATED ANIMALS,
> be now difficult to determine

(his emphasis). In his own opinion, which he believed to be supported by
naturalists, varieties arose 'by climature, soil, accident and art, under the
guidance of reason or fashion, during a succession of centuries'.[223] At the
time he was writing, Marshall could observe for himself the progressive
creation of new varieties in the hands of Bakewell and others. If, as Bakewell
seemed to claim, his own breeds had not been derived from crosses between
existing breeds, the art he had exploited to create them was of the highest
order.

## Bakewell's view of heredity

Bakewell's strength as a breeder owed much to his pragmatic attitude towards
heredity. The protection afforded by a controlled environment allowed the
perpetuation of unique combinations of traits that would never have survived
in nature. The consequent richness of domestic variation was a matter of
frequent comment. Although the nature of heredity remained a mystery, selective breeding could be successful even without such knowledge. It was on the
basis of trial and error that Bakewell had been able to discover particular
combinations of traits that seemed to produce the best progeny, classes of
character that were inherited together (e.g. small bones with quick fattening).
It was by observing the inheritance of such combinations of morphology and
physiology that a number of useful traits were 'found, in some considerable
degree at least, to be hereditary'.[224] Belief in heredity also extended to unquan-
tifiable or less readily defined traits, including resistance to bad weather,
tolerance of poor food ('hard fare') and even propensities to certain disorders.
In no case could it be doubted that heredity arose from both sexes.[225]

When a breeder spoke of his intention to concentrate good features, he
believed that changes he would bring about would somehow be represented

---

[221] Home 1776, p. 309    [222] Glacken 1967, pp. 593–6
[223] Marshall 1790, i, p. 462; Darwin (1868a, pp 186–87) reconsidered this point, for discussion of which see Wood 1973, pp 236–38
[224] Marshall 1790, i, p. 419    [225] Marshall 1790, i, p. 481

in the blood. Experience had convinced him that the blood of carefully selected domestic animals could became 'fixed' for certain characters, which made it resistant to changes in rearing conditions, even after a move to a new territory or country. Concentration of the blood thereby came to be seen as the basis of hereditary stability and thus of prepotency (i.e. the special ability of certain individuals to pass on their traits). The horse expert John Lawrence of Bury St Edmunds explains how 'The term blood has been transplanted from the horse course to the grazing ground . . . it means natural and inherent quality in a species, exhibited in certain external and visible signs.'[226]

Bakewell's actions revealed his conviction that a powerful influence on the breed must be transmitted to the lambs even when they were conceived and born away from Dishley. The old adage that 'like produces like' had taken on a new meaning. The supposed race-conditioning influence of climate and nutrition had given way to 'inherent' properties. Equally, however, Bakewell was aware of the uncertainties of heredity. To be 'well bred' from good parentage was not enough. Breeding stock had to be carefully selected on the basis of form as well as blood, in every generation, supported by wise and consistent husbandry. Only then would it be possible to recognise and 'concentrate' desired features within the breed and avoid detrimental ones, with the necessary degree of efficiency.[227] In his appreciation not only of parentage but also of *form* in a consistent environment, Bakewell avoided falling into the extreme hereditarian position taken by some racehorse breeders of his day. Lawrence explains that 'No horse, whatever may be his shape, can be fit for the purpose of the turf, unless his blood be pure and unmixed.'[228] In relation to farm animals, Bakewell would have thought this to be insufficient. Although the ancestry (blood) of the animal was of serious importance to Bakewell, particularly as his breed became progressively refined, its individual qualities (form) had also to be considered and, even more importantly, the form of its already existing progeny.

It was on this broadest possible basis that every individual animal used for breeding, female as well as male, had to be selected most carefully.[229] By following this course of action, successful breeders like Bakewell were appreciating the significance of the whole flock as the ultimate target for improvement. Mayr refers to this attitude as 'population thinking' and believes that breeders were the first group to gain an understanding of the concept.[230] Bakewell's attitude on the matter was revealed in a conversation with Young, who quoted him as saying that:[231]

[226] Lawrence 1800, p. 522    [227] Marshall 1790, i, p. 464–5
[228] Lawrence 1800, p. 522; see also Russell 1986, pp. 105–10 for a discussion of this point, referring to the eighteenth century published works of R. Wall and W. Osmer
[229] Lawrence 1809, p. 25    [230] Mayr 1971, 1972    [231] Young 1791b, p. 570

The merit of a breed cannot be supposed to depend on a few individuals of singular beauty: it is the larger number that must stamp their character on the whole mass: if the breed, by means of that greater number, is not able to establish itself, most assuredly it cannot be established by a few specimens.

To 'establish itself' fully in Bakewell's terms meant that the breed had to prove successful commercially in competition with other breeds, particularly the one it replaced.[232] Furthermore, as Marshall pointed out, a breed had to be adapted to the farmer's *climature, soil* and *system of management*, 'otherwise, if we reason from analogy, the improver appears to be setting himself up against nature; a powerful opponent'.[233] And how was opposition to nature to be avoided? Marshall had no doubt what Bakewell and his friends would answer: by carefully selecting all breeding stock against that particular environmental background.

Dealing with problems due to inbreeding required careful judgement with respect to both parents.[234] But the ram and ewe still had to related, for otherwise the characteristics of the breed would be lost. Marshall underlined this point when he wrote that 'Excellency in any species or variety of livestock, cannot be attained with any degree of *certainty*, let the male be ever so excellent, unless the females employed likewise inherit a large proportion of the genuine blood.'[235] Two further decades of Dishley success led Pitt to echo the sentiment with the words: 'The better the ewes, the better lambs may be expected . . . by this means the Dishley stock has maintained its superiority.'[236] To the end of his life Bakewell kept a lead over other Leicestershire sheep breeders, even those in the Dishley Society who had access by right to his rams. Only his ewes were different. Some years later the implications of selective breeding of both sexes were elegantly summarised by the Rev. Berry in a prize essay, when he wrote that it is not to one sex in particular but to 'high blood, or, in other words, to animals long and successfully selected, and bred with a view to particular qualifications', whether in the male or female parent, that the quality is to be ascribed. Such animals were said to 'impress their descendents'.[237]

The procedure by which Bakewell matched the parents to complement each other, 'in reference to each other's merits and defects', as Sebright put it, can be traced back to the best in racehorse breeding practice.[238] The idea of choosing an animal for breeding partly according to the observed characteristics of its proposed mating partner rested on the belief that what one parent lacked the other might provide. Later, when Bakewell had selected the kind

---

[232] Young 1791b; Young's *Tours* (1932), p. 300; Wood 1973, p. 238
[233] Marshall 1790, i, 464–5    [234] Young 1791a; Young's *Tours* (1932), p. 312
[235] Marshall 1790, i, p. 434    [236] Pitt 1809, p. 263
[237] Berry 1826, p. 30; see also a 'short account' of the Berry essay in *Prize Essays and Transactions of the Highland Society of Scotland* New Series, Vol. 1 (1829), p. 39
[238] Russell 1986, p. 107

of stock he wanted, he could employ his best rams to upgrade somewhat inferior ewes, just as the 'gentleman of the turf' might use a thoroughbred stallion. The important difference was that in Bakewell's case the male breeding stock had not been derived from noble animals imported from abroad but had been created by himself to a precise specification by selective breeding. The idea of matching male and female according to the valuable traits or points they possessed or lacked became known as 'nicking'.[239] It was sometimes observed that two comparatively inferior animals could produce superior progeny. Geneticists would come to explain the effect on the basis of epistasis (complementary genes). J.B.S. Haldane commented on this interesting aspect of breeding, noting that it is certainly in accordance with genetic theory provided that, if necessary, the progeny is inbred for a further generation.[240] This, we are led to believe, was precisely Bakewell's practice.

Practical experience was compatible with patterns of heredity in which few traits were independently inherited, most being transmitted in association with others. In cases in which a particular trait was expressed differently in each sex, there was the possibility that one form might cancel out (dominate) the other in the immediate progeny of a cross. But it was also sometimes observed that they would add to one another, or blend to produce an intermediate form. Patterns of heredity after the first generation represented a minefield of uncertainty, carrying the risk of great expense for the breeder because of wastage. As we have seen, only two possible courses of practical action were to be recommended: 'grading' or 'seizing upon the accidental varieties produced'.

When one parent was more highly bred than the other, its characteristics were more likely to dominate in the progeny,[241] a phenomenon that became known as prepotency. Usually the parent concerned would be the male, simply because it was much easier to assess the breeding value of a male.[242] The capacity of such an animal to influence its daughters was reported by Marshall in a reference to Bakewell's long-horned bull Shakespear [*sic*], 'that every cow and heifer of the Shakespeare blood is distinguishable at sight . . . an empression [*sic*] they have received with singular exactness'.[243] A causal connection was established between inbreeding and prepotency long before there was an explanation for it in terms of genetics.[244] Inbreeding was recognised as having a mysterious power to intensify and 'fix' transmitting ability to a degree far beyond what could be achieved merely by selecting and mating together similar but unrelated individiuals.[245] The practice of inbreeding continued after Bakewell's death well into the nineteenth century as 'the only known method of increasing prepotency' for the practical breeder.[246]

---

[239] Rice 1926, p. 142    [240] Haldane 1959, p. 140    [241] Berry 1826, pp. 29–30
[242] Lawrence 1809, p. 25    [243] Marshall 1790, i, pp. 326–7
[244] Winters 1954, p. 205    [245] Lush 1951, p. 501    [246] Rice *et al.* 1957, p. 324

Farmers recalled Bakewell's attitude to inbreeding by quoting his supposed dictum that 'Inbreeding produces prepotency and refinement.'[247]

Although prepotency was seen as a result of hereditary stability, and thus as a 'concentration of the blood', no breeder, Bakewell included, gave any hint of what might be its physiological basis. The phenomenon defied science. In a pamphlet published in 1812, the surgeon John Hunt, a supporter of the 'Dishley System', deplored the idea of blood being the actual vehicle of heredity as 'far exceeding the laws of nature'. He agreed with the Merino breeder, Dr Parry of Bath, that 'The word blood is nothing more than an abstract term expressive of certain external and visible forms which from experiment we infer to be separately connected with those excellencies which we most covet.' Similar sentiments were expressed by others at the time.[248] All that could be said for sure was that blood, in the hereditary sense, was divided between the parents in proportion. Thus a son mated to his mother would cause her to produce lambs with six parts of herself and only two of his father.[249] The same argument was used to calculate the proportion of high (noble) blood in grading crosses (Chapter 3). Horse breeders had pioneered the concept of proportionality in the cause of winning races. The first thoroughbred stud book was published in 1791, pre-dating the first published studbook for a farm animal by some 30 years.[250] An interest in ancestry in farm animals was a post-Bakewell phenomenon. It finds no echo in writings about Bakewell himself where the major concern is with parents, siblings and immediate progeny, compared under the same conditions. Culley's Scottish friend James Anderson, farmer and scholar, who did business with Bakewell,[251] wrote frankly of heredity as a mystery, both in 'origin and perpetuation'.[252] Marshall, who viewed breeding in terms of art as well as science, was also conscious of its mystery.[253] For practical purposes, the priority for a breeder in the Bakewell tradition was to possess breeding stock with a recognised capacity to transmit desirable traits as surely and certainly as possible, through either sex. Theoretical explanations could wait till later.

### Changes in breeding practice summarised

New ways of thinking about breeds and breeding emerged by the end of the eighteenth century, based on practical experience. These were the guidelines adopted by Bakewell and his friends, 'the new school of breeders' as Parkinson refers to them.[254] They are summarised in the following propositions:

---

[247] Lush 1951, p. 501; these actual words cannot be attributed to Bakewell directly
[248] e.g. Lawrence 1800, p. 44    [249] Sebright 1809    [250] Walton 1986
[251] Pawson 1957, p. 106: letter from Bakewell to Culley, 8 February 1787
[252] Anderson 1799, p. 87    [253] Marshall 1818, i, p. 43
[254] Parkinson 1810, title page

(1) The intrinsic nature of an animal (its breed or blood) is the most critical factor determining its form and qualities.

(2) Good husbandry practices, especially with respect to diet and—in some cases—housing, are essential for keeping animals healthy and maximising their intrinsic qualities.

(3) Transportation of valuable animals from one country to another can be worthwhile ('can more than repay the expense', as Bakewell would say), provided that the introduced breeding stock is carefully selected in every subsequent generation, and looked after well in conditions that suit it. Healthy animals in new surroundings are expected to breed more or less true to type without needing to be 'refreshed' with breeding stock from the 'original climate', as long as the breeding stock is carefully selected.

(4) Selective breeding is a powerful agent of change, even for creating new breeds.

(5) The more inbred a strain is, the more likely it will be to transmit its selected characters to its offspring.

(6) Both sexes contribute to heredity and either can be prepotent; characteristics can be transferred to the opposite sex.

(7) Progeny testing, carefully controlled, is the most efficient way to evaluate an individual's 'hereditary properties', traits it can pass to the following generation.

(8) Selective breeding can be applied to single traits or groups of traits.

(9) Visible traits may indicate hidden properties and propensities.

(10) The value of crossing as an adjunct to selection was still a matter of controversy at this time. However, the first generation from a cross was becoming accepted as particularly valuable for its hybrid vigour.

Breeders like Bakewell, who were willing to take risks, might experiment with more extensive and wider crosses. But the principal lesson they were passing on was that selection had to be accompanied by close inbreeding to be at all effective. The dangers associated with inbreeding without progeny testing were also becoming evident. It was realised that matings had to be absolutely regulated and the results carefully recorded in a written form. As pedigrees came thus to be recorded, increasing importance was attached to them, and breeders were sometimes tempted to consider ancestry sufficient in itself as a guide to being 'well bred'.[255] This was a trend that Bakewell himself deplored. His antipathy to the pedigree concept is revealed in one of his letters to Culley,[256] in which he remarks that he has seen an advertisement in a

---

[255] Sebright 1809, p. 7
[256] Pawson 1957, p. 139: letter from Bakewell to Culley, 8 May 1789

newspaper requiring a good 'Suffolk Punch' stallion but demanding that a certificate of pedigree be produced. Bakewell's scathing comment, which includes a dig at the Irish, was as follows: 'This advertisement was surely drawn up on the other side of the water or why prefer blood to form or action?"

Looking back on the Bakewell era, Young (1811) asks himself 'Did the immense changes that we all know to have been effected, originate with Bakewell?" He found the answer as follows: 'that there is not at present a Breed of Livestock in the Island that does not derive its improvement from the skill, knowledge and principles which we owe to him, and which would not in any probability have existed if Bakewell had not laid the foundation'. Young and others who wrote about Bakewell recognised that his impact on the course of events derived not only from his genuine ability and achievement but from a 'larger than life' personality and reputation, to be considered in the following chapter, together with an indication of how his ideas reached the European continent.

# 5

## *Bakewell becomes a celebrity*

He is one of the most original men to meet in the Kingdom

*Count François de la Rochefoucauld, 1784 (original in French)*

Revolutionary breeding programmes, carried out behind stable doors and high hedges, excited passionate curiosity, not only among fellow countrymen in Britain but also soon across the Channel. Experienced farmers, skilled breeders and distinguished landowners in their own countries travelled vast distances, from as far away as Russia and outlying territories of the Hapsburg monarchy, to defer to the wisdom of this most unusual farmer and, if possible, discover his secrets. At Dishley they saw a new type of sheep, renowned for its capacity to gain edible weight at a rapid rate and 'peculiarly pleasing to the eye'.[1] Amazement turned to incredulity as the visitors learned of Bakewell's claim to have designed and reshaped this fast maturing animal in his mind before bringing it into existence. It was as though an engineer had invented a new machine. Intense interest was shown in the techniques he had used to create it. Other selectively bred animals to be seen at Dishley—cattle, horses and pigs—attracted attention for the same reason. Arthur Young, who was the first to bring Bakewell's 'astonishing efforts'[2] to public notice and remained a loyal admirer all his life, was conscious that Bakewell's influence was growing with every year that passed:[3]

> That he was a very extraordinary man can admit of no doubt; for those surely may be reckoned such, who affect in their professions great changes in opinions, and in the practices of mankind.

---

[1] Marshall 1790 i, p. 389     [2] Young 1802     [3] Young 1811, pp. 3–4

English farm stock breeding at its most radical was close to becoming per-sonalised in a single individual. In the following pages we examine how it came about.

## The excitement of rising prices

An early sign to the outside world that something special was happening at Dishley was a substantial rise in the monetary value of Bakewell's breeding stock. As early as 1769 the Marquess of Rockingham commented on this in a letter to Young, recommending him to take the road to Dishley and check for himself. Other travellers followed, inspired by Young's account of his own first visit in March 1770.[4] Typical was the published comment of the Marquis de Guerchy, who braved the Channel crossing and many days on English roads in 1784: 'Near to London I saw a grazier, Mr Bakewell, hiring out (a ram) at 50 guineas (£52.50 sterling) for a single jump', he excitedly reported.[5] Such a high charge was as sensational in France as in England, being com-parable with that required for the very best racehorse stallions. There were plenty of respected opinions, coming from all directions, to back the French-man's report. The willingness of experienced farmers to meet the enormous prices demanded of them was sensational.

The reliable witness William Marshall reported that the cost of leasing one of Bakewell's rams for a season had risen from 16 shillings (less than £1) in 1760 to £10 and 10 shillings (£10.50 sterling) in 1780 and up to £315 by 1786.[6] Bakewell confirmed the elevated prices in private letters to Culley.[7] He also mentioned the substantial costs to himself that justified these prices. The prob-lem with an intensive selective breeding programme was that so many expen-sively fattened animals had to be sacrificed. Culley explained in a letter to his brother (November 1784) that Bakewell needed a regular £2000 per annum from hiring his sheep, simply to avoid going into debt.[8] On 12 September 1788, a national newspaper, *The Morning Post*, carried a report that 'Mr Bakewell had lost one of his remarkable rams. It was of such value that he had let it out for hire and received £400 for the season.' The following year Bakewell's prices were up again, according to Marshall:[9] £1260 for one lot of three rams and £2100 for a second lot of seven rams, £3360 altogether in one season.

By then Bakewell's ambitious breeding programme was becoming reflected in the prosperity of his friends, several of whom reported charges of £525 to

---

[4] Young 1771a, 1771b    [5] de Guerchy 1785
[6] Marshall 1790, i, pp. 417 and 426–7; see also *Annals of Agriculture* Vol. 8, p. 345 and Anon. (J.L.) 1800, pp. 205–6
[7] Pawson 1957, pp. 119 and 121: letters from Bakewell to Culley, 22 September 1787 and 10 November 1787
[8] Wykes, unpublished manuscript, on evidence contained in letters from George Culley to his brother Matthew in November 1784, now lodged at the Northumberland Record Office (NRO ZCU9)    [9] Marshall 1790, i, p. 427

£1050 per season. Talk of high prices fascinated Young. During a two night stay at Dishley in August 1791, he learned that his host had let three rams that year for £1050 each,[10] a sure sign of exceptional quality. Nevertheless Bakewell was at pains to act discreetly, trying if possible to keep the charge agreed with any particular hirer secret from the others. He adopted this policy for the very practical reason he explained to Culley in a letter in 1787:[11]

> More money has been bid for next [season] than I have ever known in any former season but I believe talking of great prices rather harms the cause. For, if some people cannot have the best, they will not have another . . . Therefore I think the less said of it, the better. This I said in a letter to A.Y. Esqre [Arthur Young] who has been kind enough to comply with my request.

The open trumpeting of each and every price rise was becoming something of a liability. As Marshall put it, 'So long as the fire is fanned and the cauldron is kept boiling, so long the advocates of the breed must expect to be in hot water.'[12]

## Visitors to Dishley

When Young made his first visit to Dishley in 1770, Bakewell was already a leader among his peers, with a reputation based on far more than prices. In Young's account of this experience, reported at enthusiastic length in the book *The Farmer's Tour through the East of England*,[13] he wrote:

> Mr Bakewell of Dishley, one of the most considerable farmers in the country, has in so many instances improved on the husbandry of his neighbours, that he merits particular notice. . .
>
> Nowhere have I seen works that did their author greater honour . . . Let me exhort the farmers of this kingdom in general, to take *Mr Bakewell* as a pattern in many points of great importance.

These words introduced Bakewell to a much wider public, as a consequence of which Dishley farm began to gain its international reputation. For the earnest young men who came to stay and work with him, some of whom—like two from Russia—had struggled to learn English for that purpose, the experience was unique. Knowledge of breeding was taught at Dishley in relation to anatomical variation, just as might be the practice in a progressive agricultural school. Whole carcasses of different breeds, preserved in salt, were hung side by side for comparison in the Dishley slaughter house,[14] and Bakewell converted his hall into a museum in which animal skeletons and brine-pickled joints were used to demonstrate the effects of selective breeding

---

[10] Young 1791b, Young's *Tours* (1932), p. 311
[11] Pawson 1957, p. 108: letter from Bakewell to Culley, 8 February 1787
[12] Marshall 1790, i, p. 388     [13] Young 1771, i, p. 134 (his emphasis)
[14] Holt 1793; Monk 1794, p. 28

and high nutrition.[15] Among the specimens was a bone from a neck of Norfolk mutton twice the size of one of Bakewell's own breed.[16]

His hospitality to complete strangers, eager for knowledge, was renowned.[17] John Monk, author of the first County Report on Leicestershire, who visited Dishley in 1793, wrote:[18]

> On viewing the hospitable mansion of Mr Bakewell, I was highly gratified . . . Everything at Dishley was conducted with the greatest order and regularity, and, I may add, with every politeness and attention a stranger can wish for.

This was a time when it was not uncommon for a prosperous tenant farmer to spend £10 or £12 at one entertainment.[18] Even in such company, Bakewell was considered exceptionally lavish, as John Lawrence (writing as 'Benda') reported: [20]

> It is certain that Mr Bakewell was not enriched, notwithstanding his unremitting exertions, the admirable economy of his farm, and the vast sums he obtained for his cattle. But this is to attributed to the generous style of his hospitality which he constantly maintained at Dishley where every inquisitive stranger was received and entertained with the most frank and liberal attention.

Bakewell's association with the aristocracy, both British and foreign, was a matter of frequent comment. Tradition has it that the wealthy aristocrat Thomas William Coke of Holkham in Norfolk, later created Earl of Leicester, rated Dishley Grange as 'the best inn on the road'. [21] Bakewell taught Coke to judge the quality of a ram or bull, as he did himself, with his hands. In answer to Coke's questions, he made the much quoted reply 'Mr Coke give me your hand and I will guide it.' [22]

With so much coming and going, Bakewell was forced to ration the time he spent with visitors. In his later years, he was said never to deviate from his personal schedule, always dining on his own, however distinguished the company, at a small round table in a corner, near the window,[23] on a favourite chair constructed from a willow tree grown on the farm.[24] He lived by his stated adage 'to rise with the lark and to bed with the lamb',[25] knocking out his last pipe at 10.30 p.m. 'let who would be there'.[26] Quite frequently visitors must have arrived when he was not at home, for he would spend days at a time, or even weeks, away from the farm. Such a lifestyle was bound to place

---

[15] Young 1786a, p. 478, Throsby 1791; Anon. 1795, 'Benda' 1800, p. 208
[16] Young 1786a, p. 477      [17] Anon. (J.L.) 1800, pp. 202–3 and 207
[18] Monk 1794, p. 27      [19] Mantoux 1961, p. 181
[20] Anon. (J.L.) 1800, pp. 202–3      [21] Dixon 1868, p. 341
[22] Stirling 1908 (quoted by Pawson 1957, p. 44)      [23] Dixon 1868, p. 341
[24] Housman 1894; Pawson 1957, p. 27. Bakewell's chair is now in the possession of the Royal Agricultural Society of England.
[25] Dixon 1868, p. 342 (quoting Bakewell's unpublished 'memoirs')
[26] Ernle 1888 (quoted by Wilson 1909)

a burden on other members of the household. At any time they had to be prepared for unexpected visitors, quite often foreigners, descending on Dishley in search of information and advice, or simply out of curiosity.

At this point it is necessary to consider someone of special significance at Dishley farm, the individual who devotedly kept house for Bakewell over many years, and the one who had to organise the comfortable accommodation and the lavish hospitality. This was his sister Hannah. The welcoming atmosphere at Dishley, which provided the perfect framework for Bakewell's business, could not have been created unaided by a man busy on his farm or engaged with his records and accounts, and so often away on visits to neighbours or distant journeys. He was profoundly dependent upon the loyalty and devoted support of his unmarried sister. Everything in the house was her responsibility, including no doubt the exhibits in the museum that needed to be expertly prepared and maintained in a wholesome state.

One morning in February 1785 we observe Hannah coping with the unexpected arrival of a party of Frenchman led by the Count François de la Rochefoucauld (1765–1848), heir to the Dukedom of Liancourt, while Bakewell was away for the day in Loughborough.[27] His noble visitors, accompanied by servants, had travelled from Bury St Edmunds in Suffolk, persisting along muddy, inhospitable winter roads in their carriages (cabriolets). Recalling the difficult journey, the young Count wrote later, 'The time one spends on the road in weather so awful is time entirely lost, for although it was thawing, the landscape was still snow covered.'[28] Nothing, however, was to prevent him and his party from seeing Bakewell's achievement. He recorded their appreciation of the warmth of Hannah's welcome:

> We found his sister who was very kind and offered us lunch, dinner etc, and when we declined sent for an intelligent servant to show us Mr Bakewell's farm . . . His sister's realm was indoors; the running of the house was hers. It was done with scrupulous cleanliness and simplicity.

After admiring both farmstead and land, the party returned to Loughborough to find Bakewell, who accompanied them to dinner at their 'hotel'. The next morning at 9 o'clock, arriving for the second time at the farm door, Rochefoucauld was greeted again by Hannah who arranged breakfast for him and the other guests, joined by Bakewell when he returned from walking the farm.[29]

The effort, discomfort and inconvenience experienced by visitors travelling as far as Loughborough cannot be overemphasised. The sea crossing from France was rarely comfortable and could be hazardous. To reach London

---

[27] Scarfe 1995, pp. 28–35

[28] Scarfe 1995, p. 35; see also pp. 28–9, mentioning the particular problem with coal carts forcing other traffic off the road

[29] Scarfe 1995, p. 33; Hannah had taken her own breakfast 'some hours before'

from Paris took four to five days in the 1780s[30] and was extremely expensive by the standards of continental travel. Loughborough was two days further on. Travel in England was made even more difficult by the exploitative treatment meted out to foreigners at inns and hostelries. Though warned in advanced by those who had already braved the experience, the curious army of stalwart travellers continued to arrive, some to visit model farms like Dishley, others to see the many other technological advances that England had to offer. Their reward was a wealth of practical information to carry back home.[31]

Foreign visitors who came to sit at Hannah's table and perhaps stay for one or more comfortable nights, sometimes weeks or even months, did much to advertise Bakewell's achievements across the breadth of the Continent. Among the more influential of them was Mikhail Egorovich Livonov (1751–1800), a graduate of Moscow University in philosophy, law and medicine.[32] Arriving at Dishley with an introduction from Arthur Young during a period spent in England between 1779 and 1783, he absorbed enough of Bakewell's methods of cropping and animal breeding to report them in a Russian journal and, later, two books.[33] Livanov's final years were devoted to encouraging English agricultural methods in Russia, where in 1790, at the request of Prince Potemkin, then Prime Minister, he set up a School of Practical Agriculture at Bogoiavlensk, the first of its kind in that great country.[34] To his students he always openly acknowledged his debt to England.[35] In the preface to his book, *Livestock Breeding* (1797), he declared that he was following the rules and practices of Bakewell: 'Mr Bakewell, an Englishman burning with love for his homeland, spared no expense or effort. . . [etc.]'.[36] Foreign visitors often praised Bakewell's enterprise in patriotic terms. The man himself showed no reticence about claiming that his work was 'to the good of the nation' or even 'of great importance to the honour and interest of the British Empire'.[37] In his support, the Scottish farmer and geologist James Hutton (1726–97) wrote that 'Mr Bakewell has opened the eyes of his countrymen . . . and by the successful application of his genius he has benefited his country more than if he had added to this kingdom tributary provinces.'[38]

An indication of German interest was manifest in February 1787 when Bakewell received a letter from 'Count de Bruhl', Hans Moritz (or John

---

[30] Aldcroft 1983    [31] Spiekermann 1983    [32] Cross 1980, p. 62
[33] Cross 1980, pp. 67–8    [34] Cross 1980, pp. 71–2    [35] Cross 1980, p. 85
[36] Cross 1980, p. 85    [37] Bakewell's petition, printed by Pawson 1957, pp. 181–2
[38] Hutton c.1796, unpublished manuscript. Half a century earlier a similar sentiment had been expressed by Linnaeus regarding the value of naturalising exotic species in Sweden (as Alströmer had done in the case of Merino sheep). Would not any such action by himself, he asked, 'serve the country more than if, with the sacrifice of many thousands of people, he had added a province to Sweden?' (Koerner 1999, p. 114)

Maurice) von Bruhl (1736–1809) of Martinskirchen, who lived permanently in Britain as an Envoy Extraordinary (ambassador) from the Elector of Saxony to the Court of Great Britain. On behalf of the Elector, the Count paid particular attention to political economy. He had made a tour through the remoter parts of England early in 1783 for the purpose of investigating the state of trade and agriculture.[39] His letter to Bakewell four years later[40] requesting to know 'the best methods of breeding, rearing and feeding of the best kind of stock', came at the height of development of the Saxon Merino and was undoubtedly related to it. Merino breeding in Saxony and certain other central European states and provinces was to owe much to the combination of techniques first popularised by Bakewell and reported by Culley, especially the combination of inbreeding with individual trait selection and progeny testing (Chapter 7). Inbreeding at the intensity recommended by Bakewell would be undertaken against the advice of most leading Saxon, Prussian and Hapsburg experts.

The French in particular were prominent among his visitors. The year after the Marquis de Guerchy arrived in 1784, determined to be ahead in introducing the best of English agriculture into France,[41] came Rochefoucauld with his brother Alexander, accompanied by their Polish tutor, Count Lazowski. A year later came Pierre Flandrin, Professor of Anatomy at the Royal Veterinary School at Alfort, 'who was to add his own contribution to rural economy in his direct studies of fine-wool sheep in Spain and long-wool sheep in England'.[42] According to Léonce de Lavergne, the professor returned home full of praise for Bakewell's achievements.[43] Dishley also attracted representatives of at least half a dozen other nations. One of the more mysterious was a 'native of Iceland', who presented himself unexpectedly at Bakewell's door in June 1787, stating he was acting as an agent instructed to obtain information for the King of Denmark.[44]

Probably the longest staying foreigner at Dishley was a young Russian who had been plucked out of obscurity while still a peasant boy by Prince Potemkin. Ivan Safon'kov (Saponkavich), or John Saphonkove as he was known in England, lodged with Bakewell for special training for an extended period during 1784–5, supported with money sent by the Prince. Ivan had started his English agricultural education with Arthur Young at his farm at Bradfield in Suffolk.[45] By 1785 Bakewell was ready to trust him with recording the results of a feeding experiment, to compare six rams of different breeds tethered side by side in a sheep house. Young observed the young Russian at his task on a visit to Dishley in the spring of 1785.[46] He found him chalking the results on a slate, later to be transferred to Bakewell's permanent

---

[39] Clerke 1886    [40] Pawson 1957, pp. 30 and 107    [41] Bourde 1953, p. 145
[42] Carter 1964, p. 46    [43] de Lavergne 1855
[44] Pawson 1957 p. 114: letter from Bakewell to Culley, 30 June 1787; see also Pawson 1957, p. 72    [45] Cross 1980, pp. 73–5; Betham-Edwards 1898, p 102
[46] Young 1786a, pp. 482–3

records. If this youth had not talked openly to Young, posterity would be the poorer in information on Bakewell's methods. The same individual had earlier caught the attention of Rochefoucauld,[47] who reported that he was one of two young Russians then under instruction from Bakewell, both sent by the Emperor of Russia 'to spend some years studying English agriculture'. They were in addition to seven or eight Russian boys or youths examined by Young[48] in 1784 at the end of their stay in England. After his return to Russia, Ivan was sent to Potemkin's service, together with two other youths who had been educated on different English farms. Five years later, he had become an assistant to none other than Professor Livanov at the School in Bogoiavlensk where his knowledge of the Bakewell system could be put to best use.[49]

Dishley farm was one of several attracting agriculturally motivated foreigners eager to pick up the latest techniques. Such 'guests' from abroad became a regular feature of British agricultural life in the late eighteenth century. Listed among Young's visitors in the 1770s and 1780s to his home at Bradfield near Bury St Edmunds in Suffolk were Roland de la Platière, Inspector General of Manufacturers at Lyon, M. Baert, author of a multi-volumed guide called *Tableau de la Grande Bretagne* (1800), and the Duc de Liancourt,[50] all from France. From Poland and Lithuania came Count Kalaskowski and Prince Massalski[51] and from Moravia, Count Leopold Berchtold (1759–1809).[52] Numerous Russians arrived, including Count F.V. Rostoptschin (1763–1826), who owned an estate at Voronovo near Moscow.[53] As noted, the Duc de Liancourt's two sons made the extra journey from Young's home to Dishley. It is known that Young was in the habit of recommending all such visitors to see Bakewell if possible.[54] Young's only deviation from this pattern of behaviour occurred during a critical period when Bakewell was financially embarrassed, the late 1770s and early 1780s (see later), when it was clearly tactful to keep himself and others at a distance from Dishley.[55]

Count Berchtold from Moravia, a province of the Hapsburg monarchy, proved a particularly welcome guest in Young's household, being appreciated for his character as well as his exceptional intelligence.[56] He arrived in April 1785 armed with an almost endless list of questions and had every reason to extend his enquiries to Bakewell (as will be discussed in Chapter 8). Nothing was going to deviate him from the valuable information he was determined to acquire, for he saw it as his patriotic duty.[57]

[47] Scarfe 1995, p. 33
[48] Betham Edwards 1898, pp. 102 and 125; Gazley 1973, p. 151
[49] Cross 1980, p. 75      [50] Gazley 1973, p. 155
[51] Gazley 1973, p. 194; Betham-Edwards 1898, pp. 52, 144
[52] Gazley 1973, p. 219; Betham-Edwards 1898, pp. 167–70
[53] Cross 1989, p. 89; Gazley 1973, p. 469; Betham-Edwards 1898, pp. 387, 401: Count Rostoptschin dedicated a building on his estate to Young
[54] Gazley 1973; Coke did the same (Riches 1967)      [55] Young 1786a, p. 453
[56] Gazley 1973, p. 219      [57] Betham-Edwards 1898, pp. 167–70

## Bakewell takes to the road

It would not have been unusual for an unexpected visitor to Dishley to find the farm in charge of Bakewell's sister Hannah, supported by loyal staff, with her brother not due back for days or even weeks. His risky business demanded that he should adopt an active role in seeking contacts, finding new followers and potential customers. He spent up to three months each year on the road, progressing from farm to stately residence and from horse fair to cattle market, usually on horseback. Young says that his journeys created wide interest, resulting in invitations to stay with many of the great landowners, eager to seize upon his words:[58]

> The particular merit of the Dishley stock is a matter of very small consequence, compared with the just principles which Bakewell disseminated on the many journeys, which he was always making into the various districts of these islands. . .

It was a time for radical ideas backed by practical experience. Nobody could match Bakewell in this respect.

On a tour of Suffolk in 1784, Bakewell visited a friend of Arthur Young's, the Cambridge historian Professor John Symonds, at his fine new house near Bury St Edmunds designed by Robert Adam. Numbered among the guests there was Count Rochefoucauld, who was to visit Dishley the following winter, as already described. The Count recorded his first impression of Bakewell as 'one of the most original men to meet in the Kingdom'. From the young aristocrat's journal, since published, it can be seen that he was overwhelmed with the Englishman's utter confidence about what he could achieve by selective breeding. He had proposed a bet to the Count and to Young that he could produce oxen with fat tails. Rochefoucauld remarked in his journal 'All that is astonishing. I don't really understand it, but I believe it as I believe in Religion because I have been told that you have to believe it, that everyone believes it.'[59] Bakewell's achievements were such that little or nothing seemed beyond his powers.

His personal following was never greater than in the winter and spring of 1788 when he spent two extended periods in London, 'the first about a month, the latter two months'.[60] He arrived first at the end of February, putting up at a prominent coaching inn, the Black Bear, in Piccadilly. The creation of publicity for his selected breed of heavy black farm horses seems to have been his primary purpose. He had arranged for a magnificent stallion to be brought down to coincide with his arrival. With clean (i.e. unfeathered) legs and a thicker, shorter carcass than the traditional Dutch type from which

---

[58] Young 1811, p. 6
[59] Rochefoucauld 1784 (translated and edited by Scarfe 1988, pp. 149–50)
[60] Pawson 1957, p. 172: letter from Bakewell to Young, 10 March 1788; Pawson 1957, p. 126: letter from Bakewell to Culley, 10 July 1788

Bakewell had created it,[61] the new breed was much in demand. Two years earlier, Young had stated in print that Bakewell's stallions were 'by far the finest I have seen of that breed'.[62] So favourable was the impression created that the King (George III) asked to see both the farmer and his horse at St James' Palace. They met in the courtyard in company with the Prince of Wales (the future George IV) and 'other great personages'. Bakewell hurried to report this exciting event in a letter to Young, mentioning how they had talked about breeding for nearly an hour, which the King 'seemed most pleased with and listened to what I said with great attention'.[63] The King asked to see him again later in March. Meanwhile the stallion was put on public show at Tattersall's, the London headquarters of horse dealing and betting.[64] Undoubtedly these events increased Bakewell's fame in London that year, widening the circle of those who saught his company.

After returning home for a brief period, he was in the capital again during April or May to express his support for a Bill going through Parliament, aimed at curtailing the smuggling of wool abroad by registering all wool shorn. He applauded any Government moves to restrict the free exportation of English wool, being sure that free trade in wool would inevitably increase the price of woollen goods manufactured in England. Bakewell backed the Yorkshire cloth manufacturers in lobbying Members of Parliament as they entered the House of Commons. Banks represented the Lincolnshire wool interests in opposing the Bill. Young also opposed it, on behalf of the wool growers of Suffolk, delivering his evidence to the House of Commons on 22 April.[65] The bill was passed by the Commons on 15 May and by the Lords in June, to the dismay of the landowners and farmers who wanted to see their wool gain high prices in France and the Netherlands. Bakewell expressed his opinion in support of the Bill with modest confidence: 'whether right or wrong, I believe what I said has some influence', he wrote to Culley.[66]

Bakewell was in London again the following year on an action-packed visit that occupied most of May and which illustrates further the broad scope of his activities. Arriving on the 7th, accompanied by his nephew John Honeyborn and basing himself this time at the Swan Inn in Lad Lane, his first task was to find a hirer for another of Dishley's great black farm horses. Meanwhile he was taking full opportunity to inspect the various stock for sale at the 'flesh markets'. Farm equipment was another matter for his attention, in respect to which he and his nephew called on the Rev. Mr Cook at White Lion Yard, Oxford Street, to order a seed drill.[67] On the 13 May, he

---

[61] Ernlé 1936, pp. 183–4    [62] Young 1786a, p. 486
[63] Pawson 1957, p. 172: letter from Bakewell to Young, 10 March 1788
[64] Anon. (J.L.) 1800, p. 205
[65] Betham-Edwards 1898, pp. 163–7; Gazley 1973, pp. 198 and 215
[66] Pawson 1957, p. 127: letter from Bakewell to Young, 19 July 1788
[67] Pawson 1957, pp. 138–43: letters from Bakewell to Culley, 8 and 30 May 1789

visited the House of Commons where, as he wrote to Culley, he was hopeful of persuading the 'Great Folks' to ask the King to set aside some crown lands as experimental farms. As may be imagined, even Bakewell's persuasive powers were insufficient to make palatable to the government such a radical and potentially costly enterprise. By the 25th he and his nephew were ready to set out for home, a journey he extended to five days by calling on Lord Clarendon, Earl of Essex, famous for his Berkshire and Chinese pigs,[68] the Marquess of Buckingham and the Duke of Grafton, as he proceeded northward. Their stewards were potential customers. Buckingham's steward had bred from Dishley sheep the previous season. Steward and master alike were doubtless eager to hear his latest ideas and proposals. Such visits would have been prepared by letters, of which Bakewell must have written many, although few are preserved.

## Bakewell's cause and character

Bakewell's letters to Culley and Young make clear that he viewed his work with an almost missionary zeal, a passion shared by his friends and admirers. Convinced of the rightness of their actions, they remained indifferent to the fact that their 'traffic in rams' was 'elsewhere laughed at as a visionary romance'.[69] Consistently in his letters to his friend Culley, Bakewell referred to his work as 'the cause'.[70] It was the laudable pursuit of 'free enquiry and proper investigation'. He assured his friend, 'I wish to do all in my power to serve the cause in which I have engaged'.[71] Young joined him in his passion when he wrote of 'pushing the common cause to that wonderful perfection to which it has arrived at present'.[72]

The general run of tenant sheep farmers had the reputation of being dull, like the two 'with faces bloated with beer and brandy' who shared a coach with the scholarly German pastor C.P. Moritz, as he travelled uncomfortably from Northampton towards London in 1782. They 'slept so soundly . . . and when they awoke, they talked only of their trade—sheep!' wrote the scholar, recalling his slow and tedious journey back to the metropolis.[73] How different from this stereotype was the image of 'the sagacious Mr Bakewell', 'with strong and inquisitive mind', who 'delivered himself on every occasion neatly in few words, and always to the purpose' with a fund of jokes and anecdotes to make his company doubly memorable.[74] With a happy facility to transcend normal social barriers and being, so it was claimed, 'above the sophistry of

---

[68] Young's *Tours* (1932), pp. 53–4    [69] Young 1791b; Young's *Tours* (1932), p. 288

[70] Pawson 1957, pp. 102, 108, 112, 122, 123, 132 and 137: letters from Bakewell to Culley, 11 April 1786, 8 February 1787, 30 June 1787, 20 November 1787, 22 November 1788 and 29 Dec 1788    [71] Pawson 1957, p. 112: letter from Bakewell to Culley, 30 June 1787

[72] Young 1791b; Young's *Tours* (1932), p. 297

[73] Moritz 1782 (translated by Nettel 1965, p. 180)

[74] Monk 1794, p. 191; Anon. 1795; Anderson 1799, p. 151; Anon. (J.L.) 1800, p. 209

either religious or political superstition', Bakewell became known in the widest of circles. Among the aristocracy his reception was not always without humour to match his own. 'Are you related to the Mr Bakewell who invented sheep?', asked the Countess of Oxford at being introduced socially to another man of the same name.

The practical achievements for which Bakewell became known did not even begin to express the totality of his interests. Rochefoucauld discovered the truth of this on his visit to Dishley in 1785:[77]

> After we had been shown over the whole of his farm by his servant [on instruction from Hannah] we returned to Loughborough. Mr Bakewell had to be there at four in the evening to hear a lecture on experimental physics.

This was related by Rochefoucauld as an unremarkable matter of fact. Nothing about this particular farmer was going to surprise the young Frenchman. When he invited Bakewell to join his party for supper at his hotel, they enjoyed an informative evening:[77]

> His conversion was instructive in its own way all through supper. He showed us how well, beneath a heavy and rough exterior, he had been making observations, and studying how to bring into being his fine breed of animals with as much care as one would put into the study of mathematics or any of the sciences.

It is easy to imagine how Rochefoucauld would have been primed by Young not to be surprised by the unusual breadth of this farmer's interests, nor the intensity with which he pursued every topic. Young was well aware of Bakewell's receptiveness to new ideas,[78] a theme constantly stressed by others who knew him. John Lawrence (1753–1839), the horse expert, wrote that 'he would listen to a philosophical problem with that eager curiosity and ardent desire of information peculiar to original minds alone'.[79]

Another matter for comment was the verve with which Bakewell expressed his ideas. Young spoke for many when he called attention to Bakewell's 'spirit' as well as his 'good sense and intelligence'.[80] A determination to attain the highest level of achievement in whatever Bakewell set his mind to was a dominant feature of his character. Young was ready to rank Bakewell with the best scientists and technologists of his day, like Joseph Priestly or James Watt.[81] In support of Young's judgement, we may observe Bakewell's concentrated observation of the dissection of a dead fowl carried out by his physician Dr Thomas Kirkland at the latter's home in Ashby de la Zouche. Bakewell was

[75] Humphries 1885    [76] He was a distant relative (see Trentham-Edgar 1934)
[77] Scarfe 1995, p. 33; the idea that Bakewell bred his animals scientifically and with mathematical precision may have been gained from Young.    [78] see also Cook 1942, p. 8
[79] Anon. (J.L.) 1800, p. 200. The anonymous author is J. Lawrence (see Chapter 4, note 4 and Lawrence 1809).    [80] Young 1786a, pp. 453 and 498
[81] Young 1791b, Young's *Tours* (1932), p. 269

sufficiently impressed to describe the occasion in a letter to Culley.[82] At the age of 66, inactivated by a painful condition of his legs (probably ulceration) for which Dr Kirkland was treating him as a resident patient, he gave every sign of enthusiasm for expanding his knowledge of 'the anatomy of brute creatures'.

His association with men more educated than himself gave Bakewell obvious satisfaction and not infrequently produced a corresponding response from them. The Nottinghamshire surgeon John Hunt contrasted Bakewell's approach to breeding with the inadequate words that were so often used by philosophers to explain the phenomena associated with this activity:[83]

> The experiments made by Mr Bakewell were plain, self evident and satisfactory. Whilst all the evidence of philosophy on the opposite side of the question is unnatural, mysterious and unintelligible.

Hunt, like Kirkland, was a friend of Erasmus Darwin who, in his early writings, had propounded the traditional view that the male alone provides the active nucleus from which the fetus develops, while the female merely provides the food on which it grows. It was not until he wrote his *Zoonomia* (1794–6) that he finally admitted an almost equal female contribution to the make-up of her progeny.[84] His biographer, King-Hele (1963), states that Dr Darwin came to this view after pondering over the mule, although this may not represent the complete picture. We may speculate that his theoretical conversion may have owed more than a little to the rationality of experimental selective breeding as exemplified by Bakewell and his friends.

Bakewell's business contacts extended to landowners outside the country, including some in the West Indies and the American colonies. One landowner of particular note, who ordered 'instruments of husbandry' from him, was 'that great and good man General Washington', as Bakewell refers to him in a letter to Culley.[85] The transaction took place in the year following the Declaration of Independence, with British political opinion strongly divided. Bakewell's reputation as 'one of the warmest supporters and staunchest defenders of liberty'[86] is illustrated by his support for the colonists, although it must have made him some enemies.

### Bakewell's critics

A strong element in Bakewell's status as a celebrity stemmed from the controversies that surrounded him. According to Young there were 'few persons

---

[82] Pawson 1957, p. 159: letter from Bakewell to Culley, 22 October 1791
[83] Hunt 1812, p. 12; see also Hunt 1809
[84] King-Hele 1963, p. 67 (quoting Darwin 1794–6, i, p. 490–91)
[85] Pawson 1957, pp. 107: letter from Bakewell to Culley, 8 February 1787
[86] Anon. (J.L.) 1800, p. 209

more hotly opposed than this eminent breeder'.[87] Criticism stemmed primarily from fellow farmers. Bakewell's secretive independence and disregard for traditional practices, unless confirmed by experiment, brought him into conflict with a substantial body of long established and conventional breeders. Their suspicion of him encouraged accusations of sharp practice and dishonest dealings as his growing fame brought seemingly undeserved success. Becoming ever more confident in the rightness of his cause, he disdained to deny anything. The market for his Dishley breeding stock seemed to have no limit as its quick fattening quality gained the reputation it deserved.

Major opposition came from rivals in Lincolnshire and Sussex who questioned Bakewell's willingness to have his sheep judged in free and fair competition.[88] An exchange of published letters and challenges with the Lincolnshire breeder Charles Chaplin revealed something of the revolution that Bakewell's attitude to sheep breeding represented. On the one hand there was Bakewell stressing the basic inherited characteristics of an animal's carcass and, by comparison, showing less concern about preparing sheep for inspection with a view to slaughter. His confidence in the rightness of his case was encouraged by the expanding demand for his stock for rental. On the other hand there was Chaplin, whose priority was to produce sheep in perfect condition for slaughter by applying the arts of shearing, feeding and dressing. He was happy for the butchers of Leeds, Wakefield and Smithfield (in London) to judge between his and Bakewell's sheep, in fair and open competition in accordance with traditionally agreed practice. But whereas Chaplin was confident in winning such a trial, Bakewell could not risk it. He knew full well that the 'sages of Smithfield'[89] did not favour his Dishley improvements, if only because of the loss of their profit from tallow (Chapter 4). We can understand Bakewell's reticence but also appreciate how frustrating it must have been for the conservative Chaplin to deal with a rival so dismissive of long established standards. The full acrimonious correspondence was printed by Young.[90]

Leading butchers were some of Bakewell's most vociferous critics. Supported by 'men of culinary taste', they complained of the Dishley breed's excessive fatness, which made the meat 'watery and inspid'. Bakewell had a ready answer. He confidently stated, 'I do not breed mutton for gentlemen but for the public.'[91] He was quite certain that fat mutton would continue to be in demand from growing numbers of the poorer sections of society. Young

---

[87] Young 1811, pp. 3

[88] Young 1788, pp. 566–8: open letter from Chaplin to 'The Gentleman who attended Partney Fair', dated 19th September 1788

[89] Anon. (J.L.) 1800, p. 202; 'The Sages of Smithfield before whom the fattest animals of all counties pass in hebdomadal review'

[90] Young 1788; see also Stanley 1995, pp. 34–40 and Pawson 1957, pp. 188–90

[91] Pitt 1809, p. 268

wholeheartedly agreed, having no doubt that the 'great mass of mutton eaters, which are in the towns, will for ever choose the fattest meat, and give the greatest price for it'.[92] The high value placed on mutton fat was not only for its food value but also for a variety of non-culinary purposes. Farmers would have been conscious of its qualities as a lubricant for greasing the axils of carts and the moving parts of machinery. In a domestic situation it provided the basis for the production of ointments, candles and even primitive makeup. In the context of industry, it is interesting to note its use 'by engineers in the working of part of the steam engine', to control the escape of steam.[93] Although butchers did not favour the highly bred Dishley sheep, they still sold vast quantities of the progeny of Dishley rams because of public demand.

The quality of Dishley mutton continued to be debated even after Bakewell's death, as in the columns of the *Farmers' Magazine* of 1803. A citizen of Glasgow, styling himself 'Epicurus', wrote disparagingly that Dishley mutton was 'only fit to glide down the throat of a Newcastle coal-heaver'.[94] Replying to this, an indignant 'breeder of coal-heaver's mutton' stressed the economic point of view: 'is he [Bakewell] not ... deserving of praise who produced two pounds of mutton when only one was formerly produced?', he asks.[95] This sentence, translated into the languages of several European states, became much quoted as an amazing example of selective breeding.

Unstoppable progress in supplying a growing market with cheap mutton was shared by that lively and dedicated group of fellow sheep breeders who in 1783 formed themselves into the Dishley Society. To those outside the Society, its rules and regulations to promote the breed seemed unnecessarily restrictive and illiberal, and bound to accelerate the rise in prices already being experienced.[96] Considerable opposition having been generated against it, a rival group, the 'Associated Breeders', was formed,[97] basing its practice on free and friendly interchange of stock. It was a well meaning plan which proved, in practice, unsustainable until finally, following Bakewell's example, they too became prepared to engage in selective inbreeding, 'to put those sheep together which are most perfect in shape, without regard to affinity in blood'.[98] Rivalry was nothing unusual for breeders, and Bakewell took this kind of opposition in his stride.

Not all critics of Bakewell's stock were critical of the man himself. Richard Parkinson, Sir Joseph Bank's steward, who had deep misgivings about Dishley sheep—particularly as they evolved after Bakewell's death, expressed admiration and sympathy for their creator. He could not speak more highly

---

[92] Young 1786a, p. 279; see also Rees 1819, Bakewell is quoted as saying that 'A small quantity of this fat meat, cooked over a large dish of potatoes, is a good dinner for a poor man's family; and this is what I proposed in the selection of the breed.'       [93] Johnstone 1846
[94] Epicurus 1803       [95] *Farmer's Magazine* Vol. 4 (1803), pp. 164–8
[96] Young 1811, p. 13       [97] Pitt 1809, pp. 225–56       [98] Pitt 1809, p. 267

of 'Mr Bakewell, open and candid, capable of giving, and ready to afford, every information'. His good opinion was much influenced by his first visit to Dishley when he was given a guided tour by Mr Walton, Bakewell's assistant and John Breedon, his shepherd (Breedon was Bakewell's senior herdsman for 32 years). Parkinson was sympathetic to Bakewell in his unwillingness to reveal prematurely the results of his most recent experiments before he had confirmed them. 'It was a saying of his', he reported, 'that he had a sheep in a coal pit', meaning that Bakewell had kept quiet about his more recent progress towards improvement. [99] The same was said to Young in 1791 when he asked to see Brindled Beauty, a famous daughter of the bull Shakespeare: 'Bakewell laughingly told me she was in a coal pit, by which I was to understand she was not to be seen.'[100]

Marshall was another writer ready to acknowledge Bakewell's openness. He had spent three months in Leicestershire, including a period of residence at Dishley in December 1789, when preparing his book *Rural Economy of the Midland Counties*, in which he wrote: 'I have been reputedly favoured with opportunities of making ample observations on Mr Bakewell's practice and have, as repeatedly, been favoured with liberal communications on rural subjects.'[101] This was not, however, to deny his frustrated ambition to persuade Bakewell to reveal the origin of his breeding stock (Chapter 4). Later Marshall would write of the 'art, science and *mystery* of breeding',[102] aware that aspects of breeding were mysterious even to its practitioners. It remained a matter of deep regret to him and to every serious breeder that Bakewell's records could not be examined, even after his death.[103]

Bakewell's attitude to his records is illustrated by his reaction to a visitor who came to survey his farm and publish a report about it. This was John Monk of Knightsbridge, Devon who visited him in 1794 and asked to take a copy of the records. Bakewell gave Monk an answer that surprised him,[104] offering to loan him his stock for the chance to repeat the experiment if he was prepared to remain in residence for a while. Monk had to decline the offer and did not appear to have appreciated Bakewell's motive in making it. Bakewell's offer can be seen as an action typical of an experimentalist determined to replicate his experiments by encouraging someone enthusiastic to become involved in them, under his eye of course.

An attempt at understanding Bakewell's character, after a gap of some years, was made by an anonymous author styling himself 'A Lincolnshire Grazier', responsible for the sixth edition (1833) of the well-known handbook *The Complete Grazier*. The author, William Youatt (1776–1847), was a London-based veterinarian and author of a series of books on

---

[99] Parkinson 1810, p. 283    [100] Young 1791, Young's *Tours* (1932), p. 292
[101] Marshall 1790, i, pp. 295–6    [102] Marshall 1818; p. 43 (our emphasis)
[103] Pitt 1809, pp. 255 and 269    [104] Monk 1794, p. 28

farm animals. He leapt to Bakewell's defence, convinced that the 'only mysteries he employed' rested upon his 'more than common acuteness of observation, judgement and perseverance', and that he could hardly be expected to open his practice 'to inspection by every one who sought to profit by it'.[105]

Although secrecy may thus be excused on grounds which are not unreasonable, there were other supposed faults to be mentioned. Accusations of dishonesty were levelled at him from some quarters, arising principally as an adverse reaction to his stated wish to get the maximum prices for his stock when he exhibited his animals at Dishley. The best way he found of achieving this was to display each individual separately for examination, in a way designed to enhance its best features. Most stock breeders showed rams in groups of three or four but, in Bakewell's case,[106]

> the sheep were exhibited in a small house adapted to that purpose, having two opposite doors, one for admission, the other retreat; and the inferior are always exhibited first, that the examination of the inspector might be raised by degrees to the utmost pitch at the exhibition of the last and finest.

Monk provided some further details: 'When Mr Bakewell lets his rams, he never asks a price but leaves the hirer to offer what he thinks proper. If Mr Bakewell approves of the offer he accepts it, if not, he rejects it.'[107] Another eye witness, Mr Cresswell of Ravenstone,[108] described how 'he watched the hirers drawing lots for choice, with hay slips or taking figured marbles out of a bag.' It was in 1788–9 that customers began to fix the price themselves— 'the new way'.[109] He who won the lottery had the first chance to make an offer. Marshall stated his disapproval of the practice: 'I can see no fair advantage occurring from it.'[110] Even less acceptable, if true, were what an anonymous author refers to as certain 'tricks of jockies and horse dealers', said to be practised at Dishley, such as sham contracts and the use of puffers in the crowd 'to spirit up the buyers at auctions'. He further reported that ' a young lord or gentleman, with his pockets well lined, and his senses intoxicated by the fumes of improvement, was as sure to be imposed upon by these, as by the gentry at Newmarket.'[111]

The anonymous author's allusions to Newmarket, the focus of racehorse breeding, reflect his identity as John Lawrence, whose major area of expertise was on horses. He repeated the accusation of 'jockeyship' in a book he

---

[105] 'Lincolnshire Grazier' 1833, pp. 13–14     [106] Anon. (J.L.) 1800, p. 207
[107] Monk 1794, pp. 27–8     [108] Reported by Dixon 1868, p. 341
[109] Pawson 1957, p. 157: letter from Bakewell to Culley, 6 October 1791
[110] Marshall 1790, i, pp. 421–2
[111] Anon. (J.L.) 1800, p. 202. By 'gentry at Newmarket' he refers to racehorse breeders. Bakewell's successors were accused of perpetuating 'every sort of trick' (see letter from Richard Smart, Superintendent of Royal Farms to Banks, 4 August 1809, reproduced by Carter 1979, p. 494).

published a little later,[112] although other writers were not so explicit, being perhaps less familiar with this shady world. Sham contracts were well known in racehorse dealing and worked as follows. A stallion is purchased at a high price, well publicised, to ensure that the progeny fetch correspondingly high prices. However, the actual price paid is only part of what is stated in the bill of sale. A secret agreement is made with the vendor to receive *gratis* a proportion of the offspring of the stallion in the future. If Bakewell was engaged in similar sham contracts with his sheep, it would explain his particular discomfort at Young's continual mention of the high prices. Evidence of some downward revision of the prices paid for Dishley sheep by a few percent is included in Bakewell's so-called 'memoirs'.[113] Other sources reveal that Bakewell certainly had an arrangement to receive lambs from the rams he rented out, although this is usually given a more acceptable explanation, in terms of testing the breeding quality of his rams (Chapter 4). On the sham contracts issue, Bakewell had no chance to refute Lawrence's accusation, which was not made in print until after his death. But as the action of a dedicated, single-minded breeder, it has a ring of truth about it. Bakewell certainly made no secret of favouring high prices.

## Bankruptcy reveals the extent of Bakewell's support

Nobody could claim that Bakewell was personally enriched by any of his activities, a fact proved, if proof were needed, by the issue of a Commission of Bankruptcy against him in 1776 which almost lost him his entire stock and the lease of Dishley Grange.[114] Eventually, after the sale of some of his animals in 1777,[115] he was bailed out by friends and fellow breeders who rallied round to put him back in business. A public subscription was opened, reinforced by a written appeal on 23 December 1780,[116] to allow him to reestablish his farm and stock. In it he justified his petition to continue, on the grounds of what he had achieved in overcoming difficulties and combating the prejudices of other breeders. He explained that he had had 'various experiments to make in order to ascertain which were the best kinds to breed; and that such experiments were attended with considerable expense, and more trouble than he can well convey the sense of.' He believed that he had 'brought all the different kinds of stock above mentioned [horses, cattle and sheep] to a greater degree of perfection that [had] been done by any other person, and thereby rendered important services to the country.'[116] It cannot

---

[112] Lawrence 1809, p. 54

[113] A collection of Bakewell's sayings and 'essays' in a manuscript book, together with other papers, including Mr Stubbins' sale books, kept at Dishley farm and examined by Dixon 1868.

[114] Wykes, unpublished manuscript, based on articles in the *Leicester Journal.*

[115] Young 1813, p. 402        [116] Text reprinted by Pawson 1957, pp. 181–3

be doubted that selective breeding in the Bakewell manner was a very expensive business.

Among those petitioned were the Duke of Devonshire, Lord Montagu, Lord Bangor and Sir William Gordon, who was the second husband of Bakewell's landlord, Mrs Sam Phillips. Generously they bailed him out, justified, it must be assumed, by their confidence in his outstanding ability. His problem was almost certainly one of cash flow, resulting from the lack of any fixed date for payment of hire charges. Marshall believed that it was rare for money to arrive until after the ewes had produced proof of a ram's efficiency.[117] This was the accepted procedure in Lincolnshire where the practice of hiring originated. A year's credit was given for payment, and the price could be abated if the lambs were not sufficiently good to justify the original price.[118] The abatement was generally settled on the price of hire of the next year's ram.[119] To add to Bakewell's problems, there were some Irish farmers who had rented cattle from him but never paid up at all, as he told Rochefoucauld in 1785.[120]

His shaky financial situation added an extra reason for secretiveness. Here was a self-made entrepreneur, living on his wits, whose ambitions exceeded his means to fulfil them. His elaborate experiments and costly selection programmes must have drawn him constantly towards the brink of insolvency. To maintain the Dishley style of hospitality was an additional burden.[121] The years after 1776, when he was not allowed to manage his own affairs, must have put a great strain upon his confidence. Between November 1776 when the bankruptcy was listed in the *Gentleman's Magazine* and 1778–9, a period which Culley refers to as 'the Interregnum',[122] Dishley was managed by the assignees to the bankrupt estate. There were five of them, among whom John Ashworth of Daventry, a former pupil of Bakewell's,[123] 'had the principal direction of the business'.[124] Ashworth seems to have done his best to keep the business in a viable state although Bakewell's loyal herdsman John Breedon still kept a close eye on the stock, reporting back to his master any action proposed by Ashworth of which he thought Bakewell might not approve. Ashworth could only have been at Dishley occasionally because he kept a busy inn at Daventry, nearly 40 miles away.[125] It may be supposed that he looked for a quick return by selling fat lambs in spring or early summer, or store sheep in the autumn.

---

[117] Marshall 1790, i, p. 426

[118] Letter from Bakewell to Holroyd 1772 (unpublished manuscript in Leicestershire County Record Office)          [119] Young 1813, p. 85

[120] Scarfe 1995, p. 35; Irish non-payers were hardly the major problem

[121] Anon. (J.L.) 1800, p. 202

[122] Pawson 1957, p. 194: letter from Culley to Bakewell, 1797

[123] Wykes, unpublished manuscript

[124] Pawson 1957, p. 195: Culley 1794 in answer to query by Sir John Sinclair

[125] Wykes, unpublished manuscript

The strategy of producing pedigree stock for breeding could not continue long without Bakewell's participation. Forewarned by John Breedon, he managed to block the sale of his favourite ram, 'Young A', essential to him for pedigree breeding later, when he regained control. It must have been a difficult time for all concerned, although successful sales were made, including a bull shipped to a 'gentleman in Jamaica' in 1777, which encouraged the same buyer to import five more bulls in 1785,[126] and some rams purchased by Alexander Lowe in the Scottish border country in 1778,[127] which set in motion a widespread practice, i.e. sales to Scotland, in the years following.

When Bakewell finally got the farm back under his personal control, things begin to look up again. The appeal raised more than £1000 from subscribers, who included the Duke of Rutland, the Earls of Hopetown and Middletown, and Lord Sheffield as well as some of those mentioned earlier. Bakewell's recovery also owed much to friendly help from fellow farmers, many of whom were connected to him not only by the Dishley Society but through membership of Presbyterian or Unitarian chapels.[128] A great burden was lifted from his shoulders when people paid him the money they owed him, as Rochefoucauld learned during his visit in 1784. The young Frenchman concluded that 'He is now abreast of his affairs. I am afraid he is not making much profit but there is reason to hope that he will in the future; and, in truth, I hope so with all my heart.'[129] Even as late as 1789, however, the cash flow problem was continuing, as is evident from a letter Bakewell wrote to Young in August of that year: '[I] have been so much engaged on . . . matters relating to my bankruptcy which I hope will soon be settled.'[130]

## Bakewell and the scientific establishment

A matter of great regret to Bakewell's supporters was the deeply critical attitude shown to him by a much respected figure in the scientific establishment, Sir Joseph Banks (1743–1820). As President of the Royal Society and a major sheep owner on his Lincolnshire estate at Revesby, he had substantial power of patronage. But he seemed not at all impressed by Bakewell's Dishley sheep nor by Bakewell himself, viewing him as an upstart, 'promoting his breed by cunning and impudent means, to the disadvantage of the Lincolnshire breed'. With growing suspicion of Bakewell's methods and distaste for his populism, he gradually came to suspect him of 'more art and artifice than science in his pretensions'.[131] Young's praise of Bakewell after a two day visit

---

[126] Young 1786a, p. 487
[127] Robson 1988; it is possible that they were purchased from Culley
[128] Burgess 1908        [129] Scarfe 1988, p. 35; Scarfe 1995, p. 35
[130] Pawson 1957, pp. 175–6
[131] Carter 1979, p. 573, note 21; see also H.B. Carter 1988

to Dishley in 1791 'apparently goaded Banks not a little'.[132] His opposition had been sharpened by Bakewell's support of the parliamentary lobby in favour of the 1788 Wool Bill. Bakewell's vociferously stated support for the production of cheap clothing for the British mass market made him doubly unacceptable to Banks. Added to this there was Bakewell's easy entry into society at all levels, with apparent disregard for social distinctions. Although no snob, Banks was capable of entertaining dislike of any individual failing to show sufficient deference to his authority as President of the Royal Society. It is on record that he "collected a number of 'Bakewellisms', the purport of which is that Bakewell was a shifty person and not to be trusted".[133] As the anti-Bakewell faction, championed by Banks, grew in strength, particularly after Bakewell's death, a favourite line of attack was to make a joke of him, by remarking for example, that 'his animals were too dear for anyone to buy and two fat for anyone to eat', a catch phrase still being quoted even a full century later.[134]

An influential Scotsman, Sir John Sinclair (1754–1835), the first President of the Board of Agriculture, provided a vociferous counterweight to Banks. His high opinion of Bakewell as a person of 'strong natural sagacity' who had 'in various respects most essentially benefited his country', lasted throughout his life.[135] One of the actions of the Board of Agriculture under his direction was to commission a well regarded artist and sculptor, George Garrard, to construct 'a set of models of the improved breeds of cattle— upon the exact scale from nature'. Engravings of the models (which included sheep: Figure 5.1) were published in a book in which the author paid respect to the memory of Bakewell and made the interesting statement that it was due 'to the exertions of this great man, we may trace the establishment of the Board of Agriculture'.[136] It is well known that the first formal proposal for such a Board came from Sir John Sinclair in 1792, to be set up the following year by Act of Parliament.[137] It is also recognised that both Young and Marshall had earlier suggested national surveys of land usage and development[138] which became a major function of the Board to sponsor. Garrard's commission from the Board led him to expose Bakewell's behind-the-scene influence on the setting up of the Board, a revelation which surely enjoyed Sinclair's approval as the artist's patron.

Adverse publicity did nothing substantial to reduce Bakewell's technical influence. The value of his approach to breeding was established time and again by those who copied his techniques, even including some who criticised

---

[132] Carter 1979, p. 575, note 34, referring to draft of letter from Banks to Young, 9 December 1791; see also Carter 1979, p. 575 note 35, referring to letter from Banks to Young, 20 December 1791                          [133] Hill 1966, footnote to p. 121
[134] Betham-Edwards 1898        [135] Sinclair 1832, pp. 92–3        [136] Garrard 1800
[137] Gazley 1973, pp. 312–13        [138] Gazley 1973, pp. 78–9 and 313

**Fig. 5.1.**   Plaster models of two sheep made from life to scale (2.25 inches to 1 foot, or 1:5.33) by George Garrard: (A) Old Lincoln ewe (c.1800) and (B) New Leicester ewe (1810). (Copies of photographs supplied by The Natural History Museum Trading Company Ltd (London).

him personally. No greater proof was needed of the truth of his breeding principles than that they should 'have been found to govern in a great measure the conduct of men, who had an equal dislike both to Bakewell and his stock'.[139]

Young remained Bakewell's enthusiastic supporter even as they clashed over the 1788 Wool Bill which Young had condemned 'as so atrocious an insult on the landed interest'.[140] Evidence of Young's continuing good will was revealed in an article he wrote the same year describing the pleasure of listening to Bakewell as he held forth on agricultural matters during breakfast at the Ram Inn, Smithfield on 28 April 1788.[141] Notwithstanding the presence of a 'Peer of the Realm", the central position was taken by Bakewell in all matters of debate, even on the quality of meat with Mr Goodwin, the King's butcher at Windsor. Young refers to this occasion as 'Mr Bakewell's levée'. Bakewell had truly become a monarch in his own world.

### Bakewell's influence widens

One group of farmers who accepted Bakewell's methods with particular enthusiasm were the British Shorthorn cattle breeders. The Shorthorn was a type never favoured by Bakewell, whose interest in cattle was restricted to improving the traditional long-horned variety, which he admired for its versatility, in the shafts or at the plough as well as providing milk and meat. So it was that, in other hands than Bakewell's, the Shorthorn came to represent one of the more successful examples of selective breeding in the Bakewell manner. Improvement was in the air before 1760 when a pioneer Shorthorn breeder, Mr John Change of Newton Morrow, made repeated visits to Dishley to pick up tips.[142] Progress made a big stride forward in 1782 when Charles Colling (1751–1834), later to share the credit with his brother Robert, arrived in Dishley for a period of residential instruction. No one familiar with Bakewell's methods would have been in doubt about the theme of his instruction. The great lesson Colling learned during the three weeks he spent at Dishley was the capacity of a closely inbred bull or ram to stamp its characteristics on its progeny.[142] A generation later, another great Shorthorn breeder, Thomas Bates (1775–1849), who learned the Bakewell system directly from the Colling brothers, developed one of the greatest herds of all time. In 1810 he attended Edinburgh University as a mature student where his tutors are said to have depreciated inbreeding,[143] a contrary message from the one he had come to accept.

The value of inbreeding applied with discretion was becoming widely acceptable to ambitious innovators. It was a lesson absorbed by pioneers at improving another great cattle breed, the Hereford. The most notable

---

[139] Young 1811, p. 4     [140] Young 1786b, p. 521     [141] Young 1791, p. 293
[142] Sinclair 1908     [143] Winters 1954, pp. 17 and 35

Hereford improver, Benjamin Thomkins (1745–1815), bred his herd in-and-in for at least 20 years.[144] A lesser known name to be associated with this breed, although famous in other contexts, was Thomas Andrew Knight. Later in his career, when he gained recognition for breeding new fruit varieties and improving other commercial crops, Knight recommended fellow horticulturalists to adopt the new animal breeding techniques in their plant breeding (Chapter 1), a call for action that stirred them to fresh efforts in trying to produce new hybrid varieties of superior quality. As a French observer would later write, 'the genius of Bakewell pervades all his countrymen'.[145] This was an extraordinary statement to make 60 years after a man's death, but evidently that was how it seemed to the Frenchman.

Among a growing body of foreign farmers ready to adopt Bakewell's approach to breeding, few had the opportunity to consult him personally, or even to take instruction from those breeders who had undertaken the journey themselves. Most had to make the best use they could of books and published articles by Culley, Marshall, Young, Pitt and numerous others. German and French readers were kept informed about Bakewell's achievement in their own languages. Young's *East of England* reached a German readership fairly quickly in a translation published in Leipzig in 1775.[146] The second edition of Culley's *Observations on Livestock* was published in German by F. Maurer of Berlin in 1804. Parts of the *Annals of Agriculture*, carrying various reports relating to Bakewell, were translated into German and published by Caspar Fritsch of Leipzig in three volumes.[147] Two works by William Marshall were translated,[148] the *Rural Economy of Norfolk* in 1797 and the *Rural Economy of Yorkshire* in 1800–1, both published in Berlin. At about this time information on Bakewell began to appear in original German writings, the best known of which were written by a citizen of Hannover, Albrecht Thaer, physician to King George III. His first commentary on Bakewell was included in an addendum to the German translation (1800) of Lasteyrie's book on Spanish sheep breeding (1799). Later, in the third volume of a three-volume work called *Introduction to Knowledge of English Agriculture*, published in Hannover in 1804, he covered animal breeding extensively.[149] Bakewell's influence on Thaer's ideas about variability in farm animals, inheritance and breeding has been reviewed by Zirnstein[150] (see also Chapter 7).

A large selection of Young's works were translated into French in 1801 and published in Paris in 18 volumes. French translations of articles from *Annals of Agriculture* also appeared in *Bibliotheque Britannique*, a periodical

---

144 Heath-Agnew 1983, pp. 33–5    145 Lavergne 1855, p. 25
146 Müller 1969, p. 131    147 Müller 1969, p. 133
148 Müller 1969, pp. 133–4    149 Thaer 1804, pp. 615–802
150 Zirnstein 1979, p. 54

published in Geneva from 1796. This influential work also carried articles from the *Bath and West Society Reports* and from James Anderson in Scotland. Much of Marshall's work was published in 1803 in Paris in five volumes plus an atlas. André Bourde, who has reviewed the French agronomic literature of the period, concludes that by the end of the eighteenth century Bakewell was as famous in France as in his own country.[151] Once the French were writing about him in the international language of the age, his reputation as a European figure was assured. Among a new generation of French experts in sheep breeding were P. and F. de Jotemps and F. Girod. A book on wool and sheep breeding that they published in 1824 stimulated the production within a year of two independent German translations, one by C.C. André in Prague, the other by A. Thaer in Berlin. In the Thaer version, we can read 'Did not Bakewell reduce the skeleton of his sheep by up to one half in weight while at the same time doubling the weight of flesh? Did he not change the shape of these animals in any way he wished?'[152]

The great problem for any would-be follower of Bakewell was to develop the necessary powers of discrimination in interpreting the many published words on the subject, and then to translate theory into practical reality. When it came to selecting breeding stock, the breeder's dearest wish was to match what Youatt, referred to as Bakewell's 'more than common acuteness of observation, judgement and perseverance'.[153] All commentators agreed that he possessed a phenomenal capacity for assessing the quality of an animal and its suitability for breeding. To be successful like Bakewell required single-minded persistence in moving towards an ideal type of stock already designed in the mind. The hope of emulating his success encouraged many attempts to follow his lead, even long after his death. The result was an explosion of breeding in the Bakewell manner, in the early nineteenth century over much of the civilised world.[154] Although breeders would differ in the ideal type of stock they aimed to produce, each suited to an individual requirement, the techniques they employed owed a substantial debt to the Dishley tradition.

## Bakewell's legacy

As a testament to Bakewell's skill, the Dishley sheep long remained a valuable British breed, not only at home but also on the European continent, as far away as Russia, in the former American colonies and among the dependent territories of the growing Empire. Experimental crosses were made with many other recognised breeds.[155] The *Farmer's Magazine*, published from

---

[151] Bourde 1953, p. 146    [152] Thaer 1825, p. 21
[153] 'Lincolnshire Grazier' 1833, p. 14
[154] Anon. (J.L.) 1800, p. 205–6; Young 1811, p. 17
[155] Fussell and Goodman 1930, p. 151; Trow-Smith 1959, pp. 270–74; Walton 1983; Robson 1987

1800, is full of examples. As far north as the Highlands, 'wonderful effects' were produced in crosses with black-faced, black-legged, horned, wiry ewes.[156] All the Disley characters proved dominant to the Highland ones:[156]

> All have the greater tendency to get fat . . . a round fleshy animal resting peacefully in the fields, not a single lamb is to be found with black legs or a black face, and scarcely one with any sort of horn, while all are of a very large size.

At the opposite end of the kingdom the breed made an equal mark. Half a century after the great breeder's death, the situation was nicely summarised by George Johnstone of Mereworth Castle, near Maidstone, Kent, who wrote:[157]

> The most improved breed of sheep we are in possession of at the present day is the new or improved Leicester; there is no ewe of whatever breed or description, that will not, if crossed with a pure bred New Leicester ram, produce a lamb possessing better qualities than herself.

He warns, however, that only the best breeders are advised to breed from the hybrids. To do so requires 'careful selection of animals' and 'more experience and judgement . . . then the generality of our farmers are possessed of'. That such talent was rare could not be doubted. That such talent existed was proved by the origin of several new breeds produced from hybrid crosses. Examples in the North of England included the Wensleydale, 'full of Dishley blood',[158] and the Border Leicester, derived from Leicester × Cheviot crosses.[159]

The reputation of Bakewell's quick fattening New Leicester mutton sheep overshadowed the substantial efforts of Sinclair and Banks who, in their separate ways, tried to spearhead a fresh interest in fine wool production. Sinclair formed the Society for the Improvement of British Wool, also known as the British Wool Society, in 1791. Centred in Edinburgh as an off-shoot of the Highland Society, its purpose was to promote Shetland wool and also to produce new fine-woolled breeds by selecting among the progeny of crosses to Merinos. The original plan was drawn up by Culley's friend Dr James Anderson.[160] Started with a burst of enthusiasm, it became less active when Sir John moved to London to assume the presidency of the Board of Agriculture. In fact it failed to outlast the century, being overtaken by the war with France. Later imports of Merino breeding stock, as fortunes of the allied victory, led to the formation of the Merino Society in 1811, centred in London, with Banks as President (Chapter 6). It was taken as a model for the Sheep Breeders Society founded in Brno in 1814 (Chapters 1 and 9) just at a

---

[156] Boswell 1829, p. 21   [157] Johnstone 1846, pp. 156–7   [158] Watson 1928
[159] Trow Smith 1959, pp. 272–3
[160] For a reference to Culley's friendship with Anderson, see Pawson 1957, p. 106: Bakewell to Culley, 2 August 1787; Mitchison 1962, pp. 114–115; Carter 1979, p. 185 and note 270, p. 577

time when, as fate would have it, the English Society was already declining in activity after only three years of existence. The Brno Society continued active for almost 30 years, keeping the memory of Bakewell alive throughout the Austrian monarchy and beyond. In 1829 an article was published in the journal of the Society (*Mittheilungen*) on English sheep breeding from Bakewell onwards, recounting his methods and stressing his authority.[161]

Bakewell himself had carried out some crosses with a Merino ram in the 1780s, and also a cross to some kind of Spanish sheep as early as 1772.[162] Originally he may have toyed with the idea of producing a dual purpose breed, combining the fast maturing quality of the Dishley with the fine wool of the Merino. However, it was not a concept he found acceptable: 'It is impossible for sheep to produce mutton and wool in equal ratio; by a strict attention to the one, you must in great degree let go the other', he is quoted as saying.[163] The idea of a dual-purpose sheep was shelved for the time being but gained the attention of several breeders after 1800, the most notable ones in England being Dr Parry of Bath in Somerset and John Southey, Lord Somerville with an estate in the same county. They had high hopes of producing a 'British Merino' from crosses between Spanish imports and southern British breeds like the Southdown and Ryeland. Their struggles to achieve this ambition will be described in the next chapter.

The inheritor of Bakewell's valuable stock was his nephew Robert Honeyborn (1762–1816), who had for some time been in successful charge of the arable side of the farm.[164] As the son of Bakewell's sister Rebecca, he was the closest male relative. To assist him was another 'nephew', Mr Walton, related to the husband of one of Rebecca's daughters, who had worked for Bakewell earlier and had then gained from the experience in the service of Lord Lisbon in Gloucestershire. Together they continued to try to promote the Dishley breed but made no mark as innovators. In the hands of Mr Honeyborn, the breed is said to have deteriorated,[165] along with other aspects of Dishley husbandry.[166] Writers began to characterise Dishley sheep as an example of neglected wool. Among the most critical was Richard Parkinson, infuriated that 'the Leicestershire business has been carried on with such a high hand, as if no other set of men knew anything but the breeders of that county.'[167] He and others began to deplore the damage done by continued selection for fatty meat at the expense of everything else. They warned of dire consequences at the extent to which Dishley blood was spreading into so many traditional breeds.

[161] Anon. 1828
[162] Carter 1979, p. 121: letter from Bakewell to Banks, 18 September 1787; p. 159: letter from Holroyd to Banks, 4 August 1788          [163] Stone 1794, p. 59
[164] Young 1791b, Young's *Tours* (1932) p. 302          [165] Boswell 1829, p. 20
[166] Stanley 1995, p. 46          [167] Parkinson 1810, p. 279

Those who out of loyalty to Bakewell, or for whatever reason, continued to lay stress on the superiority of his breeding methods and the quality of the Dishley sheep for crossing maintained a strong influence. For this purpose in particular, the breed had established its true value. It invariably conferred the blessing of its quick fattening quality on its progeny. The influence continued even after the more extreme barrel shape had lost favour.[168] Myths began to gather around the creator of the breed. In countries abroad he was even credited with expertise in fine wool production. This was because of the other Robert Bakewell (1768–1843), his distant relative,[169] who wrote the widely known book *Observations on the Influence of Soil and Climate upon Wool* (1808). Confusion resulted when some sheep experts on the continent assumed the two Bakewells to be one and the same person. The reputations of both probably benefited from the error.

Continental breeders continued to import Dishley sheep whenever they could afford them. Several important breeds in northern and central France, western Germany and Holland gained substantially from the crosses, two major examples of which are the Texel of Holland and the Bleu de Maine in France.[170] Within Leicestershire, Bakewell remained nothing less than a hero, although increasingly like a legendary figure of old than a real person. The attitude is revealed in the following hymn-like verses, entitled 'New Leicester sheep', published only a few years after his death, in the *Leicester Journal*.[171]

Of all the various breeds of Sheep
That Butchers skill, or Graziers keep
From which do we most comfort reap?
       New Leicester

What sheep procured the richest meat,
And in appearance look most neat
And pay best for the food they eat?
       New Leicester

Which most of all the Landscape grace
Contain most bulk in smallest space
And where they feed adorn the place?
       New Leicester

Which best our craving wants supply
To feed and keep us warm and dry
And make both cold and hunger fly?
       New Leicester

---

[168] Lawrence 1809, p. 22    [169] Trentham-Edgar 1934
[170] Trow-Smith 1967, pp. 40–41
[171] *Leicester Journal*, 5 January 1811 (quoted from Wykes, unpublished manuscript)

Which are most tractable and tame
And will, as long as sheep remain
Imortalise great Bakewell's name?
New Leicester

But with no children to remember him, tangible evidence of Bakewell largely disappeared after his death and, with it, many precious details of the reality of his life. Most of his personal papers were lost, together with his house (pulled down before 1854) and the great barn. By 1868, the church had been shut up for more than 20 years, home only to pigeons. Later it lost its roof when the landlord reclaimed the lead. As the fabric of the church deteriorated, Bakewell's grave became forgotten and remained concealed until 1919 when it was discovered by a working party of German prisoners of war, who had been given the task of tidying up. They accidentally discovered his badly cracked tombstone within the ruined church under a mass of weeds and fallen masonry.[172]

The local neglect of Bakewell contrasted with the situation abroad where his place in British history was assumed to be unassailable. The Frenchman Léonce de Lavergne spoke for many when he wrote:[173][174]

> The County (of Leicestershire) is famous for its Stilton cheese and the farm of Dishley Grange, once occupied by Bakewell, whence emanated the principle of the transformation of breeds of domestic animals, one of the most valuable conquests of human genius.

> The wealth which Bakewell has conferred upon his country is incalculable.

In the hands of some of Lavergne's countrymen and other breeders on the continent, Bakewell's methods were applied more and more widely. They were refined and more openly described by the Merino breeders, applied in directions Bakewell never travelled himself, or possibly even dreamed of. The more progressive of these fine wool breeders, in France, Saxony, Prussia and Austria, freely acknowledged their indebtedness to him and were proud to be associated with his name. Sheer puzzlement was expressed at any Englishman who criticised the universal hero.

---

[172] George Jobson, *Journal and North Mail* 9 November 1957, p. 6; Pawson 1957, p. 45
[173] de Lavergne 1855, p. 253
[174] de Lavergne 1855, p. 21

# 6

# *Merinos in Sweden, France and Great Britain*

England which in recent times has perfected every part of culture, has neglected, just at the moment, the improvement of the superfine wooled races.

*Lasteyrie, 1802, p. 118 (original in French)*

Uncertainties about the supply and price of Merino wool provided a strong
impetus for rival states to make every effort to produce the commodity themselves. Beginning with Sweden, several continental countries had made substantial progress in naturalising the breed by the third quarter of the
eighteenth century. This development paralleled Bakewell's achievements but
was at first quite separate. The private initiative, not to say act of faith, by
Jonas Alströmer in 1723 (Chapter 2) provided the first stimulus. Enjoying
strong encouragement from the Swedish crown, his project became an inspiration for other monarchs to find their own agents and experts capable of
procuring and maintaining the precious stock. From Europe, the sheep found
their way to overseas colonies and the newly independent United States. The
'trek of the golden fleece', as two American authors have described it,[1] had a
major impact on the international wool industry. Here and in the chapter
that follows we enlarge upon aspects of the dispersal most relevant to the
development of theories about heredity. In each of the countries chosen for
consideration, particular circumstances added to the body of knowledge,
because of (i) the unique environmental conditions into which the sheep were
introduced, (ii) the particular circumstances of agricultural production, or

---

[1] Burns and Moody 1935

(iii) the intellectual background that conditioned the way in which breeding records were considered in relation to theories of sexual generation. At its highest level, breeders adopted practices associated with Bakewell although directed towards a different endpoint.

## Sweden

Traditional Swedish wool suffered the reputation of being among the poorest anywhere. The native sheep, with their 'slight, thin straight and short fleeces', had almost no commercial value.[2] A primitive example that survives today, just as a remnant, is the Gutefår race on the island of Gotland (Figure 6.1). Unsurprisingly there was little demand for unrefined Swedish wool abroad. Bishop Huet, in his *Memoirs of the Dutch Trade* (1706–7), lists the many countries from which the famous Dutch weavers imported wool, stretching from Poland to Peru, among which Sweden was noticeably absent.[3] The native Swedish wool was grey and woven only as 'homespun'. From the sixteenth century onwards, successive monarchs encouraged the importation of white-wooled English sheep, or stock of English type from Friesland or Eiderstedt (western Holstein), to improve the local flocks. The resulting cloth

**Fig. 6.1.** Gutefår sheep, a traditional Swedish race, a remnant of which is still preserved (based on a photograph issued by Föreningen Gutefåret, Aspnäset c/o Edberg, 610 60 Tystberga).

---

[2] Schulzenheim 1797, pp. 319–21    [3] Smith 1747, ii, pp. 92–3; Huet c. 1706

must have been less coarse, as 'the King encouraged the wearing of it by his example', wrote John Cary, a Bristol merchant in 1717.[4] The King referred to was Charles XII, whose reign (1697–1718) was marked by a growing determination to make greater use of the domestic product. He imposed swingeing taxes on foreign cloth 'on purpose to encourage a manufacture of their own'. However, the demand by the upper classes for something softer and more refined gave Alströmer just the incentive he needed to make the Merino experiment.

Hardly four years after Alströmer introduced the first Merinos, an English writer commented that he had noticed an improvement in Swedish cloth. This was Salmon (1727) in his three-volume *Modern History, or the Present State of All Nations*.[5] Other foreign sheep continued to be imported: some from England in 1715 and others from Eiderstedt in 1726.[6] The long combing wools obtained from English and Eiderstedtish cross-breeds were found to be popular with the mass of the people. The newly introduced Merino wool was 'more profitable for gentlemen'.[7] The imported Spanish sheep were to be found almost exclusively on the estates of the aristocracy, gentry and other persons of standing; the Swedish country people of those days were not interested in them.

Beginning in 1741, premiums were distributed for the best breeding rams 'of Spanish and English race', with gratuities of 25% on the value of all wool that proved to be fine, of English, Spanish or Eiderstedtish origin, sold to the Swedish woollen manufacturers.[8] By 1746, King Frederic I was in a position to issue an edict forbidding his subjects from importing any further foreign-made cloth at all.[9] Abundant supplies from Swedish weavers justified his taking this action. However, he imposed no corresponding ban on raw wool or sheep stock, thereby revealing that the supply of Swedish fine wool was not yet assured. Most Swedish wool was still described as coarse or very coarse.

To accelerate wool improvement, breeders tried crossing Merinos with local sheep, probably mainly those derived from crosses to earlier non-Swedish imports. They hoped that Merino hybrids would inherit an extra degree of fineness in their wool while, at the same time, being better adapted to the harsh Swedish conditions. A French expert, Count C.P. Lasteyrie (1759–1849), visiting there in 1800, learned about the earlier German and English crosses.[10] Polish and Icelandic breeds had also been available for crossing.[11] A census in 1755 counted 49,350 'perfectly satisfactory sheep' and

[4] Smith 1747, ii, p. 176    [5] Smith 1747, ii, p. 217
[6] Schulzenheim 1797, p. 306    [7] Schulzenheim 1797, p. 319
[8] Schulzenheim 1797, p. 306; Schultzenheim 1804, pp. 172–3; Lasteyrie 1802
[9] Smith 1747, ii, p. 460    [10] Lasteyrie 1802
[11] Schulzenheim 1797, pp. 319–21; Karl-Henryk Suneson, personal communication, 4 January 1990

60,687 'mixed breed sheep'.[12] 'Perfectly satisfactory sheep' were not only Merinos but other pure-bred foreign breeds.[13] The 'mixed breed sheep' may have included $F_1$ hybrids with Swedish sheep although many were probably the result of several generations of intercrosses. A further census in 1764, based on returns by provincial shepherds, recorded a rather different picture,[14] given by Schulzenheim as '88,750 genuine fine-woolled sheep, without counting 23,384 good sheep of mixed breeds', and by other authors (apparently trying to isolate the Merino influence) as 23,384 of the 'mixed breed' and 65,369 of the 'pure breed'.[15] Despite differences, both sets of figures indicate that the 'pure breed', i.e. those bred exclusively from imported Merino stock, had increased greatly, both in number and also as a proportion of the total. The relative scarcity of hybrids might imply that attempts to improve native Swedish sheep had run into problems. Equally, it could reflect a growing skill in keeping the pure breed in a thriving state.

It should not be assumed that Merinos in Sweden presented no problems. Alströmer and two pioneering followers of his were forced to learn expensive lessons from their mistakes, particularly at the outset when, unaccustomed to the rigours of the long Swedish winter, the imported stock suffered from poor nutrition, neglected health and inadequate management of reproduction. Alströmer spelt out the early lessons in an instruction manual on Merino breeding published in Stockholm in 1727 (Figure 6.2).[15a] Because it was written in Swedish, its influence was not much felt outside the country, although an abstract was published in France under the title *Traité d'elevage du mouton en Suede*. Another book in Swedish, giving the benefit of six years' added experience, followed in 1733, published in Skara (Figure 6.2).[15a] Ten years later, Alströmer organised a fresh importation of Merinos, destined to be treated more skilfully in the light of experience.

News of the Swedish success filtered out through trade reports[16] and was already recognised as the most profitable branch of Swedish agriculture when a book appeared that would open the subject to wider international readership. Written by F.W. Hastfer, it was published in Sweden in 1752 and quickly issued in German translation in the same year in Leipzig and in two French editions, in Paris and Dijon, four years later. A Swiss edition appeared in Berne, with notes added by Linnaeus, and it was also translated into Danish.[16a] Hastfer emphasised principles of feeding, health care and breeding. He described the grading technique for improving Swedish sheep to the equivalent Spanish or English quality by repeated infusions of foreign blood in

[12]  Kjellberg 1943, pp. 304–5
[13]  Schulzenheim 1804, pp. 172–3; Karl-Henryk Suneson, personal communication, 4 January 1990        [14]  Schulzenheim 1797, p. 307
[15]  Culley 1807, p. 238 (quoting Lasteyrie); Rees 1819 ('Sheep' section); Martin 1849, p. 61
[15a]  Alströmer (Alström), J. 1727, 1733        [16]  Smith 1746
[16a]  Hastfer 1752a, 1752b, 1756a, 1756b; Wichmann 1795

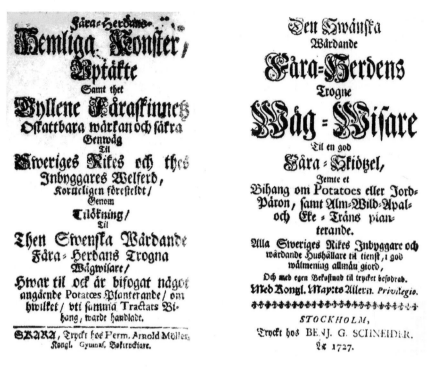

**Fig. 6.2.**   Title pages of two seminal books on Merino breeding in Sweden, written by J. Alstromer: (A) 1727, published in Stockholm, (B) 1733, published in Skara (based on photocopies supplied by Mr K.-H. Suneson).

successive generations of out-crossing. Nothing but foreign rams should be used for three generations in order to change local sheep into a new breed, able to yield the same quality wool as Spanish sheep. Hastfer's advice on this important economic matter was to be repeated many times by French and German authors later in the century, although the number of generations deemed necessary to make the transition rose from three to four and then to five. Hastfer made no proposal to select animals showing particular breed points or production traits. No potential fine-wooled breeding stock was to be wasted by Hastfer. Only undersized, unhealthy or abnormal ewes were to be culled from the stock.

On 15 June 1770, the Swedish Kommerskollegium[17] announced that the influx of pure Spanish, English and Eiderstedtish sheep, and angora goats, had been so beneficial that no further imports were to be allowed.[17a] This was the same year in which Clas Alströmer, son of Jonas, reminded the Swedish breeders about the value of Merinos in a newly published book that Carter

---

[17] A collegiate body controlling exchange of goods in and out of Sweden, on behalf of the crown          [17a] Karl-Henryk Suneson, personal communication, 4 January 1990

has described as 'the first publication . . . of any scholarly weight and substance on the subject'.[18] It was based on a discourse before the Royal Academy of Stockholm[19] and was later serialised in a French translation, to be published in 1773 as a single volume.[20] Confidence in the technique grew progressively despite the widespread belief that foreign stock would inevitably degenerate because of continuous contact with local environmental conditions. Experience was providing evidence against the supposed inevitability of degeneration, from which the Swedish Merino breeders took heart, even though the procedure of grading did not prove infallible.

In a communication to the Bath and West of England Society, a well known Swedish breeder, Count Schulz von Schulzenheim, explained that the good management of sheep in Sweden was 'carried to the furthest in the Southern part of Sweden of what is called the Kingdom of Gothland [Gotland]'.[21] Impressed with the stability of the Merino breed in this part of Sweden, he published evidence on the continuing high quality of one of his own premium flocks of nearly 50 years' standing. On six occasions he had made fresh importations of Spanish rams and ewes with the aim of further improving his stock. The imported animals were of especially high quality due to his good fortune in having a friend and relative resident in Spain. Nevertheless he could find few better sheep than those he already possessed, an opinion backed up by 'the appraisers at the public hall' in Stockholm.[22]

Only one stock imported by Schultzenheim proved superior to the sheep already established on his estate. It came from Leon and he used it to build up a separate flock kept on a different farm, noting with interest its atypical appearance, much shorter in the leg than the ones he already possessed, although breeding absolutely true for soft and abundant wool. As the flock expanded it was divided between two farms:[23]

> The Leonese ewes also lived to see five generations in lineal descent from themselves, of which none have degenerated in point of fineness of wool, upon the most accurate comparison. The wool from the above two farms, is now reckoned of the first quality of Spanish wool of home production, in the public warehouse of Stockholm.

Schulzenheim describes in some detail the plant species necessary to provide sustenance 'in the most proper pastures' for his Merinos: fallow fields in summer, stubble fields in autumn. He was clearly in favour of rich pasture land, appreciating its effect on 'the size of the sheep, their health and fitness, all of which may, in the long run, have an influence on the softness and quality of the wool'.[24] 'The idea of some people, that rich pasture is not the most proper for fine woolled sheep, is certainly ill-founded', he wrote. Regarding

[18] Carter 1979, p. 560     [19] Alströmer 1770     [20] Alströmer 1772, 1773
[21] Schulzenheim 1797, p. 310     [22] Schulzenheim 1797, p. 311
[23] Schulzenheim 1797, p. 312     [24] Schulzenheim 1797, p. 315

the housing of his sheep, he saw cleanliness, warmth, ventilation and sufficient space as the critical issues, with 'double doors in winter' and six feet square of space 'commonly reckoned for each sheep'.[25] The aim of all these provisions was to keep the sheep healthy. He did not believe, however, that circumstances of rearing had much effect on a sheep's basic characteristics. He was convinced 'beyond contradiction that the fineness of wool in sheep is principally derived from the nature of the race'.[23]

He supported the idea of races being inherently different, on the grounds 'that in the same countries, both colder and warmer, there are found both coarse and fine woolled sheep, without any particular change in the breed, *if they do not happen to couple together*' [our emphasis].[24] He was completely confident that 'The assertion of some, that housing sheep [makes] the wool [deteriorate], and that their constantly remaining in the open air, and being led from one province to another, is the principal cause of the fineness of the Spanish wool' was simply not true. He believed that he and his associates had proved the point by their 'method of uniformly feeding the sheep in the house from December to the beginning of May, by which the wool loses nothing in its fineness'.[26]

On the matter of breeding and heredity, Schulzenheim stated the 'common belief' that the best rams are 'brisk and plump' and that 'the likeness in the progeny is the nearest to the parent that is the youngest, the strongest or most spirited at the time of copulation'. He recalled, however, that the Alströmers, both father and son, had 'confirmed the opinion among naturalists, that the likeness is generally derived from the father'.[27] When the male parent came from strongly inbred stock and the female from less well bred stock, it was indeed the case that the male characters would tend to dominate in the hybrids. Breeders in Germany who were crossing Merino rams with unimproved Saxon or Prussian sheep were finding the same (Chapter 7).[28] The experience led to a holistic view of heredity in contrast to the mosaic approach favoured by Bakewell and his followers when creating a new breed based on characteristics from both sexes (Chapter 4). Each view had a logical basis, depending on the approach to breeding adopted. In the first case it was grading or the production of $F_1$ hybrids; in the second it was crossing with selection in the hybrids and their descendants.

The evident success of Swedish breeders in applying the grading technique was confirmed by foreign observers. In 1800, when the Merino expert from France, Count Lasteyrie, visited Sweden on a fact-finding tour of northern Europe, he found Swedish Merino wool to be of such excellent quality that it equalled or even excelled that from Spanish imports made five generations earlier, kept there for reference, or from Spanish sheep recently imported.[10] A

---

[25] Schulzenheim 1797, pp. 314–17      [26] Schulzenheim 1797, pp. 315–16
[27] Schulzenheim 1797, pp. 317–18      [28] Fink 1797, p. 278

few years later, when the English encyclopaedist Rees wrote about Swedish Merino sheep, he reported them to be larger and stronger than the original Spanish imports although still producing high quality, fine fleeces.[29]

Alströmer and his compatriots had thus shown that a valuable trait could be retained in a breed even under greatly changed conditions of husbandry, while at the same time other characters such as size were being modified by selective breeding. Through their experience they considerably advanced the practice of grading. Sweden had pioneered a sheep revolution that other nations could not fail to note, and would copy with even greater success.[30]

## France

Opinion about Merinos in France was for many years divided. The French possessed indigenous fine-woolled breeds and enjoyed a tradition of wool production going back to Roman times. It was commonly believed among those who were supposed to know best, that their native French sheep could be raised in quality to Spanish standard without it being necessary to introduce Spanish blood.[31] Eventually, in 1777, after a 10 year programme of expensive breeding experiments undertaken by the country's most influential expert on sheep husbandry, L.J.M. Daubenton (1716–1802), the unpalatable truth became clear, that no French breed possessed full Merino potential. Only then did France begin to make proper use of Spanish breeding stock. We shall follow the course of events.

From the seventeenth century onwards, fine woollen cloth woven in France from imported Spanish wool had become an increasingly valuable asset. For 20 years or more after 1665, various woollen manufacturers, including foreign Protestants, were 'given patents and other encouragements' by the French monarchy to manufacture 'all sorts of Spanish cloth'.[32] Their success enlarged France's reliance upon imported raw wool from Spain. Determined to give his country a measure of independence, the statesman Jean-Baptiste Colbert, chief political adviser to the King, encouraged the import of Merino sheep to grow wool on home territory.[33] Imports began in 1664[34] and, by the time of Colbert's death in 1682, the wool produced in southern provinces of France had been sufficiently improved to support manufacture. Left to their own devices, who knows what these southern farmers might have achieved? In the event, their effort was undone by the new administration who introduced Orders in Council which had the effect of encouraging traditional French breeding practices once more.[35]

Efforts made to introduce further Spanish stock were discouraged until 1751 when G.L. Chauvel de Perce, a protégé of Maréchal de Saxe, imported

---

[29] Rees 1819 ('Sheep' section)    [30] Elsner 1828    [31] Carlier 1770
[32] Smith 1747, i, pp. 215 and 281    [33] Carter 1964, p. 49
[34] Pujol 1981, p. 541    [35] Bourde 1967, p. 797

a small flock to be kept in the open air all the year round. With the approval of the King, they were located in the park of Chambord,[36] with happy results.[37] The Swedish experience of producing wool of Spanish quality was beginning to have an impact in France. The Colbertian argument was used once again: if France raises better sheep than foreigners, it will free the country from dependence upon importing wool from abroad.[38]

In 1754, the Academy of Amiens gave as the subject of a dissertation the title 'Supposing one shall not be able to do without foreign wools in France' (from original in French).[39] The prize went to M. de Blancheville, a pseudonym of l'Abbé Claude Carlier. It was published the following year in Paris under his own name. He recognised the value of good foreign breeds and gave his opinion about their care and maintenance. Two years later, Carlier organised a French translation of Hastfer's Swedish text (1756), made by M. Pohole.[40] Then in 1762, Carlier, encouraged by Daniel-Charles Trudaine, *intendent des finances* (Controller-General), published a major book of his own dedicated to the improvement of French wool.[41] An open air approach to sheep husbandry was proposed, seemingly aimed at trying to reproduce foreign conditions as much as possible. The influence of climatic pressure upon heredity was a widely discussed topic among French naturalists at the time. After that, several French authors went into print favouring import of foreign sheep for breeding in the open air manner.[42] By the 1760s there were several independent flocks of pure Merinos in France.[43]

Carlier himself preferred to take a cautious view of foreign stock. He argued that France had perfectly good, indigenous fine-wooled breeds of her own, better suited to local conditions, that could be raised to Merino standard using methods traditional to France.[41] In Carlier's major work of 1770, described by André Bourde as 'the masterpiece of these eighteenth century studies',[42] he expressed his continuing belief that in France there was a good basis for improving local races, such as those of Roussillon, to a higher level of wool production by paying closer attention to nutrition and health, particularly the prevention of diseases. Only in special cases would it be necessary to import foreign stock. He greatly preferred to select superior rams and ewes from the best local breeds.[45] He also came to the conclusion that the extreme English version of open air husbandry, *l'education sauvage*, was unsuitable for application on a general scale in France.[46] He was also against the English practice of close inbreeding because he rejected consanguineous matings as harmful. He continued to publish these views as late as 1784.[47]

---

[36] Bourde 1967, pp. 812 and 857     [37] Martin 1849, ii, p. 64
[38] Bourde 1953, p. 136; Bourde 1967, p. 745
[39] *Année Littéraire* (quoted by Bourde 1953, p. 136)     [40] Bourde 1967, p. 814
[41] Bourde 1967, p. 815     [42] Bourde 1953, pp. 137–8     [43] Bourde 1967, p. 874
[44] Bourde 1967, p. 817; Carlier 1762     [45] Bourde 1967, p. 856
[46] Bourde 1953 p. 143     [47] Bourde 1967, p. 816

In arguing that the French strains had potential for improvement, Carlier appeared to have reason on his side. The previous two centuries had already seen a considerable rise in the quality of French wool. What further advances might be possible? The valleys of the Lower Navarre on the border with Spain produced tolerably—or sometimes exceptionally—fine wool, and the sheep of the Languedoc, Roussillon and parts of Provence also gave excellent short fine-wooled fleeces.[48] As early as the seventeenth century, English commentators were warning that some French wool was as good as the best English wool. W.C., 'a servant to his King and Country', commented with concern, in a tract issued in 1671, that the French were making low priced cloth called *searge* [sic] *de Berry* 'as cheap as our northern cloth but of much better wool'.[52] It was about this time that the French began to wrest the Turkish cloth trade from England and the Netherlands, by producing *drap de Londres* (or cloth in imitation of what was sent from London to Turkey) in the province of Languedoc.[49] By the end of the century, France was supplanting England in many markets abroad: in Italy, Spain and Germany as well as Turkey. A report published in 1713 stated that France was supplying Smyrna and Constantinople with more cloth than England, or even Italy.[50] Improved local wool supplied some of this manufacture but France still imported vast quantities from Spain.

The government, in the person of Daniel-Charles Trudaine, undertook a major action relating to the further improvement of French wool, when in 1766 he sought advice from the famous naturalist Buffon. In his writings on natural history in 22 volumes (1749–89), Buffon attached central important to *degeneration*, the process by which animals vary in response to changes in local conditions either natural or artificially produced.[51] A farmer's efforts to mitigate the influence of the natural climate and native flora might slow the process of environmentally induced change but could not prevent it. For Buffon the generation of like from like was explained by means of an internal mould, the *moule interieur*, into which chemical particles are absorbed, a process of 'intussusception', thanks to 'penetrating forces'. First there was the gathering of such particles when the seminal fluids mingled. After that, many further particles were assimilated from food ingested, to generate individuals of the same or deviant form, depending on whether the conditions of the environment remain stable or become changed.[52] By comparison with their wild ancestors, all domestic animals (with the possible exception of the Arab horse) were already degenerate, according to Buffon, and therefore liable to destabilise further and revert back to the wild form.[53] Only through

---

[48] Martin 1849, pp. 64–6    [49] Smith 1747, i, p. 231
[50] Commentary by Smith 1747, ii, p. 118 on *The British Merchant*, Vol. 2 (1713)
[51] Roger 1997, pp. 299–308    [52] Roger 1997, pp. 129–31
[53] Roger 1997, pp. 299–304

human intervention, in terms of careful husbandry designed to insulate animals from wild conditions, could this be prevented. Although he recognised selective breeding as an aid to progress, Buffon, like Carlier, was flatly opposed to intensive inbreeding.

Buffon was convinced that experiments were required. He recommended the task to his associate, Daubenton, to make crosses between breeds and study aspects of physiology in their progeny. Especially to be investigated would be tolerance of reduced temperature and the influence of nutrition on wool constitution.[54] Trudaine impressed upon Daubenton the economic advantage to France of producing fine long wool from French sheep, to make the growing French wool industry as independent as possible of neighbouring states.[55] At that time, breeds in France gave either fine short wool or rough long wool. The challenge presented to Daubenton was to combine fineness with length, by selecting within the progeny of appropriate crosses.

By the autumn of 1767, Daubenton had established a special breeding farm, with 'model stables',[56] at Courtangy near Montbard in Burgundy,[54] an area he considered to be especially favourable for producing fine wool.[57] It was the district of his own birth[58] and also that of Buffon, in which the latter 'had assigned himself the part of a great landowner, and was experimenting on timber',[59] as a prelude to his later career as an ironmaster. Here at Montbard, Daubenton crossed rams from Roussillon with ewes of different French breeds, and also sheep from Flanders, England, Morocco and Tibet,[60] including those most resembling Merinos. He made individual pairings (*alliances*) and carried out a careful selection programme among the mixed progeny, concentrating on increasing both the fineness of his wool and the weights of his fleeces. During the following 10 years, the parentage of each individual lamb born was recorded so that every sheep in his flock of several hundred had a written pedigree going back three generations. By 1777, his best sheep bore fleeces superior to those of the rams of Roussillon from which they were descended but, to Daubenton's disappointment, they were incapable of producing wool equal in fineness and length to that grown on the best Spanish Merino sheep. Two government inspectors attested to the truth of this, which Daubenton reported.[61]

The significance of the disparity in quality was not lost on the new *intendant des finances*, the Marquis of Turgot. During 1776 he had personally obtained a small flock of Merino sheep directly from Spain.[62] These sheep were transferred to Montbard, placed in Daubenton's hands and successfully raised to produce wool of Spanish superfine quality. So began, in a systematic

[54] Bourde 1967, p. 857       [55] Carter 1964, p. 49       [56] Bourde 1953, p. 190
[57] Bourde 1967, pp. 858–59       [58] Martin 1849, ii, p. 64       [59] Bourde 1953, p. 206
[60] Lasteyrie 1802, p. 46; Carter 1964, p. 50; Bourde 1967, p. 875
[61] Daubenton 1782, p. 297; Carter 1964, p. 51; Bourde 1967, p. 858
[62] Youatt 1837, p. 163; Martin 1844, p. 64

way, the long history of Merino breeding in France, and its modification by cross-breeding to other races. Daubenton maintained his sheep according to Carlier's system (the *system du parc*) which included confinement in light sheds (or pens) open on all sides.[57] Daubenton wrote that[63]

> by penning flocks during the whole year, not only does one increase the pro-
> ductivity of the pasturage and the soil . . . but at the same time one makes
> the sheep more robust, in consequence of which the wool becomes more
> abundant and of better quality, and the meat more tasty. One saves the cost
> of constructing and maintaining stables, which far from being useful to
> sheep, are very harmful.

The technique can be traced back to Ellis's book of 1749,[64] favourite read-ing of many English farmers, including Bakewell, which had passed into French translation as *Guide des Bergers*. Meanwhile the Swedish breeders were forced to rely on warmly insulated stables for a large part of the year.

Within two years of the commencement of Daubenton's management, Courtangy became officially recognised as a school for shepherds.[65] His pupils were given exemption from military service, so vital was their training considered for the economy of France. Daubenton's entry into sheep breed-ing was reinforced by the publication of Carlier's *Traité* in 1770. Together they gave this branch of landed activity an entirely new dimension in France. Sheep breeding was becoming recognised as a proper occupation for men of science, rather than simply a pragmatic art practised by shepherds. Daubenton's memoir on the subject, presented to the Académie Royale des Sciences in 1777, was later incorporated into a chapter of a work of practical advice to shepherds and sheep owners, *Instruction for Shepherds and Flock Owners*, published in Paris in 1782. With its ambition to enhance wool quality by every possible means, the *Instruction* was recognised as 'one of the most celebrated treatises of the eighteenth century'. These are the words of Auguste Wichmann who translated the original French into German.[66] It also went into Italian (1787), Spanish (1790), Dutch (1791) and English (1810). Appear-ing in a period of rapid expansion of European wool manufacturing, the book had an enormous influence on sheep breeders throughout the continent. Daubenton's achievement was to provide clear, straightforward definitions of the procedures to be used. As Bourde[67] has pointed out, all the principles on which Daubenton's advice was based had already been skilfully presented by Carlier, but Daubenton spoke to farmers in terms they could absorb.

In the introduction to his *Instruction* (1782), Daubenton admitted to hav-ing drawn from the experiences of others in France and also abroad,

---

[63] Daubenton 1769 (quoted by Bourde 1967, pp. 857–8)    [64] Bourde 1967, p. 858
[65] Bourde 1967, p. 1625
[66] Editions of Wichmann's translation into German of Daubenton's *Instruction*, Published in 1784, 1795 and 1797    [67] Bourde 1967, p. 859

although he made no mention of Bakewell in this context. Daubenton's own contribution was to pay special attention to recording pedigrees and selecting for wool fineness. He also claimed to have a sound understanding of health and nutrition, although one very special English expert who travelled to France to observe for himself was far from convinced. This was Arthur Young who in 1787 stopped by to look at a farm at Charenton, managed by Daubenton for the Royal Society of Agriculture as a small replica of his *bergérie* at Montbard. In a letter to his friend Banks, Young wrote in February 1788 of the farm 'where Daubenton reads lectures on agriculture', that 'such a farm and such a flock is not to be seen in a Christian country; they fold themselves in a mud ditch and have lost scores—they are black with dirt and the ridicule of all the shepherds around.'[68]

Young's report strikes such a strongly discordant note that it begs an explanation. Was Daubenton actually in charge of Charenton at that time? Young mentions his deputy, Pierre Marie-Auguste Broussonet, although he still blames Daubenton. After his visit Young can find little good to say about a man whom everyone was praising. His *Instruction* (1782) seems to Young 'of less value than the verbal instruction of a few minutes by an old shepherd to an apprentice'.[69] Probably Young was comparing it with the wealth of detailed information he had gleaned from travelling round England, talking to such experts as Bakewell, Culley and Ellman. A single disappointing visit to Charenton may have swayed his judgement. His low opinion of Daubenton certainly did not carry weight enough to detract from the esteem in which this Frenchman was generally held, even in England. With respect to his optical measurements of wool quality in the progeny of crosses between breeds, and his precise recording of every lamb bred, he was a master. As Carter has put it: 'The European reputation of Daubenton survived the petty carping of Arthur Young.'[69] In 1788 Banks invited Daubenton to deliver an address to the Royal Society of London, introducing him as 'the greatest farmer in the world', an accolade that must have seemed like a challenge to Young and the Bakewell camp.

What Banks and many of the continental breeders appreciated about Daubenton was his zoological approach to the substantial variation to be found in the nature of wool. He was among the first sheep experts to attempt an objective assessment of wool fineness, using a microscope fitted with an attachment to create a form of micrometer. Incorporating a lamella of rock crystal, it was constructed by M. Mégnie, a mathematical instrument maker in Paris.[70] Various designs of micrometer became available subsequently, based on different optical phenomena,[71] the most famous being the English-made 'Dolland Eriometer' or 'Wool Measurer', placed on the market in 1811.

---

[68] Carter 1979, pp. 142–3; Young 1790–91    [69] Carter 1964, p. 53
[70] Bourde 1967, p. 858    [71] Carter 1979, pp. 500–503 and 586–7

Aspiring Merino breeders everywhere tried to follow Daubenton's lead in assessing wool quality in a scientific manner, classifying fleeces into recognisably defined grades, as a basis for selective breeding. Carter's (1964) claim for him as 'the first modern scientific breeder of livestock whose work has any present day relevance' reflects the high regard in which his work was held. Daubenton's pivotal importance to the history of sheep breeding is universally acknowledged. It might be said that he did for fine wool what Bakewell was doing for mutton, although—as the French writer André J. Bourde generously makes clear—Bakewell 'represents a later stage in animal research than that reached by Daubenton'.[72] The most obvious difference was that Bakewell took selective breeding important steps further than Daubenton by appreciating the potential relationship between different individual traits and the complementary qualities of each parent, and by exploiting the techniques of breeding in-and-in and progeny testing.

It is not suggested that Daubenton and other French breeders denied the value of selective breeding. In the cause of improving French sheep by crossing with foreign rams, Daubenton was prepared to use the same superior foreign sire for up to three consecutive generations, changing it only when he had a better one at his disposal. Such a mating pattern must have resulted in a concentration of the characteristics of that particular ram, accompanied by a degree of inbreeding. The procedure was not, however, as intense as the breeding in-and-in practised by Bakewell and his followers. Moreover, the French treatment of the Merino breed itself implied, in a sense, a policy of outbreeding. While sheep breeders in Spain kept the Merino *cabañas* separate and never crossed them, the French experts bred from the best individual rams and ewes they could find, whatever their origin, i.e. whatever *cabãna* they originated from. They did this with the aim of creating new types of Merino breeding stock in France. But by thus merging the *cabãnas*, they were breaking up potentially valuable gene combinations selected over many generations in Spain. Using this fresh blend of Merino stock, Daubenton had it in mind to improve several different French races.

At this stage of Daubenton's progress, we observe the entry into French sheep breeding of another major figure, l'Abbé Alexandre-Henri Tessier (1741–1837). He arrived on the scene as the King of France's choice to direct an experimental farm created on land acquired at Rambouillet, 30 miles from Paris and conveniently close to Versailles, so that experiments could be carried out under royal eyes. Tessier, who had earlier distinguished himself with an official mission to Cologne to report on the cattle plague there,[73] was a friend of Daubenton's and fellow member of the Academy of Sciences in Paris.[74] His appointment came in 1785. Within two years, during which he

---

[72] Bourde 1953, p. 145     [73] Bourde 1953, pp. 194 and 212
[74] Bourde 1967, p. 872; B-F-S (1843)

imported 367 Merinos officially from Spain with the help of the French ambassador in Madrid, he had started the Rambouillet sheep stud from those individuals that survived the long journey. Before bringing them northwards, he wintered them on the flat lands of Les Landes, South of Bordeaux, where about 60 died. Four months after reaching Rambouillet, sheep pox carried off a further 35 ewes and 60 lambs. Those of the flock that survived this intensive 'natural' selection, finally flourished, although not without Tessier experiencing considerable difficulties in keeping them healthy. Youatt, writing as 'Lincolnshire Grazier',[75] reported that, until the Merinos had been properly housed, they suffered greatly from cold, and although the wool continued to retain its fineness, it lost something in softness and length of staple.

The animals had been drawn from a large number of Spanish flocks, sufficiently distinct one from the other to show striking differences. Mating them together produced a race resembling none of those that composed the original flocks[76] (Figure 6.3). After its shaky start, this breed of multiple origin was able to establish a good reputation with spinners and weavers. Within three or four years, lambs were being distributed around the country. Lasteyrie claimed that the new breed gave wool of equal quality to that from the Spanish sheep from which it had been developed.[77] Meanwhile, additional Merino sheep were smuggled into France unofficially, mainly through the Basque country, centred in Bayonne.[78] Although these new animals were generally second or third rate, some of the best ended up at Rambouillet to go into the mix. English sheep were also kept there, particularly the Lincoln and Dishley (New Leicester) breeds.

From the beginning, the main reason for importing foreign sheep was to improve the indigenous breeds of France. Education of sheep owners on the advantages of making such improvements assumed a high priority, to which Tessier made an important contribution when he commenced publication of *Annals de l'Agriculture Française* in 1792, on the lines of Young's *Annals of Agriculture* (1784–1815) in England. His colleague, M. Gilbert, who was then teaching at the Royal Veterinary School at Alfort, purchased some of the first Spanish Merinos brought to Rambouillet and organised a practical school for shepherds under government patronage. Daubenton, the original inspiration for all this activity, maintained his premier influence through a stream of books, which his followers revised in new editions, even after his death in 1788. Editions of *Instruction* appeared in 1794, 1800 and 1802, and as late as 1810, with notes by J.B. Huzard who had written an introduction to the third edition (1800). Tessier published his own major book in 1811, which was widely studied both in France and internationally. It was translated into German the same year and published in Berlin.

---

[75] 'Lincolnshire Grazier' 1833, p. 232      [76] Bourde 1967, p. 877
[77] Quoted by Bourde 1967, p. 877        [78] Tessier (quoted by Bourde 1967, p. 1626)

**Fig. 6.3.**   Two illustrations of Merino sheep drawn by the same artist to illustrate a book on Merino sheep by C. Pictét (1808): (A) a four year old ewe with her five month old male lamb from Pictét's own flock at Lancy (near Geneva); (B) a three year old Merino ram, said to be the best from Rambouillet, imported into Lancy.

Independent of the royal *bergerie* at Rambouillet were efforts at Merino breeding by the Marquis de Guerchy in the Ile de France, which began with an intensive drive to combine Daubenton's precepts regarding racial crosses with experience de Guerchy had gained from visiting England.[79] He was enthusiastic about several aspects of English sheep husbandry, including Bakewell's techniques of breeding in-and-in, which he had learned about during his visit to Dishley in 1784 (Chapter 5). Other independent land-owners made their own arrangements for Merino breeding. Many of them were also experimenting with English sheep.[80]

In a treaty signed in Bâle in 1795, a secret clause established that Spain would allow a further 4000 ewes and 1000 rams to leave Spain and enter France.[81] Gilbert was put in charge of the first importation, used to found the national farm at Perpignan. After that, many large flocks were imported from Spain, so that by 1799 there were 6000 of the 'pure race' spread over 25 *départements*.[82] In 1802 it was calculated that there were one million 'Merino' sheep in France, either of pure stock or 'of an ameliorated mixed breed'.[29]

Knowledge of sheep husbandry in Spain, particularly the procedure for transhumance and the distinctive characteristics of different *cabañas*, was possessed by one particular Frenchman above all. This was the political economist and philanthropist Count C.P. Lasteyrie, who lived for a time in Spain. His book *Treatise on the Fine-wooled Sheep of Spain* (1799) was rich enough in new information to be translated within the year into German. Upon his return to France, and with his book published, the government assigned him the task of visiting various northern countries where Merinos were being raised and to report back on the progress being made. We have already mentioned his visit to Sweden. Points of major concern to France were whether the Merino wool degenerated after the sheep had left Spain and how far the breeders in different countries were successful in producing fine wool after crossing their local ewes with Spanish rams. The book resulting from his journey, *Histoire de l'Introduction des Moutons à Laine fine d'Espagne. . .*, published in 1802, attracted a vast international readership, being translated into several languages including German (1804) and English (1810). The author paid major attention to Sweden, Saxony, Prussia and, separately, to the Prussian province of Silesia. Owing to the state of war between France and Austria, he was unable to visit the Hapsburg provinces and therefore could say nothing from personal experience about Moravia, Bohemia, Hungary and Austria. He reported on them only briefly although he was aware that improvement of fine-wooled sheep had been encouraged there more strongly than other branches of rural economy. His book has

---

[79]  Bourde 1953, p. 145
[80]  Bourde 1953, pp. 210–12; Bourde 1967, pp. 860–72, 1620–22 and 1625*ff.*
[81]  Bourde 1967, p. 878        [82]  Bourde 1967, p. 1629

often since been taken as a standard source of information on Merino breeding in central Europe although it has this weakness: some of the better flocks existing at the time were located in Hapsburg territory, barred from his eye (Chapter 7).

In the decade following Lasteryrie's journey, further large numbers of 'original Spanish Merinos', representing all the major *cabañas* and some minor ones, reached France. During the Napoleonic era, upwards of 75,000 were transported across the Spanish frontier according to Carter.[83] In 1809 the Marquis del Campo de Alange presented six rams and 100 ewes from his Negretti flock to the Empress Josephine. Many more came as plunder of war. By 1811 France was producing almost half her own consumption of fine wool and had 10 Imperial Merino studs, generating 1200 new Merino rams a year.[84] At the same time, the French breeders were finding it more difficult than expected to introduce Merino characteristics into French breeds. In each generation of crossing, the progeny were carefully graded and only the best ewes selected to be crossed with Spanish rams. Daubenton had suggested, probably copying Hastfer, that three generations of such grading would be sufficient, although Tessier at Rambouillet was more cautious. He found it necessary to recommend five generations before a mixed breed would possess wool of the same high quality as original Spanish sheep.[85]

Various crosses were used in the construction of the Rambouillet breed, at first not altogether successfully if we can believe the opinion of one notable observer. Among visitors to Rambouillet in 1811 was Bernard Petri from Austria, about whom more will be found in Chapter 7. In a letter from this trip, later published,[86] he wrote of his disappointment with the famous *bergérie*. He found the quality of wool grown on Rambouillet sheep to be low, and the animals more suited to meat production. In an article published in the same year under the name 'Irtep' (the letters of his name reversed), he wrote that sheep breeding in France was 'incredibly retarded' (*unglaublich weit zurück*) compared with the situation in the Austrian Monarchy.[87] He had the chance to make a direct comparison with 1400 Austrian Merinos bred at the Imperial breeding establishment at Mannersdorf, near Vienna, which had been carried off as a prize of war by Napoleon three years earlier,[88] and could be viewed at Rambouillet side by side with French-bred stock. The comparison left Petri in no doubt that the Austrian sheep were greatly superior.

Petri's poor opinion apart, Rambouillet wool was still superior to most. The 'pretty flock' owned by the Empress Josephine, kept at her station at Malmaison, was descended from Rambouillet stock. As Hawksworth has pointed out, this fact 'is of particular interest, on account of the sheep having been transferred to Prussia in 1815, where they have contributed essentially

---

[83] Carter 1969, p. 18    [84] Garran and White 1985, p. 17    [85] Tessier 1811
[86] Petri 1812, p. 151    [87] 'Irtep' 1812    [88] d'Elvert 1870, p. 133

to the extension of clothing wool'.[89] Thus only four years after Austrian Merinos travelled to France as booty of war, French Merinos, possibly incorporating Austrian Merino blood, were being greatly appreciated by the Prussians. Such was the international nature of Merino migration in this era of military conquest and upheaval. An Englishman, Mr Trimmer, who inspected the Rambouillet flock in 1827, commented that the sheep were certainly the largest pure Merinos he had ever seen. On the matter of wool, he admitted it to be of varying quality, fine, 'middling' and even some 'rather indifferent'. However, he judged the whole to be 'much improved from the quality of the original Spanish Merinos'.[90]

The French themselves could be highly self-critical. This is illustrated in a publication originating from the sheep breeding station at Nancy, *New Views on Wool and Sheep*,[91] which is known in two independent German translations.[92] The authors, le Vicomte Perrault de Jotemps (Station Director), his son Fabry, and F. Girod, provided a frank assessment of problems with French wool production. They implied that French breeders had become accustomed to judging sheep by a single criterion—whether they were Spanish —and that manufacturers were lazy about evaluating wool quality. The Moravian sheep expert C.C. André, one of two who translated the book into German, recognised the manufacturers' attitude only too well. The same kind of slackness in Moravia had led him to establish the Sheep Breeder's Society in Brno in 1814 (Chapters 1 and 7). He mentioned a letter from a 'reliable correspondent' in London, written in 1821, stating his opinion that 'French alpine wool has lost all credit in England'. André's conclusion was that selection methods were still underdeveloped in France in 1825 and were much better understood in Moravia. His comparative assessment underlined what his friend Petri had written 14 years earlier.

Interpreting such value judgements raises the problem of distinguishing between reality and wishful thinking. The adverse judgement heaped on France compared with Austria may be partly because Rambouillet received more foreign visitors than Mannersdorf, thus suffering greater exposure to critical eyes. It is as well to bear in mind that the sheep of Rambouillet finally proved their quality beyond doubt. The Rambouillet breed did more than survive the criticisms of the Austrians. It thrived to be exported all over the world and still exists today. Important techniques were involved in its production which created widespread interest: the systematic measurement of wool fineness and the recording of every birth in a flock, generation by generation, with precise information on parentage. Both techniques would be vital in tracing patterns of heredity of wool quality and other characters in future years.

---

[89] Hawksworth 1920, p. 33
[90] Martin 1849, p. 65; Youatt 1837, p. 164 (quoting Trimmer 1828, p. 45)
[91] de Jotemps *et al.* 1824
[92] André 1825; Thaer 1825 (translations of de Jotemps *et al.* 1824)

## Great Britain

The contribution to sheep breeding by Bakewell was universally acknowledged. But what about the Merino breed in his native land? Had Britain got some unique expertise to offer there? Certainly this was the expectation abroad. It was even written that English expertise with Merinos was long-standing because Spanish sheep had contributed to the ancestry of certain English breeds in the long distant past (Chapter 2). But although it seemed logical to think that breeds with the finest wool might have some connection, no real evidence could be produced. If it were true that some Spanish sheep had arrived in England from time to time in earlier years, they must have been relatively few in number, now absorbed into English breeds without trace. Sir Joseph Banks discounted such introductions altogether when he wrote to Augustus Henry Fitzroy, Duke of Grafton, in August 1791, 'I have not however been able to learn from either history or tradition that the Merino sheep were ever imported into this country till the arrival of those His Majesty now possesses.'[93]

Irrespective of what may or may not have happened in the past, most of Banks' fellow countrymen showed little initial enthusiasm for keeping newly imported Merinos on home territory. Their inflated conviction regarding the superiority of British native breeds, 'the best in the universe',[94] exceeded even that of the French, producing an almost insuperable barrier for a foreign breed to overcome. Added to this, there was the anti-wool influence of the soon-to-be-famous Bakewell and his friends, preaching the economic gospel of mutton: meat first, wool second. Finally there was the rainy British climate, encouraging liver fluke and foot rot, the twin penalties of moist meadows, which were confidently expected to play havoc with the Merino's mountain-bred constitution. Many British landowners who might have afforded to make experimental introductions were convinced that the breed could not remain healthy and productive in the British climate. It was only when alerted by news of Merino breeding successes on the continent, in Sweden and Germany, and above all in France, that a few of them began to have second thoughts. Daubenton's *Instruction* (1782) was probably a major factor. Thoughts finally turned into actions as shortages developed in the supply of British wool, accompanied by an increase in the import of Spanish wool. In 1784, an Irishman William Conyngham imported a few sheep direct from Castile and others via Portugal.[95] A year later Banks obtained 'two small Spanish sheep', claimed to be of the 'pure Merino blood', for his own use. These were a ram and a ewe, from the Royal Veterinary School at Alfort (Burgundy), 'the Public Flock of M. Daubenton, by means of Dr Broussonet'.[96] After three years, at the King's command, he introduced some others under

---

[93] Carter 1979, p. 209    [94] Practicus 1802    [95] Carter 1979, pp. 155 and 554
[96] Foot 1794, p. 60; Carter 1964, p. 48; Carter 1979, p. 559

cover from Spain, via Portugal. Banks also obtained some progeny of crossed Merinos from different parts of France, including Rambouillet.

The small flock that Banks imported for the King in 1788 was conveyed secretly from Spanish proprietors in various districts.[97] Taken from different flocks, the animals were disappointingly poor in quality, failing to establish the true characteristics of the Merino breed.[98] Representations were made to the Spanish monarch for something better. In 1792, he gave permission for King George's Minister to the court, Lord Auckland (William Eden, first Baron Auckland), to select five rams and 35 ewes from one of the best and most valuable Negretti flocks.[99] The sheep which, according to Lasteyrie, were from a *cabaña* producing the largest and strongest type of Merino, and, according to Banks, with the reputation 'for purity and fineness of wool, . . . as high as any in Spain', were owned by the Marchioness del Campo d'Alange and were exchanged for eight fine English coach horses.[100] Despite their acknowledged excellence, the King thought the price too high. He passed a message through Banks in a letter to the skipper of the ship that transported them, Mr Anthony Merry (June 1792), that he wanted further sheep of the same high quality for his Hanovarian dominions but 'without running any risk of another lot of horses becoming the necessary political return'.[101]

Transported to the royal farm at Kew, West of London, the costly animals were kept as a pure flock on a rich moist soil, quite unlike their native conditions. Because of the altered environment, experts predicted that the nature of the wool would change. Among the most influential was J. Boys of Betshanger (Betteshanger) in Kent, a sheep man well known to Bakewell, having been his guest at Dishley on two visits (in 1785 and 1792). Boys shared the widespread belief that nutrition could modify the pattern of heredity. Even Bakewell was quoted as saying 'that the same kind of sheep would not produce such fine wool in rich as poor pasture, which is certainly agreeable to past experience and observation'.[102] Those responsible for the sheep were determined to prove the pessimists wrong, and for some years they claimed the production of wool to be the same or even better than in Spain.[103] Sebright summarised the situation as follows: 'Climate, food and soil, have certainly some effect on the quality of wool, but not so much as is generally

---

[97] Trow Smith 1959, p. 152. Banks saw this as his patriotic duty: 'To depend upon a Country naturally unkind to you for the Raw Material of the finest branch of your Principal Manufacture, and to be in hourly danger of the Privilege of Obtaining it being resumed, is a humiliating consideration to a great nation.' (Letter from Banks to Thomas March 1788, in Carter 1979, pp. 161–2.)    [98] Youatt 1837, pp. 177–8

[99] 'Lincolnshire Grazier' 1833, p. 233

[100] Carter 1979, p. 533 (quoting Banks 1800); 'A project for extending the breed of fine wooled Spanish sheep. . .'

[101] Carter 1979, p. 233: letter from Sir Joseph Banks to Anthony Merry, diplomatist, July 1792    [102] Redhead and Laing 1793, p. 24

[103] Sebright 1809, pp. 19–20; Youatt 1837, p. 178 (quoting *Agricultural Magazine* 1803, 1806 and 1810)

supposed.'[104] Even many years later the Chairman of the London Committee on the Wool Trade, appearing before the Committee of the House of Lords on the Wool Trade in 1828, was able to report that the price per fleece was double that of Southdown wool.[105] While the English Merino product had undoubtedly become 'more harsh', the fleece weights had increased in compensation.

Notwithstanding this evidence, most British farmers remained unconvinced that Merinos were worth serious attention. The best evaluation of British-bred stock may be gained from the sale prices, which remained disappointingly low, year after year. Despite continuing to produce an abundance of fine wool, certain aspects of the appearance of these sheep, like their hairy topknot and cheeks the fleshy protuberance behind the horns, deep wrinkles above the nose, and the hollow back (Figures 2.1 and 6.3), made the breed unacceptable to English eyes.[106] It might seem odd that such matters were so important, but 'beauty' and 'utility' were inseparable in the mind of the eighteenth century English breeder. Not only did the Merinos seem unattractive—'the body not very perfect in shape', to use one writer's words[29]—but they were also slow to mature, in sharp contrast with some of the new English breeds. Even the quality of their wool was discounted, since equally fine wool could be obtained more cheaply by importing it from abroad. At home, greater profits and a lesser insult to the eye came from breeding the impressively long-wooled sheep of Lincolnshire, or the new quick fattening varieties, like Bakewell's New Leicester or Ellman's Southdown.[107]

This is not to say that many sheepmen did not try the Merino. Back in 1792 when the King received the Negretti flock brought to him by Lord Auckland, he ordered that all the sheep obtained earlier via Portugal should be 'disposed of'.[108] They were offered as gifts to various landowners including Arthur Young, Sir Joseph Banks and the Duke of Grafton.[109] This started a trend of experimenting with Merinos in cross-breeding, which lasted about two decades. Numbers were expanded as a result of further imports from Spain, Portugal or France by private citizens. In 1802, Young was able to inform Lasteyrie that there were Merinos 'in almost every district of Great Britain'.[110]

Even so, few farmers saw good reason for keeping the exotic sheep in Britain except to cross them to native breeds. Sir Joseph Banks cross-mated one or two of his Merino rams, just arrived from France, to two ewes from Caithness supplied by Sir John Sinclair.[111] Because Bakewell had given the idea for the cross, Sir Joseph generously sent him one of the lambs, a male, which Bakewell kept side by side with one born to the same kind of ewe mated to a Dishley

---

[104] Sebright 1809, p. 25   [105] 'Lincolnshire Grazier' 1833, pp. 230–31
[106] Carter 1979, p. 534 (quoting Banks 1800); Bischoff 1828, p. 58; Youatt 1837, p. 179; see also Chapter 2   [107] Youatt 1837, p. 181
[108] Carter 1979, p. 552
[109] Carter 1979, p. 206–7: letter from Sir Joseph Banks to John Robinson, 7 August 1791
[110] Lasteyrie 1802, p. 143   [111] Mitchison 1962, p. 112

ram, to compare them. He wrote to Banks in 1787 enthusiastically offering to show him the 'striking difference' between the two hybrid animals,[112]

> which I flatter myself will throw much light on this important subject [weight gain in relation to food intake as an intrinsic property of a breed] and lead to many useful discoveries which have hitherto been almost unnoticed. If you favour me with a call at Dishley I shall be happy to communicate any knowledge I have acquired in this or any other line of business in which I have been engaged.

Although Sinclair had responded to Bank's request for Scottish ewes to match with 'Monsieur Ram' from Montbard,[113] he did not at the time want to be involved with such experiments himself. But by 1790 he had 'woken up to the subject', when he began to build up his own Merino flock at Queensferry near Edinburgh. His first 15 'Spanish sheep' came from France and were in fact of mixed origin. Their arrival stimulated considerable local interest, leading Sinclair and his Scottish friends to form their British Wool Society (Chapter 5). Eventually Sinclair put together an experimental mixed flock of 800 sheep at Queensferry to rival Banks' experimental flock at Spring Grove, Middlesex. Shearings were held annually to compare wool quality on sheep of different origins, with the intention of raising the public profile of the Merino breed in Scotland. In one year the Society imported 15 of Daubenton's best flock to compare with their own, as well as a ram and two ewes of pure Swedish stock from Lord Seton of Preston.[114] But the Society did not fulfil its intention of organising wool marketing on a large scale, and it faded out after only three years, as Sinclair spent more time in London. Appointed President of the newly formed Board of Agriculture, he was busy commissioning the famous County Reports, organised by the secretary of the Board, Arthur Young.

It was left to a few private individuals to make the Merino breed 'permanently useful'.[115] Foremost among them were two who added to their reputations through influential publications. These were Lord John Southey Somerville (1765–1819), who was the largest owner and breeder of Merinos in the country, and Dr Caleb Hillier Parry FRS (1755–1822), a general medical practitioner of Bath. With farms not far apart in the County of Somerset (Figure 4.2), the two breeders interacted closely. Together they joined with others who exchanged information in the Bath and West Society, whose Wool Committee included many clothiers as well as breeders.

Somerville was a hard-headed businessman who turned a loss-making farm into a valuable property.[116] He purchased breeding stock via Portugal

---

[112] Carter 1979, p. 121       [113] Carter 1964, p. 54
[114] Carter 1979, p. 204: letter from Lord Daer to Sir John Sinclair, 29 June 1791;
*Encyclopaedia Britannica* (1797) Vol. 18, 'Wool' section       [115] Culley 1807, p. 247
[116] Clarke 1897

and, having brought home a flock of *transhumantes* of first quality in 1801,[117] he dedicated 460 acres on his farm to sheep, both the pure Merino and crosses to English breeds. Parry, with only 60 acres for sheep, bred on a smaller scale than Somerville although no less successfully. Trained in medicine in Edinburgh, with an earlier education in geology at Warrington Academy, he recreated himself as a scientific agriculturalist at the age of 33. Obtaining rams from His Majesty and later from Somerville, he crossed them to Ryeland ewes, backcrossed them four times to Merino and then inbred.[118] Details of his progress are to be found in a long 'essay' on the Merino sheep, which he communicated to the Board of Agriculture in 1806.[119] It provided a full account in English of the grading procedure, designed to produce wool 'as good as pure Merino', in the manner of Daubenton and a host of breeders on the continent. French experience suggested that five generations of grading might be required, although Parry was able to demonstrate that excellent fine wool could be produced in a flock of Ryeland sheep after four generations, a product almost equal to the best Spanish in fineness and texture.[120] The sheep were known as Merino Ryelands, and the wool as Anglo-Merino.

Crosses to the Ryeland breed were accompanied by efforts from both Somerville and Parry at improving the fleeces of other British breeds—Southdown, Wiltshire, Dorset, New Leicester, etc.—drawing upon the enthusiasm of other Merino owners.[121] Most of the experiments were short-lived, although they were successful to the extent that appreciable Merino blood was added to the Southdown in particular, and to breeds upon which the Southdown was crossed. Typical characteristics of the Merino, the tuft on the head and loose skin hanging from the throat ('throatiness') could also be discovered in some Ryeland sheep, 'showing its origin', to an extent that a good judge, T.A. Knight, could readily 'distinguish these from the 'Archenfield or true Ryeland sort'.[121][122] Many breeders around the country possessed one or more Merino rams for crossing purposes at this time.

A much cherished aim in the British Isles was to produce a 'British Merino', a breed combining the fleece quality of Spanish sheep with the virtue of British ancestry in a quick rate of fattening for the market.[123] It was hoped thereby to create a superfine-wooled breed suitable for British conditions, with an acceptable appearance and the ability to put on weight quickly. Fine quality wool could not command a sufficiently high price in Britain to compensate for any loss arising from a deficiency in the carcass. For such was

---

[117] Culley 1807, p. 254     [118] Culley 1807, pp. 248–50
[119] Parry 1806; see also Parry 1800
[120] Parry 1806; Culley 1807, p. 248; 'Lincolnshire Grazier' 1833, p. 224; Youatt 1837, pp. 178–9     [121] Rees 1819 ('Sheep' section); Pitt 1813, pp. 28 and 217
[122] Carter 1979, p. 312: letter from T.A. Knight to J. Banks, 29 May 1799
[123] Somerville 1803; Laurence 1809, p. 402; Rees 1819 ('Sheep' section)

the demand for meat in this country and such was the price it commanded, that it remained the principal source of profit and indeed the only one that could 'meet the heavy expenses incurred in raising artificial food'.[124]

Young believed from the beginning that a breed of sheep showing all desired qualities could only be produced by the Bakewell approach, that is by mixing and matching traits. He came to this conclusion after talking to the great man while staying with him at Dishley in August 1791:[125]

> If ever it [the ideal sheep incorporating Merino blood] is successfully made, it will be owing, not inconsiderably to the unwearied attention and fortunate event of Mr Bakewell's exertions, who has so instigated mankind in this useful path, that unthought of discoveries may be the consequence.

Young considered four breeds 'most worthy of attention': New Leicester, Southdown, Hereford (= Ryeland) and Spanish. He could not conceive it possible to form the ideal sheep without the assistance of all those four breeds.[125]

T.A. Knight was among the leading breeders in this endeavour, basing his breeding programme on crosses between a Spanish ram and Ryeland ewes. In a letter to Banks in 1804, he described his technique, how he had 'found that better formed animals and indeed better fleeces may be obtained by crossing the best half-bred Ram with the best half-bred Ewes than can be obtained from the Spanish Ram in the first instance'.[126] Knight was picking out the most favourable combinations of traits from the *progeny* of hybrids (i.e. $F_2$ segregants).

Dr Parry had written to Banks several years earlier with his own plans. In a long letter dated 26 April 1800,[127] he explained how, having graded English sheep with Spanish to obtain 5/6 Spanish stock, he intended to 'select a superior form and a constitution readily disposed to fatten' by the method 'by which Mr Bakewell in Leicestershire and Mr Ellman in Sussex, have given their flocks the reputation that they ever since continued to possess'. He was intending to select for carcass quality equal to the best Southdowns.[128]

Such crosses made by Parry, Knight and others, with the aim of creating the British Merino (Anglo-Merino), proved a disappointment.[129] Although the fleeces of the first generation exceeded in fineness any of the British breeds (particularly in Southdown, Devon and Ryeland crosses), and even many Spanish sheep,[130] the hybrid animals proved very expensive to feed in relation to their gain in weight. Bakewell had been among the first to demonstrate this by experiment, weighing both the animals and the food they

---

124 Spooner 1844, p. 61     125 Young 1791b; Young's *Tours* (1932), p. 325
126 Carter 1979, p. 431: letter from T.A. Knight to J. Banks, 29 December 1804
127 Carter 1979, pp. 319–21     128 Culley 1807, pp. 251–3
129 Sebright 1809, pp. 17–19
130 Mitchison 1962, p. 112; Carter 1979, p. 369: letter from Parry to Banks, 29 August 1802

took.[131] In backcrosses to the Merino, undertaken by his successors, the form and fattening quality of the carcass deteriorated seriously; when further British blood was introduced, the quality of the wool was in danger of being lost. What had at first seemed a bright hope ultimately provided a challenge few were prepared to meet.[132] In 1809, Sebright wrote, 'This experiment has been frequently tried by others, as well as by myself but I believe, never succeeded. The first cross produces a tolerable animal but it is a breed that cannot be continued.'[133] The alternative strategy, which was to forget about fattening and to breed exclusively for wool quality, seemed more achievable.[134]

Zealous advocates of the Merino breed joined together in 1811 to establish the Merino Society, with Banks as President and Benjamin Thompson, fluent in German,[135] as Secretary. The membership, including by this time many who were economically committed to the new breed, rose to 377 by 1813.[136] The Society offered prizes for the best Merino rams but, almost at once, just at the time when German Merino breeding was beginning to flourish (Chapter 7), enthusiasm for keeping Merinos in Great Britain began to decline. Fine wool could be obtained more cheaply from abroad than it could be grown in the British Isles.[137] It was not worth breeding slow maturing sheep for fine wool when mutton production was more profitable. Here we may refer to a letter of 16 March 1812 from the Commissioners of His Majesty's Real and Personal Estate to Sir Joseph Banks, recommending a programme of slaughter, sale and castration to thin down the royal flocks to manageable numbers. Referring to Great Britain as a whole, they stated 'There are from 4–6,000 imported sheep now in the country which cannot be sold even on low terms, and the maintenance of which becomes more and more burthensome to the owners.'[138] Banks died in 1820 and, without his leadership, the Society ceased to meet by 1821.[139] A few breeders persisted with the Merino, among them Lord Western, who still retained a flock in 1844, not the pure breed but 'Merino crossed with the Leicester and the mixed breed, thus produced, perpetuated'.[140] The penalty he paid for success was a high slaughter rate, as he was obliged to practise stringent selection to remove those that had 'degenerated'.

In 1845, Low wrote that 'There scarcely exists, except in the hands of the curious, a single flock of the mixed progeny from which so much was

---

[131] Redhead and Lang 1793, p. 22     [132] Trow-Smith 1959, p. 153
[133] Sebright 1809, pp. 17–18     [134] Sebright 1809, pp. 22–5
[135] Carter 1979, p. 586
[136] Carter 1979, p. 522; among the 307 members in the year of the Society's foundation, only 14% were of the nobility (Thompson 1811; Phillips 1989, pp. 90–91)
[137] Martin c.1849, p. 717; the Spanish blockade had made Spanish wool unavailable between 1804 and 1809 (Carter 1964, Figure 5)     [138] Carter 1979, p. 513
[139] Garran and White 1985, p. 17     [140] Spooner 1844, p. 62

anticipated.'[141] William Martin summarized the position in more dramatic terms: 'Thus then, as the most extensive experiments with the Merinos were in operation, their sun suddenly went down, and has never since brightly beamed above the horizon.'[142]

But was unprofitability the true explanation for the failure of the Merino in Britain? One person who proposed a different reason was the Swiss breeder Charles Pictét, who kept a flock of 400 Merinos at Lancy near Geneva (Figure 6.3).[143] In a book published in Geneva (1802), he criticised the attempt to create the 'British Merino', as championed particularly by Parry. Pictét believed in keeping his 'Spanish flock' absolutely pure as he had done for the previous five years since importing the sheep from Rambouillet. Somerville, who was himself hopeful that the pure breed might be improved in the Bakewell manner by selecting without crossing,[144] published a translation into English of the relevant passage.[145] He also quoted a private letter in which Pictét wrote:[146]

> Is it your opinion that there will be more profit for you in crossing Spanish rams with Ryeland and Southdown ewes than in breeding the Spanish sheep in-and-in? Have you any experiments which make you suppose that a deterioration could take place by persevering exclusively in the same blood?"

Pictét was convinced that 'the Spanish breed of sheep is a true mine which the English have refused to explore till now.'[147] Lasteyrie formed the same opinion (see quotation following the title of this chapter).[148]

The outside world was unconvinced that the British were taking Merino breeding seriously. Much as there was good cause to admire Bakewell and his New Leicesters and the other newly created English breeds, the British Merino experiment could only be viewed as unproductive by comparison and unimportant from a theoretical point of view. The crosses made to create a quick fattening Merino did little to advance the understanding of heredity. A major stumbling block was created by the seeming unpredictability of trait segregation after the first hybrid generation in a cross between disparate breeds. While it was clear that the qualities of wool could be modified by selective breeding,[149] just as rates of growth and qualities of body shape had been changed, it seemed impossible to combine all these desirable characteristics within a single breed.

British fine wool was soon to be surpassed by the superior quality of German fine wools. The author of the *Complete Grazier* summed up the position:[150]

---

[141] Low 1845, pp. 147–8      [142] Martin c.1849, p. 71      [143] Somerville 1809, p. 49
[144] Somerville 1809, p. 41      [145] Somerville 1809 p. 49      [146] Somerville 1809, p. 53
[147] Pictét 1802; Somerville 1809, p. 52; see also Pictét 1808 for advice on crossing
[148] Lasteyrie 1802, p. 118      [149] Sebright 1809, pp. 24–5
[150] 'Lincolnshire Grazier' 1833, p. 277

The whole evidence before the Committee of the House of Lords, appointed in 1828 to inquire into the state of the wool trade, goes to prove that the wools of Bohemia [of which kingdom, Moravia was part] and Saxony have entirely superseded the British short wool in the greater part of our cloth manufacture; and the consequence has been that the value of the latter has fallen below the remunerating price to the grower. To this alarming fact is to be added that of the rapid increase in fine wool flocks in New South Wales, which bid fair, at no very remote period, to supply the whole demand of this country.

In defence of native British wool, it must be said that the extra long lustrous combing wool of Lincolnshire, the product of many generations of selective breeding, still dominated the European heavy worsted industry. Sales of such wool increased substantially in Britain between 1800 and 1828.[151] The number of packs of short wool registered declined to almost half, while long wool production doubled. In their class, the English long-wooled sheep were supreme and much desired as breeding stock on the continent, especially in France. As for fine wool, British interests were diversifying into colonial production. It would not be many years before Merinos bred by British colonists far overseas would supply the fine wool industries not only of Britain but of much of the world. In the meantime, breeding had been greatly expanded in the ambitious German-speaking countries, now to be considered. It was here, out of all the European regions, that the breed was most successful outside Spain, and it was here that the theoretical basis of selective breeding created greatest interest at this period of time.

---

[151] 'Lincolnshire Grazier' 1833, p. 283

# 7

# *Merinos in German-speaking countries and Australia*

Today the wool of Silesia and Moravia rivals the most beautiful electoral wool of Saxony

*M.D. de Liège, 1842 (Original in French)*

The release of Merino breeding stock beyond the borders of Spain created no greater impact than upon the countries of central Europe. Within little more than half a century, Saxony, Prussia and Austria–Hungary had become major European producers. Then, on the other side of the world, Australian colonists woke up to the breed's potential. Having imported a few of these sheep via the Cape of Good Hope or from England, they sought out the best animals for their purpose from France, Spain and central Europe. Saxon and Silesian/Moravian versions of the breed thrived particularly well in Australia, to produce first quality fine wool for export back to Europe. The development and exploitation of central European versions of the Merino, both in Europe and in territories overseas, revealed further details of the nature of heredity, especially with respect to racial crosses and the introduction of the Bakewell techniques to Merino breeding.

## Saxony

It was the opinion of Johann Heinrich Fink (Finke) (1730–1807) of Cösitz in Upper Saxony, writing to Sir John Sinclair in 1797, that although 'old and simple Germans were satisfied with any kind of wool, ... it is certain that almost one hundred years ago a higher value was placed upon fine than coarse wool'.[1] Arising from the change in taste, a market was created for fine wool imported from Spain and the way prepared for the arrival of the first Merino sheep in Saxony in 1765. They came as a flock of the famous Escurial *cabaña*, a personal gift to the Elector from his cousin King Carlos III of Spain. The 92 rams and 128 ewes that finally reached the Electorate alive, after their arduous journey, promised a new source of revenue in fine wool and cloth production, particularly in relation to their potential for improving inferior native sheep.[2] A gift of such significance to a country impoverished by the Seven Years War was most welcome. The exotic creatures, one of the smaller kinds of Merino with relatively long, curly white wool (Figure 7.1), were established in an area of pastureland considered closest to their liking, at the royal domain of Stolpen near Pirna, 30 km East of Dresden, in the approaches to the region known as 'Saxon Switzerland'[3] (Figure 7.2).

From Stolpen, sires were sent to neighbouring state farms at Lohmen, Rennesdorf and Hohenstein, to be crossed with local ewes. Generally the local sheep of Saxony and neighbouring provinces did not present a favourable impression. The most common type, described by Fink,[4] was hornless with a hairy head and almost naked belly and with 'the scrotum of the male sex . . . very short, and quite smooth, without hair or wool'. To the delight of the breeders, the $F_1$ hybrids with Spanish rams lost all these unattractive features, resembling very closely the Spanish phenotype, including their size. The imported sheep and the hybrid animals that resembled them became known as 'Electorals' (Figure 7.1). Fink assured Sinclair in his letter of 1797 that some Saxon flocks were improved to such a degree that the wool was as fine as the best sort of Spanish wool.[5] What is more, in response to a growing demand for further Electoral breeding stock, driven by the Elector's insistence that all occupiers of crown land should buy a certain number of them,[6] a second major shipment arrived from the same Escurial source in 1776–8.[7] The number of 300 ewes and 100 rams is sometimes quoted although these totals seem to have included sheep 'from Poland, Silesia and other countries'.[8] Hawkesworth mentions 89 rams and 16 ewes arriving from Spain in 1779 at a cost of £1500, 'an enormous sum in those days'.[9]

As Saxony became a new focus for Merino breeding in northern Europe, those responsible were concerned to keep technically up-to-date. We suppose that it was for this reason that in 1787 the Elector's special envoy in London, Hans Moritz Graf von Bruhl, a corresponding member of the (British)

---

[1] Fink 1797, p. 276
[2] Youatt 1837, p. 170; Cox 1936, pp. xi–xii; Baumgart 1957, p. 9
[3] *sächsische Schweiz*     [4] Fink 1797, p. 278*ff.*
[5] Fink 1797 p. 278; see also Rees 1819 ('Sheep' section)     [6] Youatt 1837, p. 170
[7] Banks 1808; Somerville 1809, p. 45; Rees 1819 ('Sheep' section)
[8] Fink 1797, p. 277     [9] Hawkesworth 1920 (no original source given)

A

B

**Fig. 7.1.** A Saxon Merino ram (A) compared with an original Spanish ram of unnamed *cabaña* (B). Both engravings are by J. Jackson, published by Youatt 1837.

**Fig. 7.2.** Map showing prominent Saxon, Prussian and other German estates for Merino sheep breeding, in relation to principal towns and cities.

Society of Arts, and long-time resident in England,[10] sought advice from the most famous breeder in Europe on methods of 'breeding, rearing and feeding the best kinds of stock'. The Englishman's answer was that the art of breeding 'was in choosing the best males to the best females and keeping them in a thriving state'.[11] It was unnecessary for Bakewell to reply in further detail, for he could recommend the newly published book by his pupil Culley, *Observations on Livestock*,[12] which reflected precisely his own methods.

Just one month after the Elector sought advice from Bakewell via von Bruhl, Sir Joseph Banks was directing an enquiry on the same subject in precisely the opposite direction. The answer, provided by Morton Eden, first Baron Henley, resident in Saxony, assured Banks that the Merinos there did not degenerate when the breed was kept pure and 'housed and better fed than German sheep and . . . not fatigued by long droving'.[13] Banks contrasted this information bitterly with his own experience of breeding Merinos in England. In a letter to Thomas March, a merchant of Lisbon, on 13 July 1789, he wrote:[14]

> The Spanish sheep which the King of Spain gave to the Elector of Saxony more than 20 years ago have not degenerated in his Electoral dominions and why should those you are so good as to send me degenerate here where sheep are so much better understood and managed?

Later reports by foreign visitors to Saxony, as for example by the French expert Lasteyrie in 1802, suggested that the problems faced by Banks were by no means unknown there.[15] The true state of Merino sheep in Saxony had been only partially revealed to Baron Henley. On a technical point, we may note that the advice that Banks got from Saxony in 1787, to keep the breed pure, was different from Bakewell's insistence upon selecting only the best individuals for breeding, both male and female.

A struggle to avoid degeneration faced every importer of Spanish sheep, whatever his nationality. Saxon breeders were experiencing the problem of degeneration, just like breeders in England or elsewhere. Recognising the risk, and eager to gain maximum value from their Merino rams while still in prime condition, they attached major importance to using them as soon as possible for upgrading their native breeds, particularly lowland flocks. The crossing system to be employed was the grading technique used by horse breeders, as described by Georg Stumpf (1785) in an influential book. Aware that the fifth generation of crosses to a Merino ram would yield progeny with 31/32 of Spanish blood, Stumpf reckoned that three or four generations

[10] Braun 1971
[11] Pawson 1957, p. 107: letter from Bakewell to Culley, 8 February 1787
[12] Culley 1786
[13] Carter 1979, p. 107: letter from Morton Eden to Banks, 12 March 1787
[14] Carter 1979, p. 173     [15] Lasteyrie 1802, pp. 26–7

would infuse a sufficient amount of it for practical purposes. The book in which he advanced this advice was considered important enough for an English translation to be issued by the Royal Dublin Society in 1800. In contrast to the breeding procedure recommended by Culley (1786), Stumpf insisted upon a different ram for every generation of crossing. He totally rejected selective inbreeding in the Bakewell manner: 'Rams should not be let to ewes of their own begetting', he wrote, 'nor should they remain in the flock more than three years.'[16] This represented a totally different attitude from that adopted by Bakewell as revealed by Culley.

Opposition to breeding in-and-in was commonplace among German writers,[17] although some took a rather softer line than Stumpf. Such a man was Christian Baumann (1739–1803), a Cistercian monk who spent most of his life in the city of Würzburg in Bavaria. In a book, published in both Frankfurt and Leipzig, Baumann claimed to offer a completely fresh look at animal breeding. The title describes its far-reaching scope: *Arrangements necessary for propagating, improving and beautifying horses, cattle, sheep, goats and other domestic animals, without degeneration, which, at the same time, are compatible with well tried expedients whereby a complete field economy is permanently established, expanded and improved.*[18] The author emphasised the importance of improved fodder crops and suitable animal housing but, above all, the value of selecting superior males to mate with the best females, to avoid degeneration. Selective breeding, he pointed out, had been brought to a high art in England, which he had learnt from reading the publications of Arthur Young.

Reviewing progress in sheep breeding in the German states, Baumann concluded that, although in general the best Merino sheep were to be found in Saxony, the master of the breeding art was located outside Saxon territory.[19] This was the talented economist von Bori (Borie) who farmed at Neuhaus, close to Bad Neustadt an der Saale about 60 km North of Würzburg. Noting with approval how von Bori had introduced Spanish rams capable of yielding the 'the most beautiful progeny', Baumann was interested to point out that 'Mr Bori hired his rams to his neighbours to evaluate their offspring following the example of Bakewell.'[20] This reference to Bakewell in the context of progeny testing and selective breeding provides early evidence of a practical impact of his methods on German sheep husbandry, even before von Bruhl's enquiry. Baumann was still a traditionalist at heart, however, which is clear from the emphasis he places on the prepotency of pure-bred Spanish Merino rams. He summarised the 'whole secret' of sheep breeding in terms of crossing Spanish rams with local ewes. Less strict than Stumpf, he allowed a particular ram up to four years in the same flock, although still

---

[16] Stumpf 1800, p. 59   [17] Rees 1819 ('Sheep' article)   [18] Baumann 1785
[19] Baumann 1785, p. 217   [20] Baumann 1785, p. 273

sharing Stumpf's conviction that inbreeding must be moderated if progeny were not to degenerate. For this reason he always had fresh rams to replace those which had reached their time. This he achieved by importing some fresh ones every two years.[21]

When Lasteyrie visited Saxony in 1800, he found 160,000 sheep there, of which 90,000 were either uncrossed Merinos (pure races) or the product of crosses to local sheep (improved races).[22] Although he found lessons to be learned from Saxon sheep husbandry, that might be copied to advantage back at home in France,[23] some of the animals he saw did not impress him.[24] He concluded that the problem of degeneration had not been solved, and attributed this to lack of care, inadequate feeding, unhealthy stabling, grazing on recently manured land and 'poorly matched couplings'.[25] Other foreign observers would repeat the criticism in the years that followed, although it was generally accepted that the fine wool exported from Saxony was a highly desirable product. 'It can be spun to a greater length than any other carded wool grown in Europe; it is also superior in fineness', wrote the author of the section on sheep in Rees's *New Cyclopaedia* (1819). At the same time it came to be appreciated, as this author himself noted, that owing to the scarcity of winter feed, the Saxon fleeces were smaller than the best Spanish ones and the wool 'not so well suited for stout clothes'.[17] Some years later, in the 1820s, the truth of the exceptional fineness of Saxon wool was witnessed by R. André from Brno in Moravia, who made close comparisons of different wool samples using the 'Dolland wool measurer', a micrometer device from England.[26] But although André accepted the fineness of Saxon wool, he was conscious that the fleeces as a whole were not as good as they might be. He was critical of Saxon breeders for failing to exploit the techniques of trait recording, individual selection and mating by 'clear affinity' (i.e. on an assortative basis), to maintain improvement and ensure a uniform product.

Standards were evidently quite variable on different Saxon farms, and quality did not necessarily correspond with the social status of the owner. A reluctance to innovate, and thus to reject the practice of their ancestors, was as strong in Saxony as elsewhere.[27] James Atkinson, in company with James Macarthur, two Britons who had settled in Australia, visited the royal Saxon farms in 1826. By this time most of the King's sheep were concentrated at Lohmen (Figure 7.2), 1100 sheep 'of all ages and descriptions'. The visitors approached the royal flock with high expectations but were shocked and disappointed by what they saw. The poor condition of the sheep was explained to them by the persons in charge, who admitted that 'for a long

[21] Baumann 1785, p. 274    [22] Lasteyrie 1802, p. 28    [23] Lasteyrie 1802, p. 32
[24] Lasteyrie 1802, pp. 23–9 and 165–76    [25] Lasteyrie 1802, pp. 26–7
[26] Teindl 1822    [27] Youatt 1837, p. 170

course of years' the flock had been 'much neglected'. Only in very recent times had a better system been adopted, the visitors were told. They could see for themselves that many of the animals did not even have the superficial appearance of Merinos. When later they inspected the flock of His Royal Highness the Grand Duke of Saxe-Weimar, which consisted of 4300 sheep 'of all ages and descriptions', they found similar cases of neglect. Enquiries revealed the surprising news that, because of inadequate supervision of breeding, dishonest shepherds had been able to enrich themselves to the detriment of the Crown by exchanging sheep of the improved flocks 'for others out of inferior flocks in the neighbourhood'.[28]

Proceeding with their tour, Atkinson and Macarthur's personal inspections of 'the most celebrated flocks in Saxony' led Atkinson to conclude 'that the pure merino race, perfectly unmixed with the original breed of the country, can scarcely be said to exist at the present time [1828] in that country, or perhaps anywhere in Germany.'[28] Nevertheless, they did come across animals of excellent quality, even though of mixed genetic background, including some that were unexpectedly large:[28]

> In the flock of General Leyser [August Wilhelm von Leysser 1771–1842; see Gorzny 1986, p. 1250] at Gersthoff, were some animals derived from the flock of Prince Libournouski [Lichnowsky] of Troppau [Opava in Moravian Silesia] which is said to be the finest in Germany. These animals were large, with long and extremely fine wool and presented more of the characters of the Negrett [*sic*] Merinos than any sheep we saw, with the exception of a few in the flock of the Grand Duke of Saxe Weimar.

The references to Lichnowsky and to sheep with Negretti characteristics (Figure 2.1) are significant in relation to the thesis to be presented below and in later chapters, that sheep stock from Moravia, and in particular its northern region known as Austrian Silesia, played a large part in reviving the Electoral reputation. The Lichnowsky family represented a major force in Silesian sheep breeding, both North and South of the Prussian/Austrian border.

These two observant and critical travellers from Australia concluded that such selection programmes as were being carried out in Saxony in 1828 were based on stock of mixed origin and relied on a compromise between two different wool quality traits, length and fineness.[29] From a different source we learn that on the better Saxon farms, selection was carried out in a most systematic manner. The technique is described by Youatt, based on an account supplied in a letter from a Mr Charles Howard:[30]

> When the lambs were weaned, each in his turn is placed upon a table so that his wool and form may be minutely observed. The finest are selected for breeding and receive a first mark. When they are one year old, prior to shearing

---

[28] Garran and White 1985 p. 134      [29] Garran and White 1985, p. 135
[30] Youatt 1837, pp. 171–2

them, another close examination on those previously marked animals takes place: Those in which no defect can be found receive a second mark, the rest are condemned. A few months afterwards, a third and last scrutiny is made; the prime rams and ewes receive a third and final mark. The slightest blemish is sufficient to cause the rejection of the animals. Each breeder of note has a seal or mark secured to the neck of his sheep, to detach or forge which is considered a high crime, and punished severely.

Youatt made no mention of compromising on wool fineness in favour of fibre length, although he quoted Howard as stressing the importance of non-wool qualities in the lambs selected.[31] The typical Saxon Merinos were smaller, more rounded and finer boned than the original Spanish breed (Figure 7.1), a form which met with Youatt's approval because it 'indicates a disposition to fatten'.[32] He noted, however, that this good opinion was not shared by another commentator, Mr Trimmer, a sheep farmer from near Kew, West of London, who had written that 'By the *constant* confinement of the sheep in winter in houses, they have degenerated into a puny weak race, producing only half the weight of wool and mutton which the present merino stock, from which they sprung yielded.'[33] Another English observer recording an unfavourable impression was William Howitt, author of *The Rural and Domestic Life in Germany. . ..*[34] He saw his first flocks between Dresden and Leipzig, and commented[34] with surprise to see

> what wretched creatures are the sheep which produce the famous Saxony wool compared our fat and comely flocks . . . In fact, it is a prevailing idea that the leaner the sheep the finer the wool . . . You may see them penning in a blazing fallow where not a trace of vegetable matter is to be seen, for the greater part of a summer's day . . . The sheep besides being lean are generally dreadfully lame with that pestilential complaint, the foot rot, and their keepers, apparently, trouble themselves very little about it.

Opinions about Saxon sheep were clearly divided, veering from one extreme to another. Henry S. Randall, late Secretary of State for the State of New York, complained that some of the worst of these 'wretched animals' were exported to the United States. Fineness of wool seemed to be 'the only test of excellence', whatever might be its quantity and however 'miserable and diminutive the carcass'. Quoting extensively from a report he had written for the New York State Agricultural Society in 1837, he proceeds to list some of the more disastrous purchases—'a most curious and motley mass'— shipped from the port of Bremen to New York by certain named Leipzig merchants during 1824–6.[35] Finally Randall had decided to take the matter into his own hands and make sure he got some really good stock. He returned to Saxony and spent the winter of 1826–7 in visiting and examining

---

[31] Youatt 1837, p. 37      [32] Youatt 1837, p. 171
[33] Trimmer 1828, p. 3; also quoted by Bischoff 1842, ii, p. 257
[34] Howitt 1842, pp. 426–7      [35] Randall 1882 (1837), p. 140

many flocks. He selected 115 breeding animals from 'the celebrated flock of Machern' and landed in New York with them on 27 June 1827. In contrast to his earlier unfavourable comments, Randall was then able to find Saxon sheep greatly to his taste at Machern (15 km North-east of Leipzig near Wurzen; Figure 7.2). He later admitted that German wool was the finest available but he re-emphasised the contrast between the high quality wool imported from Germany and the poorer quality grown in America from imported Saxon sheep:[36] 'Our Saxon wool as a whole, falls considerably below that of Germany; and I never have seen a *single lock* of the American equalling some samples given to me by a friend recently from Europe, which came from Styria, south of Vienna in Austria.'

Despite the adverse appraisals of Saxon sheep by some outside observers, a large proportion of the wool marketed through Leipzig was undoubtedly of exceptional superfine quality. How can the paradox be explained? One answer may lie in the fact that good wool could be grown on inferior looking sheep. Exceptional fineness and poor nutrition were not unrelated. Another answer is that sheep and wool labelled 'Saxon' did not always have a direct connection. The contrast between high values placed upon Saxon (Electoral) wool and low opinions expressed about some Saxon sheep can thus be explained. Much so-called 'Saxon' wool was not from Saxony at all. Wool designated 'Saxon' or 'Electoral' or 'German' in the United States or England was simply that purchased in the Leipzig market and shipped out of Hamburg or Bremen. Despite its name, such wool came not only from Saxony but also from Mecklenburg in the North and from territories to the South and East, wherever good Merino flocks had been established (Figures 7.2 and 7.4). As the Scotsman Richard Bright discovered on a visit to Hungary in 1814, 'Much of that which is sold in England, under the denomination of Saxon wool, is actually the produce of Hungary, exported in spite of the heavy duty it pays on leaving the Austrian dominions.'[37]

Merino wool was sent to Leipzig for 'assortment and valuation'.[38] The skill of the Leipzig experts in sorting the wool into different qualities gave the best of it a specially high value. What actually happened is described by James Atkinson following his visit in 1826. After shearing, the wool was put indiscriminately into large bags, without any selection or classification. In this state it was sometimes sold at home to travelling merchants, but the most general practice was to send it to wool fairs in wagons. In no case was the wool sorted by the growers themselves.[39] At first most of it ended up in Leipzig. Then, with the spread of Merinos westwards, 'a fair, open market' developed also in Hamburg.[40] Despite uneven standards of husbandry and a

---

[36] Randall 1882 (1837), pp. 141–2    [37] Bright 1818a,b    [38] Martin c.1849, ii, p. 62
[39] Atkinson 1828 (quoted by Garran and White 1985, p. 135)
[40] Youatt 1837, p. 176

conservative attitude to breeding in some if not most quarters, the marketing side of the Saxon fine wool business was supremely successful.

## Prussia including Prussian Silesia

Neighbouring Prussia was also destined to make good business out of fine wool. More than a century before Saxony had gained its deserved reputation, improvements in sheep stock in territories further East were revealed in the quality of wool and cloth exported westwards from the Baltic port of Danzig. In Tudor times, western European merchants expected wool shipped from Danzig to be coarse. Their name for it was 'Slesys', a corruption of *schlesisch* (Silesian), which has been linked to the derogatory term 'sleaze'. Then Silesian wool began to improve. The English Civil War (1642–8) brought the change to notice, when trade with England via the Baltic ports was disrupted, forcing the 'natives of Silesia and western parts of Poland' to combine the wools from the two regions to make cloth themselves.[41] Samples of Slesys cloth reaching England at this time proved surprisingly less coarse than expected; this alerted the trade to a new source of wool, which the Dutch, in particular, exploited to the full. English pride was unprepared for the shock of learning in 1655 that the Duke of Brandenburg had ordered livery for his soldiers to be made at Königsberg in East Prussia from the new, improved Silesian cloth, in preference to English cloth formerly used.[42]

When in 1763 Prussia gained a large part of Silesia from Austria, after the Seven Years War, wool production was divided: the larger part in the Prussian North with its capital at Breslau (now Wroclaw) and a smaller production in Austrian (Moravian) Silesia with its capital at Troppau (now Opava) (Figure 7.4). All the improvements in breeding stock and wool sorting, which had allowed Silesian wool to increase in value from the seventeenth century onwards, had now to be shared with Prussia. Prior to 1763 a genetic change had occurred in the best Silesian flocks through the influence of the part-Merino Paduan breed imported from Hungary (Chapter 2). The special quality of the improved Silesian sheep encouraged the Prussian monarch to bring them West to improve his Brandenburg flocks. Some arrived as early as 1748[43] to be confined and bred on the state farm at Stansdorf (Stahnsdorf, Figure 7.2) near Berlin. Some authors refer to these sheep as Merino although this is not strictly correct.

A second major focus of Prussian breeding existed on a large estate at Cösitz in the Duchy of Magdeburg (Figure 7.2) where, in 1752, a 20 year old citizen of the Kingdom of Hanover took over the leasehold. This was J.H. Fink, already mentioned as a correspondent with Sinclair. An exceptionally

[41] Smith 1747, i, p. 183, quoting *Occurrences and Ordinances*, 1641–53
[42] Smith, 1747, i, p. 196, quoting *Whitlock's Memorials*, 1643–56
[43] Fraas 1865, p. 316; Spoettel and Taenzer 1923; Baumgart 1957, p. 10

talented and enterprising individual, a supporter of the Bakewell technique of breeding in-and-in, and author of several agricultural books, he exerted a major influence on subsequent events, not only in Prussia but also in Saxony and throughout Germany as a whole. The leasehold at Cösitz, which he held for the rest of his life, was awarded to him as the King of Prussia's head bailiff. Outstandingly successful in his sheep breeding, Fink could be satisfied only with the best stock, which he drew from several sources, as Lasteyrie discovered on his visit in 1800. The construction of the flock began in 1754 with high quality Silesian rams from the royal farm at Stansdorf mated to selected Silesian ewes, known for producing finer than average wool. After the Saxon Electoral flock had been established at Stolpen, Fink brought selected 'Saxon-Spanish' rams and ewes from there to Cösitz in 1768.[44] According to Youatt[45] he later used others of the Negretti *cabaña* imported directly from Spain, the first batch arriving in 1778.

Once Fink had built up his flocks to 800 head, he was able to supply breeding animals to estates as far away as Poland.[46] Annually he produced about 300 breeding sheep for sale.[47] His reputation as the most rational of breeders kept his prices high and ensured that he was respected as the 'father' of his profession. He was an ardent exponent of 'controlled stall feeding', based on Alströmer's practice in Sweden of confining sheep in sheds in summer as well as winter, although in dry weather he would sometimes let the sheep out in the day, even in snow. Fink wrote a book about stall feeding published in 1785 in Leipzig.[47a] His growing expertise on sheep diseases (such as pox, scab and staggers) and his discovery of a practical inoculation against sheep pox formed the basis of another book.[48] In recognition of his original talent, the King of Prussia appointed him Director of the newly created School for Shepherds at Petersberg, a small estate South-west of Cösitz, just a few kilometres from Halle (Figure 7.2), where 12 shepherds attended each year to learn the practical techniques responsible for Fink's success. It was at Petersberg that Lasteyrie visited him in 1800 and formed an excellent opinion of his sheep.[49]

The King imported a further 100 rams and 200 ewes from Spain in 1786,[50] some of which were added as fresh breeding stock at the royal farm at Stansdorf, where they remained healthy. But the rest, dispatched to sheep farms around the country, mostly degenerated due to negligence.[51] Even the Stansdorf flock did not attract the admiration that Lasteyrie reserved for the sheep under Fink's direct control at Petersberg. Though smaller than original Spanish Merinos, they were not, in Lasteyrie's opinion, in any way

[44] Lasteyrie 1802, pp. 31–2; Anon 1878;     [45] Youatt 1837, p. 172
[46] Anon 1878     [47] Schrader and Hering 1863     [47a] Finke 1785
[48] Finke 1785; Fink 1799a     [49] Lasteyrie 1802, pp. 31–2 and 177–94
[50] Lasteyrie 1802, pp. 30–31; Somerville 1809, p. 4     [51] Lasteyrie 1802, p. 31

inferior to them in their wool.[52] Bearing in mind that some of Fink's breeding stock were a particularly large kind of Merino, we have to conclude that he selected for small size. Lasteyrie realised that they must have been most carefully selected to retain the Spanish wool quality. He noted that Fink had ordered a further 1000 Merinos from Spain, which he was expecting at the time of the Frenchman's visit.[53] We are told that Fink travelled personally to Estremadura in Spain to select these sheep with the greatest care.[54] Fraas later confirmed that ewes had been imported in 1800 and rams in 1802.[55]

Improvement in wool grown on indigenous races of German sheep as a result of upgrading them with Merinos or other fine-wooled stock was another matter for Lasteyrie to emphasise. Experience of crossing Merinos with local ewes had convinced Fink beyond doubt that fineness of wool was attributable more to breeding than to environmental influence.[56] This point was noted with approval by Culley in the fourth edition of his book *Observations on Livestock*.[57] Culley writes how 'M. Fink is justly of the opinion that the *fineness* of wool depends wholly on the breed of sheep, and is in no respect influenced either by climate, soil or food; but the *quantity* of wool depends entirely on the quantity and nature of the food". Bakewell had long argued in support of racial continuity being independent of environmental influence as, for example in an exchange of opinion reported by fellow farmer J. Boys from Surrey (Chapter 4).[58] We can be sure that Fink, a citizen of Hanover with its British royal connections, was well placed to be aware of Bakewell and his pupil Culley quite as soon as Culley was aware of him.

Fink underlined his success in refining coarse-wooled sheep by issuing a series of books on the subject, with new editions under different titles, from 1790 onwards, including the one Culley saw in English translation, *Various Writings and Answers Dealing with Sheep Breeding in Germany and the Improvement of Coarse Wool According to Personal Practical Experience and Facts, Collected in Spring 1799*.[59] Fink emphasised that both sexes contribute towards the characteristics of the progeny and he recommended strong selection of males according to the 'production capacity' of individual animals. Stressing the value of selecting the best rams to be used in the flock for several generations, he recommended Bakewell's method of breeding in-and-in.[60] Lasteyrie reported that 'Mr Fink has opposed the generally held opinions in Germany that races degenerate when one couples together the fathers and the mothers, the sisters and the brothers.'[61] He seems to have been the first German author to state that the practice of mating exclusively within a small

---

[52] Lasteyrie 1802, p. 32    [53] Lasteyrie 1802, footnote p. 35
[54] Martin c.1849, ii, p. 61    [55] Fraas 1865, p. 316
[56] Fink 1797, p. 278; Fink 1799a, p. 54    [57] Culley 1807, p. 260
[58] Boys and Ellman 1793    [59] Fink 1799b    [60] Fink 1799a, p. 73
[61] Lasteyrie 1802, p. 180

imported flock does not lead inevitably to degeneration although, cautiously, he still tried to get better rams from outside every two or three years if possible.[62]

Fink also made a point of stressing aspects of breeding about which his opinion was more conventional. Like Stumpf he recognised that the progeny of ewes derived from grading crosses (using imported rams on local ewes for three or four generations) grew as good wool as local Spanish sheep.[63] Both of them also optimistically stated that there was no danger of such graded stock reverting to the local type. In practice, however, it seems that Fink exceeded Stumpf's recommendation of three or four generations, making it a rule for 15 years never to cross the ewes of his original breed with any other than pure breeding rams.[64] As Culley recognised, there was much about Fink's practice that resembled his own and, thus, Bakewell's. In selecting for a small body frame like Bakewell, Fink was following an opposite course from Alströmer and his successors in Sweden whose sheep were larger than the Merino imports from which they were derived.

Lasteyrie summarised Fink's method of improving flocks in terms of harnessing Nature in order to restore a lost perfection, as follows:[61]

> The progressive march towards improvement is founded on Nature; her efforts are proved by the observation of Mr Fink and by those of the French agriculturalists. Nature can, it is true, distance herself from this progression, and she frequently does so; but she is liberal with her care and, by providing a favourable choice of individuals by which to restore perfection, she offers that which aims at racial improvement.

If it was Fink's ambition to improve his stock *beyond* Spanish quality, as seems likely, Lasteyrie's praise had its limitations, reflecting the breeding philosophy of a Daubenton rather than a Bakewell.

Meanwhile, too late for Lasteyrie's appraisal, another Hanoverian figure began to make his mark. This was Albrecht Daniel Thaer (1752–1826), the first Professor of Agricultural Sciences in Berlin, who provided a major organisational force and educational influence on the improvement of sheep in Saxony and Prussia, and made no secret at all that English practice had a profound influence upon him. For two decades, from 1784, he was engaged privately in agricultural research on his estate near Celle, while practising as a physician. There he benefited from an association with Jobst Anton von Hinüber, owner of a model farm near Hanover and a fine agricultural library. Having travelled extensively in England, Hinüber had built up a rich collection of books and pamphlets on English agriculture.[65] This happy connection was almost enough on its own to direct Thaer's thoughts towards a future in agriculture but, in addition, there was the powerful example of the

---

[62] Youatt 1837, p. 173        [63] Stumpf 1785, p. 95, Fink 1799a, p. 48
[64] Fink 1797, p. 278          [65] Braun 1971

ageing Fink. Elke Steiner has analysed the connection between Fink and Thaer in an academic dissertation submitted to Humboldt University, Berlin.[66] Both were members of an influential professional body in Germany, the Royal Electoral Society of Rural Economy established at Celle (Zelle) in 1764, which had been modelled on the (London) Society of Arts.[65] A big advantage for members lay in their Society's 'uninterrupted correspondence, and the most intimate possible connection' with the Board of Agriculture and its President, Sir John Sinclair, in London.[67] From 1789 the Society had promoted the culture of clover (with free distribution of seed), more regular rotation of crops, trials of stall feeding, establishment of fruit tree nurseries, enclosure of cultivated fields with hedges and stone walls, introduction of the system of flooding or watering of meadows, as well as the improvement of chosen flocks by the importation of Merino rams.[68] The British influence in most aspects is obvious. Zirnstein has stressed how strongly Thaer was influenced by Bakewell.[69] His easy access to English agricultural writings led him to draw up ambitious plans to introduce English breeding methods.

In 1802 Thaer began to teach farming on his estate at Celle. Two years later, King Friedrich Wilhelm III of Prussia appointed him head of the newly created State Agricultural Institute, 50 km North-east of Berlin, at Möglin (Figure 7.2), with places for 20 students drawn from all over Europe and even beyond, to undertake a two year course. It is easy to agree with Hawkesworth (1920) that the Möglin Institute 'did more for the development and propagation of Merino sheep than any other'. Thaer expressed his determination to provide a scientific rationale for agricultural techniques, which he did through extensive writings as well as the lectures at Möglin.[70] When the University of Berlin granted him a Chair in Agriculture in 1811, it was the first in this subject on the European continent. Before this appointment, agriculture was not viewed as a separate discipline in German universities, being then taught as a branch of economics.

Thaer's views on animal breeding in Britain were contained within the final volume of his major work on English agriculture (1798–1804). In response to a direct request by the Interior Ministry, he wrote a book specifically on fine-wooled sheep breeding (1811). Although modest in scope, this was the first work on the subject in Germany and it provided Thaer with a platform to reveal his passionate concern to find principles of heredity underlying animal breeding. In published articles, he sought to explain the transmission of individual parental traits to hybrids, also reporting the segregation of such traits in the progeny of hybrids.[71] He considered heredity and variability to be complementary phenomena.[72] His ideas stimulated a widespread response

---

[66] Steiner 1978, p. 23    [67] Thaer 1797, p. 376    [68] Thaer 1797, p. 379
[69] Zirnstein 1979, p. 54    [70] Klemm and Meyer 1968    [71] Steiner 1978; Thaer 1816
[72] Zirnstein 1979

in the German world, including Brno where the book was critically reviewed (Chapter 9).

By demonstrating his own substantial achievements in breeding, Thaer attempted to promote the practices that underlay his success. J.G. Elsner, a Silesian breeder with a farm at Reindorf (Gronow and Kalinowiss), who produced a retrospective overview of German Merino sheep breeding in a book published in 1857, wrote of Thaer 'that he was infinitely concerned to uphold correct breeding procedures'.[73] His experience led him to embrace the idea of predispositions (*Anlagen*) for the transmission of traits, and he had in mind that there must be laws of heredity. However, as Elke Steiner has stressed, he was aware that such laws 'are scarcely as simple as people often consider them to be'.[71] On a practical level, he recommended selection of rams and ewes for mating on a one-to-one basis and the application of Bakewell's techniques of breeding in-and-in (*Veredlung in sich selbst*) to attain constancy of inheritance of production traits. He also practised controlled crossing of his inbred strains.[74] To achieve success he found it essential to keep the breeding stock sheltered from the rain in sheep houses throughout the winter, following the example of Fink. An Englishman, William Jacob, who visited Möglin (Morgelin) in 1819, commented favourably on the fineness of the wool in the selected flock. He also noted, however, that 'the improvement of the carcase has been neglected; so that his, like all German mutton, is neglected.' Wool was obviously the major target for improvement. Jacob was impressed by the systematic way in which the professors at the Institute arranged various kinds of wool on cards for comparison and he wrote of Thaer and his staff 'discriminating with geometric exactness the fineness of that produced from different races'. Jacob commented that the finest samples on view were 'some specimens from Saxony, his [Thaer's] own was next, fine wool from Spain [Leon] is inferior to his in the proportion of 11 to 16; and the wool of New South Wales, of which he has specimens, is inferior to the Spanish.'[75]

An action typifying Thaer's desire to bring order to the practice of sheep breeding was to organise an assembly (*Konvent*) of wool growers and students in Leipzig, gathered together in 1823. Thaer's principal purpose was to define the terms used by shepherds and wool merchants when classifying different grades of fleece, in accordance with the regulations of the wool industry. By this meeting he hoped to create a scientific rationale for the procedure. Also incorporated into the programme was his fervent wish to find acceptable definitions of the 'Electoral' and 'Negretti' types of Merino sheep in the central European context. Elsner claimed that Thaer's inspiration for taking the action came from the Annual Meetings of the Sheep Breeders' Society in Brno.[76]

---

[73] Elsner 1857, p. 7    [74] Klemm and Meyer 1968, pp. 139 and 179
[75] Jacob 1820; Bischoff 1842, i, p. 355    [76] Elsner 1828, p. 120

One expert to be invited to the Leipzig assembly was Rudolf André (1792–1827) from Brno. His published criticism of Thaer's sheep expertise[77] did not deter the learned professor from exchanging letters with the younger man. André returned home with a mixed reaction to what he had seen.[78] He was disappointed to have found scant evidence of individual trait recording or the precise criteria on which to select rams and ewes for pairing. At the same time he acknowledged with admiration the exceptional fineness of Saxon wool, which could command a better price than wool produced in Moravia. Soon afterwards he imported some Saxon Electoral stock on to a farm near Brno that he was managing for Count Salm. His stated intention, as he wrote in his report of the Leipzig Assembly, was to combine the fineness of Electoral wool with the greater production capacity of Moravian Negrettis. Enthusiastically he added: 'Let us be patient and wait to see what the capable Moravian sheep breeders will achieve with these new Electoral sheep after a few generations.'[78] Tragically, the youthfully confident André died two years later, never to see his hopes fulfilled.

Rudolf's father, C.C. André, also commented on the Leipzig Wool Assembly. Writing from Stuttgart, he contrasted this once-and-for-all gathering, at which representatives of industry and commerce were noticeably absent, with the situation in Moravia where there was close co-operation between breeders, cloth manufacturers and merchants, with meetings held annually.[79] Elsner, writing from Silesia, joined André in regretting that the Assembly had not been followed up by further meetings.[80] He too stressed the superiority of the Sheep Breeders' Society in Brno with its wider membership, which met on an annual basis. The reports produced a strong response. The Moravian historian d'Elvert,[81] writing about the years following the Assembly, noted that sheep breeders from Saxony, Silesia, Braunsweig (Brunswick), Pomerania, Austria, Hungary, Poland and Russia all 'streamed into Moravia' (*nach Mähren strömten*) to avail themselves of the progeny of the noble flocks being bred there. Thus it was that in the early 1820s the wider world became aware of the Moravian achievement (see later).

Meanwhile, Thaer enjoyed privileged access to Merinos from France, Saxony, Moravia and Austria as well as Prussian-based stock, which he crossed together with the aim of producing a new ideal 'Electoral' race perfectly adapted to the local environment. The breed that finally emerged from the melting pot, known as the 'Möglin Electoral' (Figure 7.5), gained a sufficient reputation to be exported abroad in large numbers. 1505 of them entered Sweden in 1825 in a single transport.[82] Nevertheless, according to the Königsburg-born and Prussian-based expert Hermann Settegast

---

[77] R. André 1812    [78] R. André 1823    [79] C.C. André 1825
[80] Elsner 1828, p. 120    [81] d'Elvert 1870, p. 337
[82] K.-H. Suneson, personal communication, 1 April 1990

(1819–1908), Thaer's efforts never resulted in a product to rival the best of Moravian or Silesian stock. Above all there was the Hoštice breed produced by Baron Geisslern (Figure 7.5). Closer to Prussian territory, there was the newer breed developed at Kuchelna near Troppau, associated with Prince (Fürst) E.M. Lichnowsky and much influenced by Hoštice blood (see later).

A vital place in these developments was occupied by Silesia, a region given special prominence by Lasteyrie.[83] The breeder whose praises Lasteyrie sang the loudest was not centred in the heartland of Saxony nor even in Brandenberg, but in the very southern part of Prussian territory, almost on the Moravian border. His name was Count A.A. Magnis (1775–1817) and his estate was located at a place named by Lasteyrie as 'Eckersdorf' near Glatz; we believe this was probably Ebersdorf (now Domaszkov). To reach this far southern corner of Prussian Upper Silesia (Oberschlesien) meant a journey of more than 400 km from Lower Saxony where Fink operated. Unsurprisingly there were differences in the methods of Magnis and Fink, noted by Lasteyrie.[84] Count Magnis had the benefit of the fine-wooled Paduan race established in Silesia in the previous century, with which to cross the imported Merinos. He developed a more heavily built type of Merino than Fink. This was because the original Merinos he imported had been selected for large size in territories to the South, collectively referred to by Lasteyrie as 'Hungary' but certainly coming under Moravian influence.

Carrying more than 9000 head of sheep,[85] the Magnis estate justified being described as a major breeding centre on the basis of size alone. Lasteyrie's visit convinced him that the wool produced there by the young Count was the best he had come across during his whole tour. He was made aware that Magnis, who had begun breeding with a mixture of small Silesian and large Hungarian sheep, later had access to the most beautiful sheep he could find in all of Germany. The result of his skilfully crossing these various blood lines was a type of animal which carried an abundance of relatively long fine wool.[86] Lasteyrie wrote enthusiastically about it: 'The wool of most of his flocks equals the most beautiful Spanish wool; and the individuals surpass, in strength, in grandeur and in form, the most beautiful of continental flocks.'[87] He noted the care taken over selective breeding, particularly the registration of traits in each generation, the use of progeny testing and the recording of pedigrees. The technical sophistication Magnis displayed persuaded John Lawrence to write of him as 'the most renowned tup breeder on the continent'.[88]

Lasteyrie was unable to proceed further South from the region of Glatz into Moravia because of the state of the Napoleonic war 'that [was]

---

[83] Lasteyrie 1802, pp. 32–5 and 194–206
[84] Lasteyrie 1802, pp. 32–3; Youatt 1837, p. 174     [85] Youatt 1837, p. 174
[86] Lasteyrie 1802, p. 33     [87] Lasteyrie 1802, p. 34     [88] Lawrence 1805, p. 441

devastating Europe', as he remarked.[89] He could only write of what he learned in Silesia, that 'the best wool from the Austrian lands is from Moravia and Bohemia' and that 'the importance of fine wooled sheep has been encouraged in Austria more strongly than other practices of rural economy.'[90] The approach to pedigree and trait recording he observed in Silesia on Magnis's estate seems, according to his description, strongly reminiscent of the best Moravian practice, using methods associated with the name of Baron Geisslern.

As soon as political circumstances allowed, the Silesians began again to import the best of stock from Moravia, Hungary (present-day Slovakia) and Austria, paying high prices for rams mainly of the Negretti type, as much as 10,000 guilders (about £1250) per individual.[91] When Elsner had reviewed European sheep breeding in the late 1820s,[92] Magnis was still a force to mention although by then the best fine-wooled sheep in Silesia were associated with a different name. Prince Eduard Maria von Lichnowsky (1789–1845), educated in the sciences at Göttingen and Leipzig, was a member of a family with estates both in Prussian territory and also over the border in Moravian Silesia, in the Hapsburg monarchy. It was a family well known to the world of music for generously supporting the composer L. van Beethoven. Prince Eduard's sheep breeding activity was centred principally at Kuchelna (Chuchelna),[92] near Opava (Troppau) in Austrian territory. His approach to breeding seems to have been much like that of Magnis, although we have a few more details in Lichnowsky's case. An early report of his achievement was published by André in *ONV*.[93] In 1799, he introduced sheep from the 'famous Geisslern' at Hoštice in Moravia, to be followed by stock from Holič (Holitsch) in Hungary (now Slovakia) and Mannersdorf in Austria.[93] He crossed these large and heavily wooled, so-called 'Negretti sheep', with smaller 'Electorals' he imported from Fink's flocks at Cösitz or Petersberg. By selective breeding from the hybrids he managed to produce his own race of Merinos which he built up to an excess of 20,000 which he later exported throughout Europe and even to Australia and America.[92]

The richly wooled Negrettis (Figure 2.3) had a closer, shorter, duller staple than the small, sometimes delicate Escurials of Saxony (Figure 7.1). Silesian owners appreciated valuable features in both types. By the 1820s a so-called *vollblut* ('full blood' or 'thoroughbred') race of mixed Escurial and Negretti stock had been established both at Kuchelna and Glatz (in Prussian territory) by the Lichnowskys.[94] This information comes primarily from Elsner who himself became a major force in Silesian sheep breeding. In his book of 1827, *My Practical Experience of Superior Sheep Breeding*, published in Stuttgart

[89] Lasteyrie 1802, p. 36    [90] Lasteyrie 1802, p. 210    [91] Elsner 1857, p. 3
[92] Elsner 1828    [93] Lichnowski [sic] 1811
[94] Elsner 1828, p. 50, Settegast 1861; d'Elvert 1870, p. 338

and Tübingen, he wrote that every year 'more than 30,000 sheep pass through my hands'. He corresponded with the Brno Sheep Breeders' Society, supplying information and opinion to C.C. André, and is remembered for his graphic evaluation of sheep as 'the levers' (*die Hebel*) of all economy.[95]

Evidence of Negretti blood began to appear increasingly in German Merino flocks as the influence of the larger, stockier, shorter-wooled type moved northward and westward. Atkinson observed the influence at Gersthoff in Saxony in 1828 (see above). German Merino sheep were becoming generally more rounded with an increase in neck folds. The wool continued to be known as 'Electoral' in Saxony and in countries importing the wool, such as England. Progressive changes in the body form of the sheep, as well as wool quality, were achieved by selective breeding, borrowing techniques associated with the Bakewell tradition in England.

## Austrian Empire

Factory weaving of woollen cloth in the Austrian provinces of Bohemia and Moravia began before Merino sheep arrived there, when little fine wool was produced locally. At that time the quality of wool produced in Moravia was infinitely inferior to that of the Spanish Merino. The Hapsburg Empire offered a huge market for coarse cloth.[96] Russia was another major importer of coarse textiles where, during the seventeenth century, Moravian cloth was more popular than Silesian.[97] Supplies of coarse wool to be spun and woven in Moravia could be obtained locally, whereas fine wool had to be imported almost exclusively from abroad, mainly direct from Spain, starting in 1732 with the peak of importation in 1773–4.[98] A fine-cloth factory had been set up in 1751 at Kladruby in Bohemia but was disadvantaged by poor communications.[99] After the shock of losing most of Silesia to the Prussians, following the Seven Years War, the main production was moved to Brno in 1764.[100] The cloth from these factories satisfied Austrian needs and was also exported eastwards, mainly along the Danube to Constantinople and the Levant. Austria was then beginning to rival France for this lucrative trade, in a market which demanded light, fine cloths for a warmer climate. By the 1780s much of Brno's production was in the form of so-called *Londrins Seconds* cloths, many of which were sold to Constantinople. This was light-weight material on the model of that manufactured predominantly in Languedoc (France) in imitation of similar cloth woven in England.[101] At the Brno fine-cloth factories, the number of looms was increased to 119 by 1788, employing a maximum labour force of 2000. By this time, the price of Spanish fine wool had risen sharply, by as much as 30%

---

[95] Elsner 1827, p. 207      [96] F.W. Carter 1988, p. 45      [97] Carter 1988, p. 47
[98] Freudenberger 1977, pp. 42 and 112      [99] Freudenberger 1977, pp. xiv, 19 and 32
[100] Freudenberger 1977, p. xiv      [101] Freudenberger 1977, p. 78

after 1774.[102] It was becoming more profitable to produce fine wool locally in Moravia, or in surrounding Austrian controlled territories, where 'the efforts of the government in improving the breed of sheep had paid off so well that the local manufacturers paid only half as much for wool as the Belgians for Spanish wool of commensurate quality.'[103] So began an era of Moravian fine-wool production that would rank it among the highest quality in Europe.

How was the transformation from relative mediocrity to excellence achieved? The native sheep of Moravia were no more prepossessing than those elsewhere in the region. According to an English sheep expert, W.C.L. Martin, the rough-wooled native Moravian sheep was 'closely allied to the Wallachian breed',[104] the fullest description of which is given by his contemporary Youatt.[105] Characterised by unusually long spiral horns, which in the males sprang perpendicularly from the head (Figure 7.3), it carried a substantial amount of hair among its wool. It was acknowledged to be a most vigorous type of sheep, which had spread from Wallachia (southern Romania) into Hungary in the sixteenth century, and then onwards as far as Bohemia. Its ultimate origin can be traced to a hilly region of the Volga-Kama crook, far to the East in Russia, from where it was carried westwards by the original Magyar tribes to all the territories they conquered.[106] It is a type of sheep often referred to as Zackel, which is common even today in South-eastern Europe.[107] The unimproved modern Zackel in Greece has very coarse wool of the carpet type. It is known as Racka in Hungary and Cápovice in Moravia.

In addition to the Zackel type, there existed in Romania and Hungary a relatively fine-wooled breed, which is probably equivalent to the modern Tsigai (Tzigaya) breed (Cigaj in Czech),[108] derived from crosses between Zackel sheep and a fine-wooled Turkish breed.[109] A substantial introduction of Turkish sheep, the best of which had their origin in Persia, was a natural consequence of the Turkish colonisation of that part of Europe. An English account of trade published in 1727 states that 'Persia furnishes most of Romania and Anatolia with sheep.'[110] It appears that an axis of trade in fine-wooled sheep stretched from Persia, through Turkey and the Black Sea to Romania. Neighbouring Bulgaria also had fine-wooled sheep in the sixteenth and seventeenth centuries, when wool merchants stimulated wool improvement by standardising the Bulgarian production into three grades.[111] The further migration of the Tsigai breed into Hungary and from thence into Moravia suggests that the Persian influence was probably felt even beyond the limits of Turkish conquest.

---

[102] Freudenberger 1977, p. 112
[103] Freudenberger 1977 p. 152, based on the Brno state archives
[104] Martin c.1849, ii    [105] Youatt 1837, pp. 138–9    [106] Gaál and Gunst 1977
[107] Ryder 1983, p. 337    [108] Bökönyi 1974, p. 188    [109] Ryder 1983, pp. 336–8
[110] Smith 1747, ii, quoting the *Atlas General* 1727, p. 178    [111] Ryder 1983, p. 329

**Fig. 7.3.** The Wallachian breed of sheep of Southern Romania, closely related to a type of sheep found in Hungary and Moravia that was extensively improved by Merino crosses (from an engraving published by Martin c.1849).

A second favourable influence on central European sheep, evident in Hungary as early as the seventeenth century, came from the silk-wooled Paduan breed from northern Italy. Information from Hungarian sources has revealed that the first importations of such stock were organised by Archbishop G. Szelepcsényi and Prince Eugene of Savoy (Chapter 2).[112] Gaining access through the ports of Trieste and Rijeka, the influence of these Paduan animals spread northwards through Hungarian or Austrian territory into Moravia and from there further northwards into Silesia, which was then the monarchy's major centre for wool production (Figure 7.4). The best of the fine wool from Silesia was exported for manufacture in western Europe,[113] where the weavers were ready to admit that 'the sheep of the Paduan afford a good sort of wool, little inferior to that of England.'[114] Fresh introductions of the breed continued to reach Hungary well into the eighteenth century, although the impression gained is that these superior sheep from Italy were few and highly localised, available mostly to owners closest to the Adriatic

---

[112] Gaál and Gunst 1977, p. 258 (quoting late nineteenth sources in Hungarian written by J. Rodiczky)   [113] d'Elvert 1870, p. 133

[114] Smith 1747, ii, p. 218 (quoting Thomas Salmon 1739, p. 350)

**Fig. 7.4.** Map showing prominent Austro-Hungarian and Prussian Silesian estates for Merino sheep breeding, in relation to principal towns and cities.

ports. Their spread northwards was only in the hands of specialist fine-wool breeders. In central Europe as a whole, most sheep were native, unimproved stock and most animal breeding was left to chance. Jerome Blum has written generally about continental Europe in that period:[115]

> Only rarely was there an effort to improve the breed by selective mating, and then it was done almost exclusively by those exceptional landowners who interested themselves in agricultural improvements, and who kept their animals apart from the village herd.

Most sheep were in the hands of the peasants and unenlightened land-owners; they were thin, leggy creatures kept for milk and wool or slaughtered for their meat, pelts, tallow and soft fat, the latter having a multiplicity of uses, from greasing the wheels of farm carts to making crude ointments and soap.

Paduan sheep were in a category altogether different, valued above all for their wool. They were specially well looked after on breeding farms within easy reach of a factory or market where the wool would be properly priced. After 1764, it became important to establish or expand Paduan breeding farms on estates with access to the new factories being built in Brno and surrounding villages. Representatives of industry there are known to have encouraged local estate owners to upgrade their sheep. By 1774, when the price of Spanish wool rose sharply, it was fortunate that domestic wool of quite good quality was already being produced in Moravia. Over one third came from the estates of Prince Johann Liechtenstein 'whose interest in improving the breed of sheep in the Hapsburg Monarchy predates the efforts of the government itself.'[116] Those who had already improved their flocks with Paduan crosses would have pressed for access to Merinos direct from Spain and be well placed to make the most effective use of these treasured animals, once they began to arrive. Writing about this matter, J.K. Nestler, Professor of Agriculture at the University of Olomouc, noted that even a modest improvement from Paduan crosses gave Moravian breeders an important advantage: 'In this way the sheep were better prepared for cross-ing with Merinos.'[117] But it was probably not until after 1800 that enough wool of really superior quality could be procured from surrounding estates and other Austro-Hungarian lands, to reduce dependence on foreign supplies completely.[118]

By the time Youatt came to write about Moravian sheep in 1837, the orig-inal Racka type was no longer widely distributed in the Austrian provinces, except on the estates of religious establishments. Many had been simply killed off; others had been improved by crosses.[119] The old breed had come

---

[115] Blum 1978, p. 151    [116] Freudenberger 1977, p. 112
[117] Nestler 1829, section 16    [118] Freudenberger 1977, p. 113
[119] Loudon 1844, p. 99; Gaál and Gunst 1977, p. 259

through quite a series of changes since the sixteenth century with the Persian/Anatolian crosses in Romania, the Paduan crosses in Hungary, Austria and Bohemia/Moravia, and finally the all-pervading influence of the Merino from the late eighteenth century onwards. As Youatt describes the Moravian sheep before Merino improvement, it was larger than its closest relative, the Wallachian, with a small head, back somewhat bowed and carrying long wool, finer than the Wallachian.[120] Basically, however, according to Martin, the two breeds were 'closely allied'.[121] We cannot suppose that the Wallachian-related type was the only 'native' breed of sheep in Moravia at that time, but it was the one which came to the notice of Martin and Youatt, the breed which they reported to have been improved by Merino crosses and therefore that which concerns us here. In summary, it was a type of sheep ripe for raising to fine wool status by Spanish imports, having been already partly improved by crosses to fine-wooled stock of Persian and/or Italian origin.

The earliest Merino arrivals were housed on the Imperial estates at Mannersdorf near Vienna and at Holič (Holitsch), in Hungarian territory, now Slovakia, close to the Moravian border (Figure 7.4). The flocks at Holič were joined in 1768 by 325 sheep imported directly from Spain.[122] Some of the newly introduced stock went to two prominent aristocratic landowners in Hungary, Count P. Festetics and Count D. Ehernal. The rest were distributed to the Imperial farms at Mannersdorf in Austria and Pavlovice in Moravia. Most Merinos entering Austria or Hungary by the Adriatic route arrived through the port of Trieste direct from Alicante. Trade with Trieste was actively encouraged by the Brno-based Moravian Loan Bank, founded in 1751, which had a strong interest in building up the port as a major centre for the exchange and distribution of Hapsburg goods.[123]

Importations were organised by Count Dominik A. Kaunitz (1739–1812), Austrian Ambassador in Madrid, who housed the precious sheep not far from Trieste, on the Imperial estate of Mercopail (Mrkopalj), just off the road from Rijeka (Fiume) to Zagreb (Figure 7.4). A breeding farm had been established there in 1772 for Paduan imports, associated with a training school for shepherds.[124] Lasteyrie referred to 300 Merinos arriving there in 1775 as the first shipment, then 300–400 in a second shipment (in about 1780) and 400–500 in a third, but nothing else before 1800, at which time an Austrian agent in Spain was negotiating for 800–900 more.[125] Kaunitz's estate manager, A. Hitschmann, wrote about a large importation in 1786.[126] Whatever the precise numbers and dates, there is no doubt that a stream of Merino sheep flowed through Mercopail, some to chosen private landowners,

[120] Youatt 1837, p. 139   [121] Martin c.1849, ii, p. 49
[122] d'Elvert 1870, pp. 69–70   [123] Freudenberger 1977, p. 61
[124] d'Elvert 1870, p. 68–9; Gaál and Gunst 1977, p. 259
[125] Lasteyrie 1802, pp. 36–7   [126] Hitschmann 1812

including Kaunitz himself who kept a small flock on his estate at Jaroměřice in southern Moravia, but the majority to the Imperial estates at Holič and Mannersdorf. Merinos coming from other directions into Austria included Saxon Merinos in 1788 and stock from Roussillon (in France) in 1790.[127] The major point to make about the importations is that the Merino gene pool available in Austria–Hungary became exceptionally rich.

Kaunitz's involvement in this activity was in a family tradition. His grand-father, Count Andreas Dominik Kaunitz, had established a woollen textile factory in 1701 to weave coarse cloth and stockings on his estate at Slavkov (Austerlitz).[128] In those early days, Kaunitz's manager had experimented with crossing Paduan rams with ewes from Bohemia. He noted that the progeny far surpassed lambs fathered by Bohemian rams, introduced earlier.[129] The availability of locally grown fine wool to the manufacturers of Brno led to an increase in demand and a substantial rise in its market price (from 20–70 guilders/100 kg to more than 80 guilders/100 kg between 1772 and 1782). The Brno woollen manufacturer, Wilhelm Mundy, a rugged individualist and self made man, who arrived in Brno from the Rhineland in 1772, is credited with a key role in the Merino expansion. 'He is held responsible for having got the noble landowners in Moravia to refine the breed of sheep so that the quality of the wool approached more nearly that of the Spanish Merino sheep.'[130]

On behalf of the Imperial Government, Baron A.V. Kaschnitz von Weinberg (1744–1812), a prominent economist, was entrusted with the administration of State estates (*Kameraladministrator der Staatsgüter*).[131] It was his responsibility to organise a rapid expansion of flocks and their dis-tribution throughout the Hapsburg monarchy and to upgrade local sheep over several generations. He and others also made great efforts to encourage the cultivation of clover introduced from Flanders, to provide a valuable extra source of nutriment for the expanding flocks throughout the whole year. Disengaging from his official appointment in 1785, he purchased an estate at Zdounky (Figure 7.4), about 65 km East of Brno, to practise sheep improvement on his own account. The Emperor Joseph II presented him with 50 ewes and 10 rams from Holič, out of the original shipment from Spain, which would become the nucleus of the famous Moravian flocks.[132] On the basis of his experience, Kaschnitz wrote a manual on sheep improvement, published in 1805, entitled *Practical Notes and Introduction to the Improve-ment of Sheep Breeding*. Touching upon breeding methods, he drew attention to one of his neighbours Baron F. Geisslern at nearby Hoštice (Hoschtitz), about 6 km from Zdounky. According to Kaschnitz, Geisslern was a 'true

[127] Lasteyrie 1802, p. 37    [128] Freudenberger 1977, p. 12
[129] Hitschmann 1812    [130] Freudenberger 1977, p. 169    [131] Balcárek 1977
[132] Nestler 1837

researcher' who had created, during the previous 20 years, such a superior race of fine-wooled sheep 'that it cannot be surpassed without great effort'.[133]

With the turn of the century a much-travelled Austrian, Bernard Petri (Bernhard Petry) (1767–1853), created a fresh insight into sheep breeding throughout Europe by his lively and frequent publications. The son of an adviser on economic matters to the Duke (later King) of Bavaria, Petri became similarly skilled. His first job was to manage the gardens of the Duke of Deuxpoints near Mannheim. From there, on 12 June 1787, he wrote to Sir Joseph Banks in England offering his services. He believed he could be useful in plantations abroad and he proposed botanical visits to Tahiti and Jamaica.[134] For four years after that he lived in England, where he managed to associate himself with Bank's interests. Upon returning home he was appointed Economic Adviser to the King of Bavaria but when Napoleon's army occupied that country, he astutely moved to Austria, where he entered the service of Prince Johann Liechtenstein. Petri's enthusiasm in applying the most up to date agricultural and economic principles led to widespread cultivation of clover and increased production of higher quality wool. Under his influence, sheep were bred selectively for fine wool on the Imperial estate at Mannersdorf. Although centring his activities in the region of Vienna, he co-operated closely with breeders in Moravia and with their associates in Hungary and Silesia.

In 1802 Petri was sent secretly to Spain to buy the best Merino sheep for Prince Liechtenstein. He travelled incognito under dramatic conditions, disguising himself as a dumb Spaniard accompanied by an interpreter from Segovia, of German origin.[135] His adventures were recounted graphically in a series of 16 letters written to the Princess of Liechtenstein, which were serialised in *PTB* under the title *Extracts from letters about an agricultural trip to Spain, mainly on sheep breeding*.[136] He was successful in obtaining valuable animals from the Paular, Guadeloupe, Infantado and Negretti *cabañas* (Figure 2.1), personally selected from about 60,000 animals. In absolute secrecy, doubtless employing considerable bribery, he managed to smuggle 416 sheep through Barcelona. Most ended up finally on the Prince's farms at Loosdorf and Hagendorf, villages near Vienna, although Petri kept some for himself on his estate at Theresianfeld near Vienna Neustadt (Figure 7.4). The cost of bringing in these sheep was very high, estimated to be 24,000 guilders (silver coins known as gulden) when the exchange rate was 7–9 guilders per pound sterling. By the following year, 143 rams and 92 ewes were sold for 32,214 guilders, illustrating the high profit that could be made. A man who could achieve such monetary gains was destined to have consid-

---

[133] Kaschnitz 1805     [134] Dawson 1958, p. 668     [135] Criste 1905, pp. 240–46
[136] Petri 1812

erable influence as a financial adviser. Petri presented his ideas to a wider public in a book with the optimistic title *Everything about Sheep Rearing* published in Vienna in 1815. His reputation was further enhanced by the rising standard of wool from the Prince's estates, and the particular quality of his own flocks at Theresianfeld. Flocks of fine-wooled sheep based on Petri-bred stock became established throughout the Province of Austria and beyond.

Of all the countries of central Europe, Hungary possessed by far the largest number of sheep, with an impressive history of fine wool production. As a centre for wool growing, it outlasted everywhere else in that region.[137] Writing about their unique kingdom, the Hungarian experts Gaál and Gunst are sure that Merino sheep arrived there some time before the first 'Saxon' Merinos reached Austria (i.e. before 1765).[138] They emphasise how the first Merinos entering Hungarian territory, housed at Mercopail, were of the Escurial *cabaña* ('overbred, small bodied, thin wooled') later to be replaced by the 'easier to acclimatise' Negretti type (hardy, thick-set, wrinkled). After 1772, sheep were being bred at Mercopail, as well as merely housed there temporarily. These authors mention that the Negretti type at Mercopail was called 'infantando' [*sic*]. The use of this name suggests that two Spanish *cabañãs*—Negretti itself and also Infantado—may have been involved in formation of this type of Merino sheep. The same suggestion arises with respect to Merino sheep in Austria. A modern German expert H. Baumgart has written about how it was reported at Thaer's Leipzig Wool Assembly in 1823 that the Austrian Merino sheep 'was named Infantado and later Negretti'.[139] This is not to suggest that Austrians confused the two *cabañas*. Indeed, Petri in his book distinguished four races he was then keeping separately on his estate at Theresianfeld, which he illustrates by clear engraving to show their differences. In fact, the Negretti and Infantado seem, from the examples given, remarkably similar (Figure 2.1).[140] It is no surprise that later Austian breeders came to doubt a real distinction between them, a judgement that was accepted by the reliable Mr Carr, 'a large sheep owner in Germany' who corresponded with the English expert W.C. Spooner,[141] reported in the *Agricultural Journal* and quoted by Martin.[142] As Martin remarked, 'We may observe that in Spain a distinction is made between Infantado and Negretti, which Mr Carr overlooks.'[143] It is clear, however, that Mr Carr had not overlooked it but was simply recognising that the designation 'Negretti' had come to be used in a different sense in Austria–Hungary than in Spain. It was no longer restricted to sheep from a single *cabaña* but was used as a morphological description of a robust, heavily wrinkled breed of mixed Spanish origin, selectively bred to the requirements of the wool market.

---

[137] Jenkins and Ponting 1987, p. 44      [138] Gaál and Gunst 1977, p. 258
[139] Baumgart 1957, p. 10      [140] Petri 1815, 1825      [141] Spooner 1844, p. 58
[142] Martin c.1849, ii, pp. 62–3      [143] Martin c.1849, ii, p. 65

Negretti sheep in the new sense were to be found in large numbers at the official state farms at Mannesdorf and Holič. However, the height of selective breeding of this type of sheep was taking place elsewhere. Independent evidence comes from the Scottish traveller Richard Bright, writing in 1818 about a visit to Hungary in 1814, who was sure that the best sheep he had seen there were Moravian, derived from Geisslern's Hoštice stock.[144] Bright was particularly impressed with the great flocks he saw on the estate of Graf Hunyadi at Tarrany (now part of Slovakia).[144]

> It is about fourteen years since the first Spanish sheep were introduced upon the Hunyadi estates, from Moravia, where Baron Geissler [*sic*] had been many years employed improving the breed. Since that time the Graf has exercised unwearied assiduity in crossing and recrossing, and introducing new and more perfect Merinos. By keeping the most accurate registers of the pedigree of each sheep, he has been enabled to proceed, with a degree of mathematical precision, in the regular and progressive improvement of his whole stock. Out of the seventeen thousand sheep composing his flock, there is not one whose whole family he cannot trace by reference to his books; and he regulates his yearly sales by these registers.

The methods used by Graf Hunyadi were precisely those of Baron Geisslern himself. Spooner took this for granted and also commented on the similarity to the practice in Saxony. Spooner knew about the Saxon system from Mr Carr, and he noted that each breeder 'adopts pretty nearly the same system'.[145] Several important landowners in Hungary depended on Geisslern's breeding stock and received practical instruction during the year 1805 from the well known Moravian estate manager Martin Köller (1779–1838), who was closely associated with Geisslern.[146] Handwritten instructions on selective breeding of sheep in Moravia, composed by Köller, dated 1827, are lodged in the state archive in Brno. Köller prepared these when he returned to Moravia to become senior manager of estates belonging to the rich and politically powerful Count A.K. Kounic (Kaunitz).[147]

Hungarian Merinos of the quality owned by Count Hunjadi attracted very high prices. Bright wrote about a sale at Holič on 12 April 1810, at which many rams were sold for prices between 2000 and 7000 florins. These are equivalent to Geisslern's Hoštice prices of six years earlier,[148] at a time when an ordinary ram fetched 25–50 florins. However one three year old wedder that Bright saw sold at Holič, surpassing all others in the superiority of its wool, fetched 16,200 florins.[149] At this time there were 20 florins to the pound sterling, according to Bright. The inflated prices of Geisslern's rams and also

---

[144] Bright 1818a, p. 124; Bright 1818b, p. 295
[145] Spooner 1844, pp. 58–9; see also Youatt 1837, pp. 171–2     [146] Orel 1977
[147] Balcárek 1977     [148] Köcker 1809, Chapter 8     [149] Bright 1818a, pp. 411–12

those from the Imperial farm at Holič were later highlighted by d'Elvert in his *History*.[150] Bright was in no doubt that the prices asked reflected exceptional quality compared with those he knew from Saxony: 'In Saxony, much attention had been paid to the Spanish sheep, but the economy of the sheep farm is there less understood than in Hungary.'[151]

The sheer volume of sheep in Hungary was a matter for comment in itself. Ten million was the estimate, of which six million were supposed to be 'improved' i.e. brought up more or less to Merino standard by grading crosses. This figure compared with 2.5 million in all the rest of Austrian lands put together, including 438,501 in Moravia and Austrian Silesia combined, according to an 'authentic statement published in 1813' that Bright quoted.[152] Important as Hungarian wool production had grown by 1814, further advances were made in the 20 years that followed. Gaál and Gunst state that the Merino character in Hungary was spread mainly by the 'Electoral–Negretti' breed 'later called felt-woolled sheep'.[138] Settegast makes clear that the 'Electoral–Negretti' breed was developed in Austrian Silesia by breeders such as von Lichnowsky on the basis of crosses between Geisslern's Hoštice Negretti and the Saxon Electoral (compared in Figure 7.5).[153] Allowing that Gaál and Gunst are referring to a stage in Hungarian Merino development when the direct influence of Geisslern, who died in 1824, had ceased, we may state the order of establishment of Merino sheep in Hungary to have been as follows: Paduan (from Italy in the seventeenth and early eighteenth centuries), Escurial (from Spain or Saxony in the 1760s or possibly earlier), Infantado–Negretti (from Spain in the 1760s–1780s), Hoštice Negretti (from Moravia in the period 1790–1824) and Electoral–Negretti (from Prussian or Austrian Silesia after 1820).

What Moravia lacked in numbers, compared with Hungary, it made up for in exceptional quality. This was underlined by J.G. Elsner in his book *My Practical Experience of Superior Sheep Breeding* (1827). In the final chapter, headed 'A comparative review of superior sheep breeding in Germany", this Silesian landowner compared Moravia, Silesia and other German states, as follows:[154]

> Moravia has launched itself vigorously into superior (*höheren*) sheep breeding during the past two decades. The Sheep Breeders' Association in Brno has brought together a great many agriculturalists, enthusiastically united in their ideas, who have raised sheep breeding in the province to a very high level. Eternal credit must be given to the founder of this association [the much esteemed adviser, André] not only within that province but in the whole of Germany, because of his having brought about so extensive an exchange of ideas, both by word and letter.

[150] d'Elvert 1870, pp. 133–4     [151] Bright 1818a, p. 140
[152] Bright 1818a, p. 413     [153] Settegast 1861     [154] Elsner 1827 pp. 202–5

Moravia and Silesia are now zealous competitors, and neither can relax if the one is not to be overcome by the other. In both lands, sheep breeding is carried out with passion and untiring attention, of which beforehand there was not the slightest hint. Continued success has come through a most rational and systematic approach, and will probably continue. A specifically superior feature of both provinces is that the landed gentry themselves take personal care of their estates, most of them supervising their own sheep

**Fig. 7.5.**   Merino rams of the 'Negretti' (A) and 'Electoral' (B) types drawn in 1847. They were reproduced by Burns and Moody 1935, where (A) is described as an Infantado Negretti Merino ram from Hoštice sheepfold and (B) as a ram from Thaer's sheepfold at Möglin. In the original engraving of the Hoštice ram, 'drawn from nature', it is shown against a background of the model farm at Schleissheim in Bavaria. The artist was C.C. Fleischmann, a draftsman with the United States Patent Office. In this version a different background has been substituted.

breeding farms. *Should Saxony fail to consider these lands, and not assume control of its own sheep farms, with the same attention and vision, it runs the risk of losing its premier position in the world market and suffering a big reduction in profits.* Silesia and Moravia, by founding their flocks on an *independent noble blood stock*, have made themselves independent of Saxony. In the course of history, it would not be the first time for a colony to overtake the mother country. [our emphasis]

Three decades later, in a retrospective review of sheep breeding, The same Prussion expert divided progress in sheep husbandry in central Europe into two phases, before and after 1820.[155] He considered that up to 1820, efforts to improve sheep breeding in the German states had been a struggle against ignorance and prejudice. As early as 1756, a German writer, Schlettwein, had defended the principle of inbreeding in a controversial academic thesis on sheep breeding, *De Lana Ovium Emendanda*, at the University of Jena. This is mentioned by C.A. (Auguste) Wichmann in the introduction to his German translation of Daubenton's major work.[156] It seemed to Wichmann that the self-same ideas and principles 'as Mr Daubenton brings forward' were detailed by Schlettwein. Clearly it was not as intensive a programme of inbreeding as Bakewell and Geisslern were prepared to engage in. Nevertheless, Elsner reported that there was intense opposition to what Schlettwein defended in 1756, which he put down to 'ignorance and prejudice'. Wichmann attributed the missed opportunity of following Schlettwein's lead to lack of official patronage. He compared the conservative attitude of the Saxon authorities unfavourably with the enlightened support Trudaine had given Daubenton in France.

Lending support to Elsner's high opinion of Moravian breeding stock is the published report by the Scotsman Bright of his travels through Lower Hungary in 1818, which includes the telling statement, already quoted, that the economy of the sheep farm was better understood in Hungary, under Geisslern's influence, than in Saxony.[157] There is also an account from a French author, M.D. de Liège, in his book *Les Animaux Domestiques* published in 1842, in which he wrote:[158]

> Poland, Moravia and Silesia also received strains of Merinos which rapidly multiplied and prospered, thanks to intelligent care. Today the wool of Silesia and Moravia rivals the most beautiful electoral wool of Saxony.

Reports by various foreign observers, already quoted, suggested that Saxon flocks were far from uniformly impressive and that a large proportion of wool marketed through Leipzig, as 'Saxon', had its origin outside the boundaries of Saxony.

---

[155] Elsner 1857   [156] Daubenton 1784, p. 43   [157] Bright 1818a,b
[158] de Liège 1842, p. 586

Intensive selection for particular wool characteristics carried out on Saxon sheep, to the exclusion of other qualities, led to a deterioration in their constitution and state of health, so that by the late 1830s the type of Spanish Merino originally imported into Saxony had almost disappeared. Ambitious breeders began to look for better rams. The orthodox adherents of the 'pure' Merino races, either Electoral or Negretti, came into conflict making claim and counterclaim. Hermann Settegast, the influential Prussian expert, describes the conflict in his book of 1861. It is illustrated with a diagram of a family tree representing the major and minor branches of the Merino breed as it had spread from Spain into other European countries (Figure 7.6). Settegast awards the branches of the tree varying status, according to their origin and degree of selection. The title '*vollblut*' is awarded to only three 'races' of Merino outside Spain: the 'Negretti' of Geisslern in southern Moravia, the 'Electoral–Negretti' of Prince Lichnowsky at Kuchelna in Austrian Silesia and certain 'Merino–Mestiz' flocks in three regions of France, based on stock of mixed origin from Tessier's farm. In the mind of

**Fig. 7.6.**   Family tree of Merino breeds in continental Europe, taken from Settegast 1861, indicating the three races outside Spain considered by this stage of selective breeding to have attained the status of *vollblut* (thoroughbred). (An enlarged version of this figure appears on page ii).

Settegast, *vollblut* referred to a consistently superior stock, breeding true to type. It evidently did not disturb him that these might have a mixed origin. Selection was the key to pure breeding rather than a 'noble' ancestry. It was a point of view by no means shared by every breeder although perfectly logical to those in the Bakewell and Geisslern tradition.

Settegast could find no Electoral flock which had proved consistent enough to be awarded *vollblut* status. The Rambouillet race in France, even then still developing, was also too variable. While the Rambouillet flock had been kept 'pure Merino' (although derived from several different *cabañas*), Merino–Mestiz (*métis*) flocks were half-bloods,[159] out of native ewes served by Merino rams. According to a later source,[160] flocks of such hybrids remained remarkably constant over generations and were valuable breeding animals, surpassing the Rambouillet in body mass. This can only have been the case if the progeny of crossed ewes were backcrossed to Merino rams in every generation. For the Merino–Mestiz flocks, the term *vollblut* was used by Settegast in the knowledge that it was not 'pure blooded' in the old-fashioned sense (Chapter 10). He was using the term in the sense that 'thoroughbred' was understood by English racehorse breeders.

Geisslern's 'race' might not be as mixed as the French Merino–Mestiz flocks, but it certainly did not owe its origin to a single bloodstock. Bearing in mind that sheep of at least five Spanish *cabañas* (Escurial, Infantado, Negretti, Paular and Guadeloupe) were reported to have entered Austria at different times, a mixed origin for the so-called 'Negretti' flocks of Geisslern and his friends cannot reasonably be doubted. In central Europe the name 'Negretti' described a type of sheep with thickly matted wool growing to the eyes and down the legs to its feet; it was a large sheep with a heavy fleece and wrinkled skin, the body being characteristically rounded, with head and neck short and broad and the nose short and somewhat turned up (Figure 7.7). Such sheep were inevitably contrasted with the smaller Electoral (or Escurial) type of Merino, with longer legs and a spare neck and head, with very little wool on the latter, and less wool in general, although with a finer, shorter, softer character to the fleece (Figures 7.1 and 7.5). We have descriptions from Martin in about 1849, based on the expert opinion of Mr Carr.[144] The fleeces of the Negretti sheep were about twice the weight of those from the Escurial type, the difference depending on degree of improvement. It must have taken considerable breeding skill to create the new Escurial–Negretti *vollblut* race developed in Silesia. As we have seen, the Prussian-based Settegast gives the credit to Prince Eduard Maria von Lichnowsky on the Austrian side of the border.

Central European Merino wool found a ready market in the west. Demand encouraged a rapid increase in production. In 1810 the quantity entering

---

[159] Martin c.1849, ii, p. 63
[160] Spoettel and Taenzer 1923 (quoted by Burns and Moody 1935)

England from 'Germany', either from the North Sea or Baltic ports, was little more than 10% of total imports, but within six years it represented half of a much greater total. The increase occurred after the Spanish flocks had been ravaged by war. Wool labelled 'Saxon' got the highest price. This can be attributed in the first instance to the skill of the Leipzig wool graders, and secondly to the introduction into Germany of better stock from territories to the South-east. In fact, more wool was imported into England from the

**Fig. 7.7.** A Hoštice Merino ram (A), as illustrated by Janke 1867, compared with an Australian Merino ram (B). The source of the Australian photograph, believed to have been taken before 1910, is not known, but it bears a close resemblance to the Peppin Wanganella stud ram photographed in 1874, presented by Garran and White 1985 (Fig. 50).

Hapsburg territories via Silesia than from Saxony although, it must be said, at a slightly lower price.[161]

By 1830, the vast majority of Merino wool imported into England was shipped either from the ports of Hamburg and Bremen (i.e. from 'Saxony') or from Stettin and Danzig (i.e. from 'Silesia'). Although much of the wool coming to Britain under either label would have originated in territories further away, little would have come from Moravia whose own fine-wool cloth industry in Brno and surroundings absorbed most of the local production. The amount of wool reaching England from Germany as a whole increased from 412,000 lb in 1800 to 778,000 lb in 1810, then to more than 5,000,000 lb by 1820 and to a maximum of more than 26,000,000 lb in 1830. [162] Then, more quickly than the trade had grown, it declined. By 1850, German wool imported into Britain had been reduced to a token because of an overwhelming influx of Merino wool from the British colonies. The major supplier was Australia, able to produce it at a highly competitive price.[162]

Moravian agriculture moved on to other crops, particularly wheat but also fruit, vines and cattle. An efficient agriculture was essential to support expanding industrial production. In 1844, the Scotsman J.C. Loudon was able to write of Moravia that 'With the exception of some districts in Netherlands, scarcely any part of the continent is so well cultivated. It bears too a larger proportion of wheat than any other district in the east of Europe.'[163]

## Australia

It may seem strange to include Australia in the present context, but it was from central Europe that a large proportion of the Merinos entering the new British territory originated, and not from Britain, despite the fact that this country had access to many thousands by 1810, as a result of the devastation to the Spanish economy caused by the activities of Napoleon and the Peninsular War that finally excluded him from Spain in 1814.[164] Little use was made of the woolly booty for breeding purposes, either at home or in colonies overseas. Partly this was because many of these unfortunate animals, shipped in hurried circumstances, died in transit or arrived in such a wretched condition that they had to be destroyed. For whatever reason, the opportunity for Merinos to become established in British hands was not widely exploited until quite late in the breed's history (see also Chapter 6).

The enormous significance of Merino sheep in Australia has inspired thorough research into their origin there; this has revealed very little true Merino breeding before the 1820s.[165] After the first so-called 'Merinos',

---

[161] 'Lincolnshire Grazier' 1833, p. 284      [162] Baines 1970 (1858), p. 79
[163] Loudon 1844 (1831), p. 98
[164] Carter 1964, p. 325; Carter 1969; Carter 1979, p. 487: letter from A.C. Johnstone (army officer and adventurer, later Member of Parliament) to Sir John Sinclair, 9 May 1809
[165] Garran and White 1985; Massy 1990

probably cross-breeds, were landed in Australia from the Cape of Good Hope in 1797, about five rams and seven or eight ewes,[166] followed by five Negretti rams and one ewe from Kew in 1805,[167] fewer than 30 of the many thousands then flooding into Britain from Spain ended in the new territory.[168] The sheep that helped to form the foundation of the Australian Merino were in fact of very mixed origin. Early crossing breeds came mainly from Bengal, later Ireland and England, the latter being more or less improved by Dishley blood.[169] Genuine Merinos arrived from different sources. Britain provided some at first from the few flocks that still existed in the home country, like those of Joshua Trimmer, Lord Weston and the Marquis of Londonderry. Others were imported from Spain. But after 1825, imports were mainly from France or central Europe.[170] Merino breeding was established both in New South Wales and Tasmania, the latter being said to compare closely with Saxony for both climate and pasture.[171]

Examples of shipments included eight 'Saxon' sheep in 1826, among which was the ram Kayser (*Kaiser*) on lease from Albrecht Thaer at Möglin, who was 'the author of the best work extant on sheep culture', according to the *Hobart Town Gazette*.[172] Further sheep from Möglin arrived in 1828. Meanwhile, 103 rams and 61 ewes from Saxony had been unshipped in 1827, after which there were three further shipments from the same source (firstly 14 rams and 291 ewes; then 13 rams and 359 ewes; and finally 45 rams and 100 ewes) in 1829.[173] Two hundred ewes from Ochatz, in northern Moravia (Austrian Silesia) (classified misleadingly as 'Saxon Negretti') were brought in by J. Bettington in 1829, and 200 Silesian sheep (also referred to as 'Saxon') were shipped from Stettin via London in 1827 and purchased by Alexander Riley. A principal importer of Möglin sheep was William Hampden Dutton (b. 1805), an English pupil of Albrecht Thaer who had studied at the Agricultural Institute at Möglin in 1822–3,[174] before moving to Australia in 1826.[175] Dutton got other sheep from flocks at Warsen, Dresden and Anhalt (probably from Fink's stock at Petersberg (Figure 7.2)) and also from Kuchelna in Austrian Silesia (northern Moravia) at the breeding farm of Prince Lichnowsky (Figure 7.4). A total of more than 5000 Merino sheep entered New South Wales before 1831 and over 1500 reached Tasmania.[175]

Australian settlers were deeply shocked and disappointed by some of the shipments they received. So-called Merino sheep reaching them from unnamed German flocks could be poor, ill-made specimens and often diseased. Significant exceptions included those from the Lichnowsky estate at

---

[166] Massy 1990, p. 27    [167] Massy 1990, p. 309; Bischoff 1828, pp. 60–62
[168] Garran and White 1985, pp. 35–6    [169] Massy 1990, pp. 926–9
[170] Massy 1990, p. 66    [171] Massy 1990, p. 55    [172] Massy 1990, p. 58
[173] Hawkesworth 1920, p. 37    [174] Garran and White 1985, p. 143, 150
[175] Massy 1990, p. 70

Kuchelna. So much is clear from a report by Dutton to the Australian Agricultural Company in London in 1826:[176]

> I found the sheep which I purchased from Prince Lychnowsky [*sic*] in Austria both better able to endure fatigue and, upon the same allowance of food, kept themselves in much higher condition than the others, and in the case of a future purchase I would strongly recommend its being made wholly out of his flocks.

Dutton correctly refers to the Lichnowsky sheep as from Austria, i.e. from the Austrian part of what had been Silesia and was then incorporated into Moravia. More typically the Lichnowsky sheep in Australia were incorrectly referred to as Saxons![177]

The contrast between the excellent sheep imported personally by Dutton and the 'job lots' sent by agents encouraged further Australian wool growers to make the journey to Europe themselves. One of these was 'the redoubtable Mrs Eliza Forlonge',[178] the wife of a Glasgow merchant; her determined efforts in droving her precious Saxon purchases from the famous Kliphausen stud (Figure 7.2) 300 miles across Germany and 100 miles more to Liverpool, with the help of her two sons, have been described graphically by Garran and White. She made her epic journey in 1829, and by 1835 the Forlonge flock had increased to 8000.

James Atkinson, who farmed in New South Wales, was another sheep breeder who travelled personally to Europe. He made a tour of German states in 1826 and, upon returning home, reported that the Lichnowsky sheep at Troppau (Opava) were the finest in 'Germany':[179]

> These animals were large, with long and extremely fine wool and possessed more of the character of the Negretti Merino than any sheep we saw, except for a few in the flock of the Grand Duke of Saxe-Weimar, and a few in a flock near Leipzic [*sic*], both of which had been lately obtained from the National Farm at Rambouillet in France; these were evidently of the pure breed, and, in fact, were the only sheep we met with, which we were decidedly of the opinion were so.

The kind of sheep Atkinson was describing at Troppau was closely akin to the Hoštice Merino. As we have noted, the Lichnowskys began their breeding work with sheep from Geisslern. Co-operating closely with the Brno Sheep Breeders' Society, the Prince and his associates learned the selection methods used by Geisslern. With stock of such quality, Australian sheep breeding and wool production flourished, and enthusiasm spread throughout the region. Complex programmes of crossing and selective breeding were undertaken, as described in detail by Massy (1990). By the final quarter of

---

[176] Dutton 1826 (quoted by Garran and White 1985, p. 152)
[177] Garran and White 1985, p. 153    [178] Garran and White 1985, p. 174
[179] Garran and White 1985, p. 134

the nineteenth century, Australian rams were being produced of the type shown in Figure 7.7, compared with an engraving of a Hoštice Merino ram published in 1867 for comparison. It is easy to agree with Hawksworth that 'In no country were the improvements so successful as in Germany and a portion of Austria [i.e. Moravia] excepting Australia and Tasmania.'[180]

When reviewing the major trends in sheep breeding, it thus proves impossible not to look towards the heart of Europe and in particular to enquire about Geisslern in the Moravian village of Hoštice. What can be said about that shadowy figure? We need to ask whether he really did have the powerful influence on the course of events that was claimed for him. It is also of intriguing importance to know whether his success in breeding provided extra insight into heredity, as Bakewell's achievement had done. Was the name 'Moravian Bakewell' justified in the sense of a real similarity between the two breeders or an equality of achievement, or was it simply used to enhance Geisslern's reputation in a local context? In the following chapter we introduce Geisslern and evaluate his achievement.

---

[180] Hawkesworth 1920, p. 33

# 8

# *Ferdinand Geisslern, the Moravian Bakewell*

Sir, you have transformed the wilderness at Hoštice into a place of pilgrimage for our agriculturalists. Your action has spoken volumes; something inestimable has spread out from here with the brilliance of electricity, truly and rapidly.

*R. André, addressing his patron Baron Geisslern (1816) (translation from the German)*

It took unusual and even conflicting qualities of personality to be successful at breeding Merinos outside Spain. Open-mindedness was required to adapt and combine the best ideas in sheep management into a suitable breeding programme, single-mindedness to carry the programme through to a successful conclusion. Among the few individuals whose efforts stood the test of time was Ferdinand Geisslern, second son of a Moravian judge; his abundant talent in this direction emerged unexpectedly from provincial obscurity in the 1790s. His farm at Hoštice in eastern Moravia became renowned for its superior flocks of fine-wooled sheep and acted as a Mecca for other breeders. Respected rivals with official positions as directors of Imperial Hapsburg breeding farms soon had reason for surprise, even discomfort, that a self-taught farmer of relatively slender means should excel them. For not only was his sheep stock unsurpassed, and therefore much in demand at high prices, but he seemed to improve it with every generation under his care, making it famous far outside the boundaries of Moravia or even of the Imperial Empire itself.

Geisslern's accomplishments were reviewed by J.K. Nestler, Professor of Natural History and Agricultural Science at the University of Olomouc

(Olmütz). Remarking upon the favourable impact of Geisslern's breeding activities on the Moravian economy, Nestler wrote: '70 years ago it was a matter for comment when the annual profit from 1,000 sheep was 100 guilders; now we have more than a few sheep breeding farms on which 100 sheep bring a profit of 1,000 guilders.'[1]

The claim to excellence that supported these high profits lay in Geisslern's way of co-ordinating the improvement of wool and body-form together, using the most advanced techniques. By this means he produced an exceptionally valuable strain based on imported original Merino breeding stock, modified to local requirements. More remarkably, he was able to upgrade the relatively coarse-wooled native sheep of Moravia (Chapter 7) to a condition in which they yielded some of the most consistently fine fleeces in Europe. Others used his Merino stock in their own territories, as far apart as Hungary and Pomerania. Recognised as the 'Moravian Bakewell' by the value placed upon his rams, he resembled the Englishman in a surprising number of ways, as we shall show.

In 1779, Geisslern's father, an ambitious individual who had worked his way up to the office of judge in the southern Moravian town of Znojmo, bought the neglected estate of Hoštice, about 60 km East of Brno (Brünn) (Figure 7.4), a stretch of rolling countryside covering about 300 hectares (about 730 acres). Geisslern was then aged 28. There is no record of his background except that both he and his younger brother Johann Nepomuk studied at a superior *gymnasium* in Vienna, the Theresianum, established for the education of young men for high civil or military service. Commissioned into the Austrian army, they both saw active service before returning to civilian life. By 1782, Geisslern was managing the Hoštice estate while his brother held a position in Vienna in the state administration. The father and, later, both brothers were awarded the title of Baron (Freiherr) Geisslern for meritorious public service. Before that they used the name Geissler. Ferdinand Geisslern never married and, when he died at Hoštice on 7 September 1824, the estate passed to his brother. The distinguished reputation of Hoštice sheep continued into the 1860s.[2]

Geisslern's neighbours, who competed with outsiders to give elevated prices for his breeding animals, were free with their praise, inspiring glowing reports in the agricultural press and literature. More difficult to find are comments about his life and personality, although several writers bear witness to his professionalism and his friendly openness in sharing ideas, a rare quality in the competitive world of breeding. For personal information we are restricted to a brief obituary notice published in Brno in 1824, reprinted later by d'Elvert:[3]

---

[1] Nestler 1841, p. 216 (quoted also by Fraas 1852, p. 607 and d'Elvert 1870. i, p. 340)
[2] Settegast 1861; Körte 1862; Janke, 1867          [3] d'Elvert 1870, ii, pp. 137–8

Here at Hoštice he [Geisslern] was engaged in agricultural activities of the highest merit at a time when nothing progressive was being done in agriculture and when there existed a distinct prejudice in higher circles against a landowner taking personal care of his estate, an activity considered to be demeaning. His inner strength and determination enabled him to surmount the prejudice, leading quickly to success in terms of greater profits both from field crops and animal husbandry. Above all he improved his sheep to an extent, and with such precision that his achievements became an object lesson to others. Hoštice emerged as the focus of attention for every progressive agriculturist in the Austrian Empire. Anyone was free to visit Hoštice where he would receive a friendly reception, and discover much to discuss and emulate on this model estate. The owner willingly shared his knowledge with others who, on following the advice given and adopting his principles as their own, achieved similar success. This was especially true with respect to sheep improvement. We may conclude with certainty that the breeding of noble sheep had its origin in the experience and correct deductions of Baron Geisslern.

Although Geisslern's personal papers and records have not survived, except for a few letters, there is no shortage of information on his farm and its management. The principal credit for this must go to C.C. André, who acted for Geisslern in much the same way as Arthur Young had done for Bakewell, adopting the role of enthusiastic publicist. The first mention of Geisslern appeared in André's journal *Patriotisches Tagesblatt* (hereafter abbreviated to *PTB*) in 1802, in the form of a visitor's report on the farm, most probably written by André himself.[4] The account begins with a Latin quotation from Horace: '*Omne tulit punctum qui miscuit utili dulce*' ('He who has mingled the useful with the pleasant has gained every point [i.e. full marks]', suggesting from the outset that Geisslern's claim to achievement rested upon his capacity to combine beauty with utility. André extended the idea by remarking on how Geisslern's farm presented a 'harmonious unity', greatly admired by visitors. In a detailed description, he commented upon the high productivity of the soil, the development of new varieties of clover and fruit trees and of how Geisslern was able to 'improve upon Nature'. Reserving special praise for the sheep, he noted the resemblance of Hoštice to an agricultural school where sheep breeding and other farming skills could be studied. His reference to 'harmonious unity' and his likening of Geisslern's farm to a school are obvious echoes of Bakewell, recalling remarks in similar terms about Dishley made by Marshall.[5]

Further details of the farm are to be found in an account of a visit there, entitled 'The Estate at Hoštice in Moravia'.[6] The author, once again probably André himself, comments upon the value of the estate which had increased

---

[4] C.C. André 1802, Orel and Wood 1981
[5] Marshall 1790, i, pp. 287 and 297; 'Benda' (1800); Housman (1894) (Chapter 5)
[6] C.C. André, 1804a

from 43,000 guilders in 1779 to 100,000 by 1804. The credit for this is attributed mainly to the owner's breeding skills, although other aspects of husbandry also attract the writer's praise, including the planting of special aromatic herbs in the meadows to augment the sheeps' diet. Much is made of Geisslern's good relationship with his estate manager and also with prominent members of the village: the local physician, the enlightened village priest and the teacher. Geisslern's practice of paying his shepherds a good wage is another matter for favourable comment. Their living conditions and those of other villagers impressed André as being unusually high. He notes that when the family had bought the estate, all the houses were built of unfired clay bricks, but by 1800 most of them had been rebuilt with properly fired ones. Living conditions for farm workers could be very bad in central Europe at that time. Joseph Marshall, an English traveller who passed through neighbouring Bohemia in 1772, was clearly shocked by the condition of the peasants there, forced to live in 'hovels of the worst sort'.[7]

M. Köcker, a Silesian economist who visited Geisslern in 1809, wrote an account in a letter for André to publish. Fresh from a stimulating experience, Köcker was full of enthusiasm. 'Geisslern, sheep breeder of great esteem, may justly be called the 'Moravian Bakewell', he wrote.[8] In the next few lines he mentioned some of the techniques associated with sheep breeding on Geisslern's farm, how he numbered every one of his animals and recorded their parentage and progeny in order to select the best rams and put them to the best ewes. He noted with approval the well ventilated stables for segregating sheep in small groups and the use of artificial irrigation to increase fodder production. Geisslern was the first landowner in Moravia to establish water meadows artificially, a technique in which Bakewell was a renowned expert.

As a practising economist, Köcker was deeply impressed with Geisslern's claim to be self-taught, especially when he discovered that his 400–500 sheep produced an annual profit of about 30,000 guilders (approximately £4000 sterling). Hoštice ewes fetched 80 guilders each, a one year old ram could be sold for 300 guilders and a two year old ram for 1000 or, exceptionally, even 2000–3000 guilders (see also Chapter 7). The price of an ordinary ram was about 15 guilders at that time. The financial evidence underlined the conclusion that here indeed was a new Bakewell.

Köcker may have originated the description of Geisslern as the 'Moravian Bakewell' in 1809, but it was already implicit in the writings of André from 1802 onwards because of the frequent parallels drawn between the two breeders. After 1809, the term 'Moravian Bakewell' appeared repeatedly in the pages of journals published in the Austrian territories. The first volume

---

[7] Marshall 1772, iii, p. 307        [8] Köcker 1809

of *ONV* carried a short article deploring the delay in introducing English breeding methods to the Austrian provinces of Kärnten and Steyermark.[9] The Vienna-based author contrasts local backwardness with the progressive attitude of Geisslern, referred to here as the 'Austrian Bakewell'. He points to him as an example of the financial advantage of applying English methods, enabling large profits to be made out of wool production.

Among the stream of visitors coming to learn about sheep breeding at Hoštice, one of the more significant was Count F.A. Magnis whose flocks over the border in Prussian Silesia, in the county of Glatz, were made famous internationally by the praise of the French expert C.P. Lasteyrie.[10] Other important 'disciples' included Prince Lichnowsky with an estate at Kuchelna in Moravian Silesia, Count Emmerich Festetics of Hungary, F. Maas from Alt Kenzlin in Mecklenberg and J.G. Elsner, a Prussian landowner with estates at Reindorf. Count Festetics and fellow breeder Count David Ehernal each attracted the nickname 'Hungarian Geisslern'.[11]

Within Geisslern's own locality, one of the first to come under his influence was Martin Köller, who acted as his ambassador, travelling widely on his behalf. In Hungary Köller introduced the great man's methods on to the estates of Counts Ehernal and Hunyadi.[12] Another regular visitor to Hoštice was J. Petersburg (1757–1839), friend and fellow Moravian breeder, who had married one of Köller's sisters. There was also Rudolf André, C.C. André's son, who became Geisslern's most outspoken pupil. After studying his methods as a resident on the farm, the young André wrote a book called *Instructions for Improvement of Sheep* (1816),[13] claiming to give the most detailed and precise practical information then available on selective breeding.

## Breeding methods

Köller gave his own account of sheep breeding at Hoštice anonymously under the pseudonym 'K in Mähren' (1811). The title of his article took the form of the double question: 'Is it necessary always to acquire original Spanish (Merino) rams in order to maintain a noble flock of sheep, and does the stock stay true to type when related blood is mixed?' He was influenced by a controversy proceeding in the agricultural press between two well known writers on sheep breeding, the Englishman C.H. Parry and the Swiss C. Pictét about whether or not Merinos should always be crossed with local improved ewes (Chapter 6). Recognising Geisslern's breed as a constant race with a high yield of fine wool, Köller looked for an explanation of its consistent quality. He found his answer in the practice of inbreeding, coupled with directional selection for defined production traits. He knew that most writers on the subject opposed inbreeding on religious grounds, believing it to be

---

[9] Anon. 1811b    [10] Lasteyrie 1802, pp. 32–5 and 194–206    [11] Anon. 1818b
[12] Bright 1818a,b    [13] R. André 1816

contrary to the laws against incest, but he felt impelled to identify himself with the technique. In support he recalled an inscription placed in 1796 above the entrance of a newly built farm owned by the Archbishop of Olomouc. It read as follows:

| | |
|---|---|
| *Nicht von Vorurtheil beirrt-* | Not by prejudice misled, |
| *wird von Müttern, die zu Gatten* | Mates for Mothers |
| *Brüder, Söhne, Väter hatten* | Come from Fathers, Sons and Brothers. |
| *gesunde und edle Lämmer ziehen* | Healthy and noble lambs are bred, |
| *die durch seine Klugheit blühen.* | Which flourish through such prudence. |
| | *(free translation)* |

If the Archbishop was prepared to countenance such close inbreeding in his sheep, why should anyone else not do so? The Brno historian d'Elvert, recalling these events many years later,[14] was sure that Köller's acceptance of inbreeding in sheep was a precise reflection of Geisslern's practice. However, the sequence of events points to an early influence also from Petersburg who, as manager of the Archbishop's farm, was responsible for the inscription. He had written a strongly worded pamphlet carrying the quoted verse, circulated in 1797. It was later published, in major part, by C.C. André in *PTB*[15] as a protest against authors who rejected inbreeding. These included Professor F. Fuss (1785) in Prague, the Cistercian monk C. Baumann (1785, 1803) in Bavaria (later in Moravia) and Stumpf (1785) in Saxony. Above all there was Buffon, considered to be in serious error by both Petersburg and Köller for his 'pernicious views'. They challenged this so-called expert 'who penetrated the heads of trusting farmers with the idea that animal improvement could come from crossing animals raised under different conditions'[16] (Chapter 3).

When prudently applied, inbreeding represented for Köller the only way of guaranteeing 'a completely noble flock without hereditary defect [*Erbfehler*]', yielding the finest wool and a well flavoured carcass.[17] Such flocks did not require the regular introduction of Merino rams from Spain to upgrade them, he stressed. After the question 'Where can such unblemished flocks be found?', he answered 'More than once did our Moravian Bakewell exhibit them at Hoštice.' Köller defined these as 'noble flocks' on the basis that they did not degenerate even when no new foreign rams were introduced. He insisted that 'Noble sheep without hereditary defects crossed with ewes without hereditary defects produce offspring also without these defects.' This was no more than 'the law and process of Nature' (*das Gesetz und Gang der Natur*); only people with ingrained, old fashioned prejudices would say that it was not so. He was convinced that rejection of inbreeding offered the lazy

---

[14] d'Elvert 1870, ii, pp. 148–9    [15] C.C. André 1804b, p. 1314; see also Petersberg 1815
[16] Fraas 1852 p. 556; d'Elvert 1870, ii, pp. 148–52; Buffon 1795 (German translation)
[17] 'K in Mahren' 1811

breeder a way of avoiding inconvenience. However, 'The general laws of nature cannot be changed just because it is convenient to do so.'[17]

Reading Köller's account, one can agree with Fraas[18] regarding the close parallel to be observed with the views of Bakewell on inbreeding, as expressed for example by Culley.[19] Köller never claimed a whole flock of noble sheep to be absolutely perfect, always stressing the need to keep them under permanent selection as Bakewell had done. Pospíšil, the skilled manager of Geisslern's, farm might truthfully state that while there were some sheep he could see no way of improving further, there were others that failed to meet his high standard. The better the breeder the more likely he was to recognise the limits of his skill, just as Bakewell had done (Chapter 4).

An anonymous author, probably C.C. André, had written particularly about English sheep breeding in the 1809 volume of *Hesperus*,[20] describing the improvement of local flocks through crossing with imported Merino rams. Köller expanded on this topic in his 'K in Mähren' article, making clear that he accepted the idea of crossing only under strict conditions. Simply to cross was not enough. In order to match Geisslern's achievement, a breeder must select his stock with particular prudence and intelligence, choosing both sexes with desired production traits and free from hereditary imperfections. Only by analysing traits within inbred families could such faults be revealed. Several generations of crossing would be necessary and the breeder must have detailed knowledge of individual animals and their ancestry in order to make the most effective crosses and eliminate the hereditary defects. Each animal had to be considered in its entirety. Köller was describing the procedure of grading crossing coupled with assortative directional selection and progeny testing.[17]

C.C. André, who commissioned Köller's account, was intrigued to know even more about Geisslern's methods. It was for this reason that his son Rudolf eventually found himself there as a pupil. The great breeder had agreed to teach him. At the age of 23, Rudolf lived at Hoštice for several months in 1815 with direct access to the master and to his estate manager Pospíšil, being at the same time in contact with Köller, then managing Count Lamberg's estate at Kvasice (Figure 7.4), about 15 km from Hoštice. Köller, whom the historian d'Elvert was to describe as R. André's 'spiritual father',[21] helped him to interpret the new ideas in breeding. Inspired by his visit, the young man was quick to draft out a book that revealed details of Geisslern's methods, which his father helped him to complete and get onto the market within a few months. In the introduction, R. André wrote, 'I was most impressed with what I saw at Hoštice, above all with the magnificence of the sheep. Since the time when I was lucky enough to stay at Hoštice, I have

[18] Fraas 1852, p. 556    [19] Culley 1786, 1804    [20] Anon. 1809
[21] d'Elvert 1870, ii, p. 183

devoted myself wholeheartedly to sheep breeding.' Dedicating his work to 'the benefactor Baron Geisslern', André gave many more practical details than Köller had been able to include. In the introduction, he explained how Merino sheep in Moravia were still insufficiently improved *as a whole*, despite the efforts of leading landowners and skilled breeders. There were still farm managers who, when entrusted with the valuable stock, proved ignorant of their basic needs exposing them to moist pastures more suited to cattle than to sheep, and disastrous for Merinos. An even greater obstacle to Merino improvement, according to André, was sheer ignorance of the principles of breeding. Even the new book by Petri (1815) did not describe the breeding business as thoroughly and convincingly as André wished. With no practical handbook on the subject, André found it hardly surprising that farmers and farm managers had made expensive errors. He intended his own book to remedy the situation.

Much of the book was devoted to Geisslern's procedure for improving local 'common' sheep (*gemeine Racen*) by upgrading them with imported rams (*original Racevieh*) or with stock of equivalent quality and purity. André made a distinction between 'pure noble breeding stock' (*reines edles Racevieh*) and 'noble sheep' (*edles Schafvieh*). The difference between the two categories rested upon the consistency of racial traits, their 'certain stability' (*gewisse Beständigkeit*) as he referred to it. The traits of the pure breeding stock 'remain constant even when external conditions are unfavourable for their preservation'.[22] Noble sheep were typically the product of grading crossing between local ewes resembling those shown in Figure 7.3, more or less already improved by Paduan crosses, and imported Merino rams, as shown in Figure 2.1. Although they might demonstrate 'rare, excellent qualities of body-build and wool', their progeny would be variable, particularly in unfavourable conditions. André explained the instability between generations in terms of degeneration (*Ausartung*). By contrast, the pure noble breeding stock degenerated only very slowly: 'After many generations one will still be able to detect some almost unquenchable qualities.' As well as showing the desired wool and body characteristics preserved in every generation, such animals were 'distinguished by their endurance (*Ausdauer*), vitality and suitability for breeding as a result of a balanced, firmly established organisation (*Organismus*)'. They were propagated consanguineously due to the complete exclusion of outside rams. To ensure uniformity of production traits, André recommended close mating of the best rams with their daughters and their mothers, each ram forming a 'sire's family'.

The value of selective breeding could not be overestimated. The inconsistencies of noble sheep between generations when management and other

---

[22] R. André 1816, pp. 6–7

external conditions were not absolutely favourable to the excellent traits of the breed, did not prevent their being finally raised to the status of the pure race.

> With care and attention a merely noble flock can be raised to the pure race [*Racevieh*] if one refrains from intermixing alien bloods and, through an appropriate control of pairings, brings together specific characteristics with regard to the body build and wool of these animals, which will then be transmitted to the progeny and preserved in the same degree. In this way something constantly unique arises, something fixed in the organisation of these animals, something derived totally and exclusively from pure blood relatives [*aus lauter Blutsverwandten hergeleitete*], which is characteristic of the lineage.
>
> R. André [22]

This then was the secret, to match the parents for their traits, to practise rigorous selection and to fix the type by close inbreeding. Individually controlled matings (*Sprung aus der Hand*) were the answer. It is no wonder that Geisslern was seen as a second Bakewell. Selective breeding was required from time to time even in the pure race, because events were to show that racial stability was never absolute. Probably neither Geisslern, nor indeed Bakewell, believed that it would be. Both, however, were happy to exploit the impressive influence of 'the blood' as much as possible.

André's designation 'pure noble race' (*reines edles Racevieh*) does not correspond precisely with what Köller in 1811 had called the 'pure original race' (*reine original Rasse*) because André's category was not restricted simply to original sheep from Spain. It also embraced their pure bred, highly selected descendants. André even appears to have believed that something like the original race could be recreated from the progeny of grading crosses to local sheep, if the lambs were sufficiently well selected, over a large number of generations. Such a Bakewellian belief in the power of selection was a big advance in the Moravian context. André stressed that the purity of a group of inbred animals could only be proved by establishing constancy between generations. The breeding value of each ram had to be assessed. If a ram produced consistently good progeny from a series of females, then it was considered most suitable as the instrument of improvement, either to maintain the pure noble race or to improve a local race. André was convinced that the greatest obstacle to strain improvement arose from exchanging rams between flocks:[23] 'It is absolutely worthless and highly injurious to a thorough improvement if one takes breeding rams sometimes from one source, sometimes from another.'

Having established the theoretical basis of breeding, André carefully described the details of Geisslern's procedure: the best way of housing sheep,

---

[23] R. André 1816, p. 9

separating the breeding rams, young rams, stock ewes and young ewes, and how to number and mark individual animals as a prelude to pedigree and trait recording and individual selection. The details are given precisely. First he briefly described 'twelve advantageous traits of the body', illustrated by quantitative data taken from a table published by Petri (1815). Then, in 12 pages, he enlarged upon the traits. He stated that wool samples should be taken from four parts of the body, cut with shears close to the skin; the first should be taken from the head, the second from the middle of one shoulder, the third from the back and the fourth from the hind leg, 'not too low'. Every sample was to be carefully and separately packed in a paper bag and described, all samples of a given animal being placed together in a cloth sack with the description on each sample recorded absolutely clearly to avoid mix-ups. André included a set of specimen forms for recording each different aspect of wool quality and body shape. On the basis of an analysis of all traits in both sexes, breeding animals were to be further selected for their compatibility with one another.

The matching of ram and ewe on the basis of complementary traits was exactly the mating policy adopted by Bakewell as reported by William Marshall.[24] But whereas Marshall was referring to qualities affecting the ability of a sheep to put on edible weight quickly, this being Bakewell's single-minded interest, André's concern, reflecting Geisslern's practice, was broader. Describing in detail how to evaluate the fleece in various parts of body with respect to its fatness and fibre characteristics, he stressed that noble sheep should also combine optimal characteristics of body conformation. In thus defining the ideal Merino sheep towards which the breeder should aim, he reflected the complexity of the task Geisslern had set himself, to upgrade a local race to the highest degree of 'nobility' possible.

André noted that high quality wool could be sorted into five degrees of fineness. Two separate categories were average wool and coarse wool. His task of classifying individual fleeces into grades was more difficult because every sheep carried wool of varying quality on different parts of the body, in different proportions. André described a scheme to classify fleeces into 28 major grades, most of which were subdivided, to give 82 grades altogether. These were defined on the basis of total weight of washed fleece, uniformity of fleece, proportions of fine wool in different grades, proportion of average wool and coarse wool and the greasiness of the fleece. In an intensely practical way, he ranked each grade of fleece by the price to be gained from a hundred of such fleeces: from 650 florins for grade 1 to 79 florins for grade 82. A sample of these grades is illustrated in Table 8.1, abstracted from the complete table in André's book. The stress André placed on the economic

---

[24] Marshall 1790, i, p. 422, footnote

**Table 8.1.** Characteristics of a selection of the 82 fleece grades recognised by R. André, based on a scheme originally drawn up by J. Petersburg.

| Grade of fleece | Wool yield (washed), in lb | Index of uniformity of fleece | Fineness of wool (%) | | | Greasiness of fleece | Price/100 head (florins) |
|---|---|---|---|---|---|---|---|
| | | | Fine | Average | Coarse | | |
| 1 | >6 | 3 | 80 (quality 2) | 15 | 5 | significant | >650 |
| 3 | 5¾ | 3 | 80 (quality 1) | 15 | 5 | significant | 600 |
| 6a | 6 | 3 | 80 (quality 2) | 15 | 5 | low | 543¼ |
| 9b | 5¼ | 3 | 80 (quality 1) | 15 | 5 | low | 510 |
| 28a | 5½ | 3 | 80 (quality 4) | 15 | 5 | low | 397 |
| 28b | 5 | 3 | 80 (quality 3) | 15 | 5 | low | 397 |
| 28c | 5 | 4 | 60 (quality 4) | 30 | 10 | significant | 397 |
| 80 | 1½ | 4 | 60 (quality 5) | 30 | 10 | low | 97 |
| 81 | 1½ | 2 | | 60 | 40 | significant | 93 |
| 82 | 1½ | 2 | | 60 | 40 | low | 79 |

significance of fleece quality is clearly evident. Furthermore, he equated this directly to the breeding value. At the meeting of the Sheep Breeders' Society in 1816, Petersburg[25] drew attention to the fact that André had taken over the 82 grade fleece scheme from him, without acknowledging the fact in the book. By that time Petersburg himself had come to realise that the scheme was too elaborate, being tiresome to use and inappropriate for practical breeding. He was now recommending a classification into seven grades.

Selection to maintain or attempt to attain 'pure noble race' status was not based only on these features; it also took into account body form, health and fertility. The selection procedure was clearly a complex and time-consuming business, requiring strictly accurate records in every respect.

At the root of the selection procedure was Geisslern's insistence that individual pairings should be based not only on visible qualities but also on existing offspring: 'If one has correctly interpreted these characteristics [body build and wool] and finds them constant in the offspring in full measure after many years of precise observation and investigation – only then may one describe such an animal as *original Racevieh*', wrote André.[26] Geisslern, like Bakewell, identified and classified traits separately. Then, by following the inheritance of each trait in the progeny, he was able to evaluate the parents in all important respects.

André stressed how essential it is to have detailed knowledge of every individual sheep in a flock.[27] Only then it is possible 'through an appropriate distribution of rams to improve the flock in its offspring, not only from the viewpoint of enhancing the quality of particular individuals, according to the proposed goal but also of spreading the characteristics of these individuals among the others as much as possible'. By considering the quality of a flock as a whole, he was adopting the true 'population approach' to

---

[25] Petersburg 1816, p. 113      [26] R. André 1816, p. 7      [27] R. André 1816, p. 35

Abtheilungs-Schema
des
Schafviehs
nach seinem Ertragswerth
in LXXXII Stufen.

| Stufe. | Wolter- trag nach Pfund. | Sortirung des Bliefes. | Qualität der Wolle in Rück- sicht auf Feinheit. | | | Fettigkeit des Bliefes. | Wollnutzung von 100 St. Schafen je der Stuf., nach Conve ti- ons Geld- Werth. |
|---|---|---|---|---|---|---|---|
| | | | An feiner, 1. 2. 3. 4. oder 5ter Qualität. | Am Mit wolle. | An rober. | | |
| L. | Ueber 6. | 3 fach | 80 Pct. 2Qual. | 15Prct. | 5 Prct. | bebeutend | Ueb.650 fl. |
| II. | 6. | 3  s | 80 s 2 s | 15 s | 5 s | bedeutend | 6 ,8 s |
| III. | 5½ | 3 s | 80 s 1 s | 15 s | 5 s | bedeutend | 600 s |
| IV. [a, [b, | 6. 5½ | 3 s 3 s | 80 s 3 s 80 s 2 s | 15 s 15 s | 5 s 5 s | bedeutend bedeutend | }556½ s |
| V. | 6. | 4 s | 60 s 2 s | 30 s | 10 s | bedeutend | 552 s |
| VI. [a, [b, | 6. 5. | 3 s 3 s | 80 s 2 s 80 s 1 s | 15 s 15 s | 5 s 5 s | gering bedeutend | }543½ s |
| VII. | 5½ | 4 s | 60 s 1 s | 30 s | 10 s | bedeutend | 538 s |
| VIII. | 6. | 4 s | 60 s 3 s | 30 s | 10 s | bedeutend | 516 s |
| IX. [a, [b, [c, | 6. 5½ 5½ | 3 s 3 s 3 s | 80 s 4 s 80 s 1 s 80 s 3 s | 15 s 15 s 15 s | 5 s 5 s 5 s | bedeutend gering bedeutend | }510 s |
| X. [a, [b, | 5½ 5. | 4 s 3 s | 60 s 2 s 80 s 2 s | 30 s 15 s | 10 s 5 s | bedeutend bedeutend | }503½ s |
| XI. | 5. | 4 s | 60 s 1 s | 30 s | 10 s | bedeutend | 490 s |
| XII. | 6. | 3 s | 80 s 5 s | 15 s | 5 s | bedeutend | 436 s |
| XIII. [a, [b, | 6. 4½ | 4 s 3 s | 60 s 4 s 80 s 1 s | 30 s 15 s | 10 s 5 s | bedeutend bedeutend | }480 s |

**Fig 8.1.** One page of the grading scheme for sheep stock published by R. André 1816.

breeding, which Bakewell had embraced several decades earlier.[28] There was a clear practical reason for it: to produce the highly uniform product appreciated by the cloth manufacturer. As André wrote: 'No animal should be perceptively better or worse than the others, particularly in wool. In this way one should work from the beginning of the improvement programme.'[29]

When André came to consider inbreeding, he recognised there were 'no firm principles beyond argument'. He could do no better than summarise the

---

[28] Mayr 1971, 1972; Wood 1973    [29] R. André 1816, p. 37

conclusions reached by 'prominent breeders', reflecting, above all, Geisslern's own views on the subject.[30]

(1) In a noble race, consanguineous mating is the only available means by which to propagate valuable traits in a pure state to the progeny and thus avoid degeneration. 'I improve such animals [the noble race] purely and simply in and through each other', wrote André.[31]

(2) In flocks yet to be improved, close consanguineous mating should be avoided because it does not lead to advantageous results. The use of inbreeding could result in retardation or suppression (*Hemmung*) of improvement, or even degeneration. For improvement purposes it is necessary to use rams from a noble race.

(3) In local races of sheep, consanguineous mating has no advantage at all.

Thus, on the benefits of inbreeding and outbreeding under different circumstances, André's account gives more defined advice than could be found elsewhere, in any language.

A striking feature of André's account is his expressed conviction that it is possible to improve even the noble race. He criticised the many Merino owners who believed that when they had bought 'pure Spanish original stock', they had done everything necessary. For a breeder like Geisslern, 'perfection' was not an absolute state, for his aim was always to achieve a 'higher perfection'.

> The whole art lies in keeping animals pure and unmixed and particularly to show skill and judgement in handling individual pairings in a masterly way . . . Such animals possess the natural capacity, the potential [*Anlagen*] for a higher perfection . . . and one should only assist Nature to develop the extra potential towards perfection, and thereby victory is achieved!
>
> *R. André[32]*

Not everyone was pleased to see the book. Baron J.M. Ehrenfels, an Austrian sheep breeder, accused André of nothing less than treason in betraying the Moravian secret of successful breeding to foreigners. The same charge was directed against Köller. It is recorded that both men heartily denied the accusation. On the contrary, they were pleased 'that the world outside has learned something from us: because we certainly acknowledge openly what we have learned from our German neighbours, and even now continue to learn from them'.[33] The view of Köller, and evidently of André and Geisslern too, about secrets in science was represented as follows: "Progress in knowledge and skills is not to be achieved in the manner of one who lays his empty hand palm down upon the grass and calls to the children,

[30] R. André 1816, pp. 41–2    [31] R. André 1816, p. 94    [32] R. André 1816, pp. 95–6
[33] d'Elvert 1870, ii, p. 184 (quoting Nestler's (1838a) extended obituary of Köller)

'I have something under here but I am not going to show it to you.' "

Most, if not all, of the so-called 'secrets' came from abroad, although it is fair to say that Geisslern and his school showed genuine originality in combining information from different sources to create their own particularly successful breeding programme. They rejected ideas that did not appear useful; they defined precisely the procedures to be adopted and produced clear and comprehensive schedules for recording and interpreting the breeding results. By their rigorous approach, they produced what was widely acknowledged to be one of the best breeding stocks of fine-wooled sheep in Europe.

## Geisslern's reputation

The Frenchman Lasteyrie, who went to see for himself how Merinos were faring in different countries,[34] returned with particular praise for the flocks he saw in Silesia on the estate of Count Magnis, close to the Moravian border. Magnis had married the sister of Geisslern's fellow landowner, Count Leopold Berchtold (1759–1809). Letters from Mrs Magnis to Geisslern (preserved in the state archives in Brno) illustrate the friendly relations that existed between Geisslern and the Magnis family. The race that Magnis developed at his estate near Glatz was largely derived from it, and came to rival it for influence on Merino development, providing founder stock for flocks in Saxony and further West (Chapter 7). Lasteyrie may have been unaware that the animals which so impressed him on his visit to Magnis were based on Geisslern's stock and bred according to his methods. His omission of any reference to Geisslern can be largely explained by his exclusion from Moravia because of the war situation. This unfortunate omission has misled generations of writers using his account as their source. A different picture emerges from the accounts of Baumann and Bright, who had the opportunity to see the situation for themselves.

Christian Baumann, a Cistercian monk who lived for much of his time in Bavaria, wrote extensively on agricultural matters. Towards the end of his life, he moved to Moravia where he produced a new textbook on agriculture published in 1803. Outlining the development of fine-wool production in central Europe, he concluded that the two best breeding farms were both in the Imperial Empire: one at Holič in western Slovakia (then part of Hungary) (Figure 7.4), and the other on the estate of 'the noble Ferdinand Geisslern' at Hoštice in Moravia.[35] Baumann informed his readers that everything at Hoštice was in the most perfect state of improvement, an inspiring example to other sheep breeders. He noted that despite the smallness of Geisslern's property, his stock was exported widely, beyond the borders of the Empire, to Silesia and Poland and even as far as Courland (Latvia).

---

[34] Lasteyrie 1802     [35] Baumann 1803, p. 706

Among the few westerners who penetrated into the Austrian Empire with an agricultural interest was one who reported at some length, and in most favourable terms, about Geisslern's sheep. The Scotsman Richard Bright made his assessment in 1814 and reported the experiences in a book,[36] summarised in the *Farmers' Magazine*.[37] It was soon reviewed by C.C. André senior, who noted Bright's obvious acceptance of Geisslern's prominence alongside 'those who were among the first zealous improvers of the Merino breed'.[38] Bright's experience of Geisslern's sheep and of other flocks bred according to his methods had been gained not at Hoštice but on Hungarian estates over the border in what is now South-western Slovakia (Figure 7.4). A few years earlier, in 1805, Geisslern had sent his former pupil Martin Köller on an extended visit into Hungary. The result had been a massive expansion of Hoštice influence there. Bright's favourable impression was reinforced after meeting a 'Saxon nobleman', who told him that sheep breeding in that country, under Geisslern's influence, was more advanced than that in Saxony. André was obviously delighted at the prominence given to Geisslern by this 'Englishman'[38] (for further details of Bright's visit see Chapter 7).

The Scotman's account was passed over by most writers, in preference to the more widely recognised expertise of Lasteyrie. The bias which resulted is now clear. Among the many nineteenth century agricultural writers in English, we can find only two, Loudon and Spooner, who quoted Bright.[39] Ignorance of Geisslern in the West meant that his precious stock and the techniques used to produce it remained for a time the property of a select group of friends, when leading sheep breeders in Moravia collaborated mainly among themselves. One with whom Geisslern was bound to interact was Baron Kaschnitz, who organised the distribution of officially imported Spanish sheep (Chapter 7). From his farm at Zdounky, a few kilometres from Hoštice (Figure 7.4), it was easy for him to swap both stock and information with Geisslern. He did the same with two other landowners, later to become his sons-in-law, Baron F.S. von Vockel (1772–1829) and Baron W.F. von Podstatsky Thonsern (1781–1833), each with an estate close at hand, at Zdislavice and Litenčice respectively (Figure 7.4). According to d'Elvert,[40] the 'researcher Geisslern' collaborated particularly closely with Baron Vockel, encouraging him to pay special attention to improving the sheep's diet (by introducing new fodder plants) and building ventilated stables. A little further away was the estate of Count Berchtold at Buchlovice (Figure 7.4); he maintained a friendly association with Geisslern, encouraged by the breadth of their common interests as landowners, including fruit tree breeding. Eventually, as Geisslern's reputation travelled abroad with his sheep, he

[36] Bright 1818a    [37] Bright 1818b    [38] C.C. André 1819
[39] Loudon 1844 (1831), p. 99; Spooner 1844, p. 58    [40] d'Elvert 1870, ii, p. 158

exchanged information with more distant landowners. The result was a chain of progressive sheep owners which extended into Hungary. They were creating 'a spontaneous, *de facto* sheep breeding society, not to be rivalled for liveliness and passion for research by any other region of Austria or Germany', wrote Professor Nestler of Olomouc University in 1838.[41]

In time the exchange of stock would be widened into a network of contacts arising from the formal establishment of the Brno Sheep Breeders' Society by C.C. André in 1814, the first such organisation in central Europe and the most influential.[42] The Society retained its strong influence until the close of the 1830s. This was when wool began to pour into Europe from Australia and other British colonies, causing home production to become less economic. By then Geisslern had been dead for several years, although national pride kept his memory alive. At the Fourth Congress of German Agriculturalists and Foresters, held in Brno in 1840, Professor Nestler described Moravia under Geisslern's influence as the 'fountain head of present day rational, scientific breeding'.[43] In the same breath he praised Rudolf André, who had exposed the great breeder's methods to the world.

Fresh information about the spread of Geisslern's influence was contained in a report to the Moravian and Silesian Agricultural Society by J. Waniek, who had represented the Society at the 1842 Congress of German Agriculturalists and Foresters held in the Mecklenburg spa town of Doberan.[44] On his way back home, he visited the estate of S.F. Maas (1787–1864) at Alt Kenzlin, near Demmin in Mecklenburg (Figure 7.2), where he was surprised to see 'excellent sheep, clearly of the Negretti race, bred to a high level', quite unexpected in that country. On enquiry he was told that, many years earlier, Maas had visited Geisslern, who welcomed him with great kindness and instructed him in sheep breeding techniques. Much impressed, Maas had purchased a small flock from the Hoštice estate and, having persuaded an experienced 'honest shepherd, from there to supervise them, returned with him to Kenzlin. Thanks to the shepherd, Waniek was able to see the descendants of this sample of Geisslern's Negretti flock flourishing in a quite different environment. The naturalisation of these sheep on the northern fringe of Germany was no mean enterprise considering the distance and climatic contrast. The shepherd was still working at Kenzlin in 1842 when he showed animals to Waniek that seemed to him indistinguishable from those at Hoštice. Waniek was delighted by the praise Maas heaped upon Geisslern, both as breeder and teacher. Maas showed him a picture of the great man in his workroom at his table, looking, according to Waniek, 'just as I remember him'. This picture, the only known one of Geisslern to be painted, has now unfortunately disappeared, probably destroyed in the Second World War.

---

[41] d'Elvert 1870, ii, pp. 158–64    [42] Elsner 1857, p. 77    [43] Nestler 1841, p. 337
[44] Waniek 1842

For his achievement in sheep breeding, Maas was awarded the Order of the Red Eagle by the King of Prussia. His breeding stock attracted exceptionally high prices, greater than any other of his countrymen. Flourishing flocks of the Hoštice race were established throughout Mecklenberg and also in Pomerania, based on the Kenzlin stock. The brothers Friedrich and Eduard Kunitz were also very active in this region, drawing their stock directly from Hoštice.[45] Waniek was proud to report that the spirit of Geisslern was still alive in those distant places: 'Thus it is that chance carries a seed to a remote field where it flourishes in good soil', he wrote.

The Prussian breeder Hermann Settegast came across the name of Geisslern when investigating what he referred to as the 'battle of the golden fleece' (*Kampf um das goldene Vliess*). This was the term he coined for the disagreement about which of the supposed two original Merino *cabañas*, Escurial or Negretti, was the superior. The controversy is featured in the introduction to Settegast's book, *Breeding of Negretti Sheep in Mecklenburg*, published in 1861. Visits to sheep farms in Pomerania, as well as Mecklenburg, had convinced Settegast of substantial differences between the Escurial and Negretti races, both in wool and also in body characteristics. 'But what was the reason for it?', he asked himself. Was it due to a different ancestry in Spain, as most people assumed, or was it due to selective breeding? To find the answer he needed an objective scale of assessment, not only for wool but also for other economic characters.

He devised a way of assessing individual animals on the basis of a scale with a range of 100 points, 20 of which dealt with body conformation, the rest with wool. It was something like André's (Petersburg's) scheme although with wider terms of reference. At two farms in Mecklenburg he evaluated the qualities of a flock of Negretti sheep originating from the French breeding farm at Rambouillet, obtained from Tessier, giving them a score of 63–64 points, which fell a long way short of the Geisslern-derived sheep at Kenzlin (also called Negretti), which were given 92 points. Having found them so different, Settegast could not but conclude that the designation 'Negretti race' had nothing to do with the Negretti *cabaña* in Spain. He recommended replacing the term 'race' with 'lineage' (*Stamm*).[46]

Upon discovering that some of the Negretti sheep of Mecklenburg had been introduced originally from Austrian territories, Settegast was eager to visit the breeding farms at Holič and Hoštice to discover their source, which he did in 1857. He was surprised to find that the manager at Holič believed in the old-fashioned dogma of racial constancy promulgated by the German horse breeder Justinus (1815). He could not answer Settegast's question about the best sheep for making improvements because he considered all

---

[45] Janke 1867     [46] Settegast 1861, pp. 12–13

animals 'noble' if they had a 'noble origin' from imported Spanish sheep. According to his scale, Settegast was able to classify the sheep at Holič no higher than 55 points, even lower than the Mecklenburg flocks of French origin . But after moving on to Hoštice, Settegast was delighted to find everything he hoped for. The sheep he inspected there were of the same high quality as those on Maas's farm at Kenzlin. Settegast concluded as follows:[47]

> Thus the achievements of Geisslern have not been buried but further developed by his worthy pupil and transferred into a wider orbit. We have to acknowledge with gratitude the achievement of Maas at Kenzlin.

In another place he wrote:[48]

> German Merino breeding was fortunate in attracting a series of experienced and distinguished individuals, possessing the kind of character indispensable for animal breeding: devotion to the object of their study, strength of purpose in pursuing their goals, a keen eye and power of observation and a continuing willingness to learn. Among these individuals to whom honour is due, the greatest of all was Freiherr v. Geisslern of Hoštice. From the same material kept at Holič he created one of the most beautiful flocks that Germany has up to now possessed. His approach to breeding depended not on purity of race but on the capacity of prominent animals to achieve what he wanted. To pick out such exceptional individuals from the mass, to evaluate them for their wool value and to use them decisively in reaching the designated goal, these were the means used to attain something unusual in animal breeding, which von Geisslern set in motion.

'How is it possible that Geisslern's achievements could be so widely forgotten?', Settegast asked, and we may echo his words. International rivalry was probably close to the root of it. A few years afterwards, even Settegast seems to have been affected. In his *Textbook of Animal Husbandry* published in 1868, he omitted all mention of Geisslern, even in the section on breeding. Can it possibly be that he had already forgotten his impressive experiences at Kenzlin and Hoštice? Or was he under pressure to write the later book with a Prussian bias? The country under Bismarck was then in an expansionist mood and at war with Austria. Settegast concentrated attention on German sheep although he deplored the lack of information available in Germany on early developments in breeding when he wrote: 'It is justifiable to complain that historical information arising from German Merino breeding, on which a critical review of methods of breeding might be based, is up to now very scanty.'[49]

His words express his concern at not finding detailed information about what pioneer Merino breeders in Saxony and Prussia had been doing. There seems to have been no published account to compare with the deep analysis

---

[47] Settegast 1861 p. 21    [48] Settegast 1861, p. 19    [49] Settegast 1868

of Geisslern's techniques made by R. André in Moravia. Settegast wanted more than a broad outline of Merino history, which he had described in 1861. In that book he included a family tree illustrating the origin of Merinos in Spain and of the relationships between the major Merino lines disseminated into other European countries (Figure 7.6). The three main branches of the tree were indicated as Negretti in Austria, Electoral in Saxony/West Prussia and Negretti in France. Among 'German' flocks, the description 'full-blooded' (*vollblut*) was given only to those at Hoštice in Moravia and those on Lichnowsky's breeding farm at Kuchelna in Moravian Silesia. Settegast indicates how the latter had its origin in a blend of racial influences, including Hoštice. His Merino family tree also shows how the gene pool ('blood') of the Hoštice breed diffused into nearly all the German states.

The special qualities of Geisslern's stock were further illuminated by A. Körte, the Prussian author of *The German Merino Sheep, its Wool, Breeding, Nutrition and Care*, published in Breslau in 1862. He explains how Prussian sheep breeders experienced problems of disease in their flocks, which they tried to overcome by obtaining big-boned stock from breeders who had 'kept faith with former Austrian breeding principles'.[50] The sheep they wanted were those bred originally at Hoštice. In earlier times many Silesian breeding farms had their descent from them, Körte claimed. He added some extra information about the origin of the Hoštice flock, stating that the first two original Spanish rams and 16 ewes to reach there had arrived in 1775. We believe this to be a misprint (probably for 1785), the earlier date being four years before Geisslern's father purchased the estate. Körte adds that the Geisslern family obtained additional original Merino stock from Baron Kaschnitz, Count Lamberg and from the Imperial Austrian breeding farm at Mannersdorf near Vienna. He also mentions that the breeding stock at Hoštice was kept 'without blending' with any newly introduced foreign sheep, according to strict Austrian rules.

The continuing value placed on the Hoštice breed caused it to feature prominently in Heinrich Janke's *The Principles of Merino Breeding with Special Reference to German Merino Breeding*, published in Berlin in 1867. The frontispiece to this widely read work is an illustration of a Hoštice ram (Figure 7.7), winner of the first prize of 2000 guilders at the Vienna Exhibition of 1866. While praising the Hoštice stock, Janke left no room for doubt that the Saxon Electoral sheep had declined in significance by comparison, even by the 1820s. He was emphatic that the most successful breeding methods for wool production had been elaborated by Geisslern in Moravia. He had no doubt that they were the best type of sheep available in the 1860s, reinforcing Settegast's opinion expressed in 1861 and Körte's in 1862, without quoting them.

---

[50] Körte 1862, p. 182

Janke took the same line as Settegast on another matter, that the difference between 'Electoral' and 'Negretti' sheep was largely a matter of selective breeding. He agreed that Spanish sheep of the same origin had been introduced into both Saxony and Moravia in the 1760s.[51] Selective breeding in Saxony had then been directed exclusively towards fine-wool production, whereas the Austrians had been determined to improve body size as well as wool quality. Janke adds further to the information on Geisslern's sheep which he claims were the true foundation of the Austrian Negretti race and 'had a decided influence on so many famous German thoroughbred herds of our time'. Geisslern's two original rams and 16 ewes were all pure-bred Spanish Merinos that came as a gift from the Empress Maria Theresia, to which 100 sheep were later added from the Emperor's Mannersdorf breeding stock. As a result of selective breeding, the Hoštice race became quite distinct from the original Merino, lying midway between the Austrian 'Infantado' and Saxon Electoral. Janke wrote:[51]

> The herd remained unnoticed for a long time but within the course of a decade a particular fleece quality and structure, and character of the body, was established through heredity. Baron Geisslern named the herd 'Negretti' although it had not been refined from the Conde-Negretti cabaña kept in Spain

(see also Chapter 7). Before writing the book, Janke had visited the British Isles, sent there by the Prussian Ministry of Agriculture to study methods of animal breeding. His experience is reflected in attempts to define the terms 'pure blood' and 'heredity' and in the conclusion he reached that reliance on 'pure blood' alone has limited value in breeding. He added to the chorus of breeders who stressed how essential it was for sheep to be carefully selected and matched, to protect them from degeneration. Illustrating this principle with the example of Geisslern, he was happy to restate the parallel with the great achievement of Bakewell in creating the New Leicestershire sheep for meat production. He stressed the similarity of the two 'Bakewells', just as Moravian authors and Settegast had done.

Janke's book contains the last mention of Geisslern we have been able to find until a fleeting reference to Hoštice appeared in the *Journal of Heredity* of 1935. This was in a review of Merino differentiation worldwide entitled 'The trek of the golden fleece', written by R.H. Burns and E.H. Moody from the Wool Department of Wyoming State University in the USA. Although they referred to Austria only briefly, one of the illustrations to the article was of a ram entitled 'Infandando [*sic*]–Negretti Merino from Hoschtitz sheepfold in Moravia (1847)'. The figure of the Hoštice ram (Figure 7.5) is taken from a drawing by C.C. Fleischmann, a draftsman with the United States

---

Patent Office.[52] In the background of the original drawing (omitted in the reproduction by Burns and Moody) is the Royal Agricultural School and model farm at Schleissheim in Bavaria, providing evidence of yet another destination for the influential Hoštice breeding sheep.

## Transmission of Bakewell's ideas to Geisslern

Geisslern's resemblance to Bakewell was not only in his breeding methods. Contemporary writers pointed to a host of other similarities, including advanced designs of buildings for animal housing and use of artificial irrigation. Also stressed was the manner in which each of them exploited his farm as a kind of agricultural school. It is difficult to avoid the conclusion that Geisslern was consciously modelling himself on Bakewell. The latter had reached the height of his fame in the mid 1780s, at precisely the time when Geisslern was taking his first steps in sheep breeding. If the resemblance in methodology was truly as great as suggested by André and others, conscious imitation seems a feasible explanation. But did André and the other commentators exaggerate the resemblance? Were Bakewell and Geisslern simply responding to comparable economic climates of expansion and revolution? Let us examine the evidence and determine what opportunities for influence existed.

Germans had the chance to read about Bakewell in their own language after 1775 when a translation of Young's *Eastern England* was published in Leipzig. This book was well circulated among progressive landowners in Moravia, appearing in a number of castle libraries. Among other relevant English works published in German editions, the most significant was *Observations on Livestock* by George Culley.[53] The early part of *Annals of Agriculture and Other Useful Arts* (1790–96), edited by A. Young for the Board of Agriculture in London, was also translated, published in Leipzig from 1791. The first German author to mention Bakewell was probably C. Baumann in his pioneering book on farm animal improvement 'without degeneration', published in 1785. As an example of English methods applied in Germany, Baumann singles out von Borie of Neustadt an dem Saale in Franconia (Franken) who adopted Bakewell's practice of leasing rams to his neighbours to evaluate their progeny (Chapter 7). Baumann's other major book, an extensive work on modernising agriculture, was published in Brno in 1803. The printed list of subscribers at the beginning of this book includes the name Ferdinand Geisslern, down for three copies. Geisslern's subscription

---

[52] The origin of the figure has been researched for us by Professor Donald R. Wood of the Department of Agronomy, Colorado State University, Fort Collins. In the library of the late Dr Burns, he found a letter sent from Germany in 1847 written by Fleischmann. It included extensive comments about the wool industry, sheep husbandry and early meetings of wool growers (Ex.Doc. 54, 30th Congress—first session, House of Representatives).

[53] 2nd edn, 1794; German version in 1804

points to his being familiar with earlier works by the author and therefore having access to at least one source of information on Bakewell. His request for multiple copies indicates that his interest in the book was not only for himself. Here we may note the astonishment recorded by Köcker (1809) on seeing several books on sheep husbandry in one room of a shepherd's house in Hoštice. Perhaps one of them was the book by Baumann referring to Bakewell.

Albrecht Thaer's contribution to agricultural improvement in German-speaking countries after 1800 has been reviewed in Chapter 7. Thaer represented a major force for progress in Germany although not a primary influence on Moravian breeding. By the time he published his account of Bakewell's techniques of 'breeding-in-and-in' (1804), Geisslern and fellow breeders Petersburg and Köller were already convinced of its value and ready to teach others how best to apply it. In Prussia it seems to have been Fink who first recommended the practice,[54] while in Moravia it was Petersburg in his pamphlet circulated anonymously in 1797 and finally published, still anonymously, in 1804.[55] The point to note is that Geisslern and his collaborators had established the value of the technique in practice quite a few years before it became generally written about in German agricultural literature.

In 1815, Köller commended to the Sheep Breeders' Society a list of authors who had provided Hoštice with 'the proper principles of sheep breeding'.[56] These were Hastfer, Daubenton, Tessier, Albigard, Kaschnitz, Thaer and Petri, among whom only Hastfer and Daubenton wrote their books early enough to have provided a primary stimulus to Geisslern. The book by Hastfer published in German translation in 1756 was the first account in German of sheep grading, as practised in Sweden. Techniques for measuring wool quality had been developed with precision in France, particularly by Daubenton, and were widely copied. It is noteworthy, however, that Geisslern was never known as the Moravian Hastfer or the Moravian Daubenton but only, and repeatedly, as the Moravian Bakewell. We have therefore to discover how he came upon and exploited the three major techniques associated with Bakewell: systematic selection for clearly defined production traits, progeny testing and breeding in-and-in.

Books and articles were only one source of potential information, and possibly no more than an inspirational spur to Geisslern. Up-to-date knowledge about the latest ideas in breeding was more readily obtained by word of mouth. Geisslern was fortunate in being on friendly terms with a wealthy and well informed neighbour, Count Leopold Berchtold, who had made it his mission in life to obtain the latest information on all economic matters. Berchtold, who styled himself a 'patriotic traveller', was a frequent visitor to

---

[54] Fink 1799a      [55] C.C. André 1804b      [56] Köller 1815

England and other countries during the 1780s, with no other admitted intention than to gather information of benefit to his countrymen. His perception of the new rationalisation in industry and agriculture was enhanced by an extraordinary linguistic ability, enabling him to grasp a new language in a matter of weeks.[57]

Count Berchtold's great estate at Buchlovice was about 30 km from Geisslern's more modest acres. He was owner of one of the first Merino flocks in Moravia and turned to Geisslern for breeding stock to improve it. He also commissioned the services of Köller, Geisslern's manager. Geisslern responded by cultivating fruit trees from Berchtold's nursery. In both activities they attracted the co-operation of neighbours, such as Baron von Vockel. The association between Berchtold and Geisslern seems to have been quite close enough for a regular exchange of information on agricultural matters. Berchtold passed on his enthusiasm about Geisslern's breeding stock to his brother-in-law, Count Magnis in Silesia, made famous as owner of the best sheep Lasteyrie saw on his tour in 1800. Magnis visited Geisslern and took back not only sheep stock but also valuable practical information on how to breed them using the Geisslern method.

The widely travelled Berchtold visited England five times in the period between 1788 and 1800. In search of information of value to his countrymen, he was recognised as a true philanthropist. The Humane Society, established in London in 1774, publicly acknowledged their regard for him by his election as an honorary member of the Society. During his first visit in 1788, he wrote and arranged publication of a book, *An Essay to direct and extend the Inquiries of Patriotic Travellers to which is annexed a list of English and foreign works, intended for the instruction of travellers.*[58] Written in English, the book was quickly translated into French by Lasteyrie.[59] In it Berchtold listed more than 4000 questions that a patriotic traveller ought to ask on every possible aspect of economics. There were more than 400 in a section on agriculture, cattle and sheep. One of them was particularly relevant to Geisslern's interest: 'What means have been used to prevent the degeneration of foreign fine woolled sheep and what was the consequence of it?' This economically sensitive subject[60] was nowhere debated more keenly than in Moravia. Berchtold's reference to it in 1789 occurs just at the time when Geisslern was developing his breeding programme.

Berchtold's first action on reaching England that year was to make for Suffolk to contact Young, who invited him to stay in his home at Bradfield. Every subsequent day seemed to strengthen their regard for one another. Young admired the Moravian aristocrat for his Spartan simplicity, his gift for languages and his enlightened interest in the Austrian peasants: 'he talked

---

[57] Betham-Edwards 1898       [58] Berchtold 1789a; Betham-Edwards 1898, p. 172
[59] Berchtold 1789b       [60] Baumann 1785

English like a native, walked like a giant, and of all the multitude of foreigners who frequented my house, he was the most persevering and most intelligent', wrote Young.[61] Berchtold's attitude to Young was equally admiring. In the dedication of his book, he enthuses: 'Your unparalleled zeal, ability and patriotic labour, have caused your name to be respected and pronounced with reverence in both hemispheres.' 'I cannot ... omit to give this public testimony of gratitude for the important service your labours have rendered to my native country.'[62] We may imagine Berchtold asking Young to enlighten him on sheep breeding and Young referring him without hesitation to the achievements of Bakewell. How could he have avoided doing so at that particular time? Bakewell was at the height of his fame in 1788–9, renowned not only as the most famous stockbreeder in the country but also for his feeding experiments, his use of irrigation ('watering') to increase yield of grass from meadows and his advanced designs of animal houses and stalls, all subjects of great interest to Berchtold, as revealed by his published questions.

It is tantalisingly uncertain whether Berchtold actually exchanged words directly with Bakewell. Young regularly took foreign visitors to Dishley[63] and many others found their way there privately on his recommendation. Bakewell's farm bordered a main coach road, within two days' journey of London, and he was renowned for his hospitality. It is also true, however, that Berchtold did not need to visit Dishley to meet Bakewell. More than one opportunity was presented for their paths to cross in London during 1788 and 1789. Bakewell visited the capital from early February to the beginning of March 1788 and was there again for much of April and May. During the first visit, he enjoyed the enthusiastic attention of the King (Chapter 5), which added to his notoriety. Berchtold, who was there himself in 1788, preparing his book, would surely have found this example of royal interest in animal breeding intriguing, a fine example of patriotism as he defined it. In the following year, both Berchtold and Bakewell were in London during the month of May. Berchtold can be placed there because of a letter he wrote to Young on 25 May.[64] Bakewell's presence is revealed by letters to Culley dated 8 and 30 May[65] (Chapter 5). One of Bakewell's reasons for visiting London at that particular time was to lobby Members of Parliament, to put pressure on the Government to set aside some crown land for experimental farming (Chapter 5). Here was another example of patriotism to stimulate Berchtold's curiosity. He had both motive and opportunity to meet and talk with Bakewell if he wished, although there is no evidence that he did so, and no reference to Bakewell in his diary for the period. However, even if the opportunity for a meeting was missed, Bakewell's activities and opinions

---

[61] Betham-Edwards 1898, p. 170      [62] Berchtold 1789a      [63] Gazley 1973
[64] Berchtold 1789a      [65] Pawson 1957, pp. 138–43

were too obvious to be overlooked by Berchtold. Young would have been sure to introduce him fully to the concepts of breeding-in-and-in, selection of individual traits and progeny testing. It was very much in their mutual interest to do so.

What Berchtold would have heard about English breeding, and about Bakewell in particular, contrasted strongly with the practice of most continental breeders at that time. The widely accepted continental view, stated explicitly by Hastfer and others,[66] was that in order to maintain quality in a fine-woolled flock, fresh pure Merino rams must be reimported from Spain every few years. Only a minority of breeders, Berchtold's neighbour Geisslern among them, were beginning to doubt the need for repeated importation. The secret that Berchtold would have learned from Young, or perhaps from Bakewell himself, was to ensure that the original imports were high enough in quality and were properly fed and housed and, above all, that only the very best of the progeny were used for breeding. Bakewell's technique of breeding-in-and-in must have seemed to Berchtold most interesting and significant. Included in his *Instructions* is a list of books a patriotic traveller should read. Five works by Young are included, one of which is the German edition of *Eastern England*, containing the first published account of Bakewell's husbandry.

During Berchtold's second visit to London in May 1789, he wrote a long letter to Young giving him details of economic observations made during his recent visits to Spain.[67] There he had met Lasteyrie who had given him access to the unpublished manuscript of a book on Spanish sheep. This was real privilege for him and for his friends like Geisslern back home. In his diary he wrote a German translation of an extensive extract of the book, the original of which did not appear in print until 10 years later.[68] Having thus gained a more accurate knowledge of the Merino breed in all its variability, he was able to apply the lessons he had learned on selective breeding in England within a context especially relevant to Moravia.

What, then, can be concluded about possible cause-and-effect relationship between Bakewell's achievement and Geisslern's? Kaschnitz tells us that Geisslern had perfected his noble flock during the previous 20 years, i.e. from approximately 1785 onwards.[69] This agrees with the information from the obituary notice published by d'Elvert that Geisslern was managing the estate by 1782.[70] However, both dates are before Berchtold's visits to England. The only published information about Bakewell then available in German was contained in the translation of Young's *Eastern England* (1775). It took a further 10 years before Culley's book was published. And the new century had arrived before Geisslern was recognised as the Moravian Bakewell.

---

[66] Hastfer 1752b; Alströmer 1770; Baumann 1785; Stumpf 1785   [67] Berchtold 1790
[68] Lasteyrie 1799   [69] Kaschnitz 1805   [70] d'Elvert 1870

It seems probable that Geisslern grew to resemble Bakewell more as time passed and he learned more about him. Even so, there were quite obvious differences between the two breeders. Geisslern showed unique qualities expressed under the particular circumstances of Moravia. In the first place there was consistency and openness in his breeding practice, to a degree that allowed R. André and others to describe the methods he used in detail. The contrast with Bakewell's more guarded manner and selective response to questions is obvious. Another big difference lay in the particular traits that interested each of them. Finally, there was the unique Sheep Breeders' Society in Moravia, which encouraged close and free interaction between breeders, manufacturers and academics, thereby making progress particularly rapid. A small, relatively backward territory managed to overtake much more powerful states in this particular aspect of breeding, during a few critical years. Geisslern was called by right the 'Moravian Bakewell' but he was much more than that. By creating a synthesis of techniques in a highly regulated manner, he raised the art of sheep breeding to new heights.

# 9

## *From breeding principles to genetic laws*

Climate, nutrition and generation remain the levers of Nature in the formation of matter. In the interaction of these three potentials, generation, the
genetic force, is the most powerful.

*Ehrenfels, 1831 (translation from the German)*

The startling changes wrought by selective breeding in the eighteenth and
early nineteenth centuries demanded an explanation from the world of
science. Wealth and national prestige could come from unlocking the door to
a deeper understanding of heredity. Interest encompassed a variety of animals
and plants although species of high economic value provided the primary
stimulus. In Moravia, where major attention became concentrated on fine
wool production, certain aspects of breeding were considered of particular
significance: the improvement and stable maintenance between generations
of traits determining quality and quantity of wool; the role of sex in the
transmission of traits; and the influence of climate and nutrition on health,
fertility and growth rates. Following the success of breeders like Geisslern
and Petersburg, the challenge of interpretation was taken up first and foremost by C.C. André, teacher and writer on science and economics (Figure
9.1). With a growing knowledge, soon to be encyclopaedic, of developments
in sheep husbandry throughout Europe, he published information on the latest
literature and encouraged experiments and scientific discussions on breeding
by every means open to him. For 22 years, up to 1820, his activities were
centred in Brno.

**Fig. 9.1.**   Portrait of C.C. André, agricultural expert and journalist, printed from a negative in the archive of the Mendelianum in Brno.

This energetic and talented citizen came into the world in 1763 and grew up in the town of Hildburghausen in Thuringia.[1] After reading sciences at the University of Jena, he made his name first as a teacher at the Philanthropinum school at Schnepfenthal, which was founded by G. Salzmann (1744–1811) on the principle of the discovery of nature as a new force in education. The school was based on that established by J.B. Basedow at Dessau where Salzmann had taught. Early on, André specialised in mineralogy and became a founding member of the

---

[1] Wilhelm 1867; Franke and Orel 1983

Mineralogical Society of Jena. Then, to supplement his income from the school, he edited a series of encyclopaedic volumes, reviewing different fields of scientific endeavour, an ambitious work undertaken with a co-author, the teacher and natural scientist J.M. Bechstein (1751–1822). In the introduction to the volume on zoology, published in 1795, André drew attention to the then much discussed topic of 'developmental history' (*Entwicklungsgeschichte*), a term used by the physiologist C.F. Wolff (1734–94).[2]

As an early promoter of the idea of epigenetics, Wolff believed the embryo to be initially formless but carrying a hidden entity transmitted from parent to offspring in the process of generation, which expressed itself in a some-what variable form, depending on the guiding influence of climate and nutrition. New physical variants came into being as a result of environmental influences acting on the seed of the parents, both egg and semen.[3] Having considered the particular case of Ethiopeans and Europeans, Wolff concluded that 'with climate it is the heat and its digestive force that have a strong effect on the formation of variations'.[4] 'Constancies appear in structures', Wolff asserted, 'only because environmental conditions are in common or remain the same.'[5]

Comparing the theories of preformation and epigenesis, André came down in favour of the latter, the gradual organisation of parts from disorganised matter. He supported his argument with examples from both animal and plant crosses, illustrating the transmission of traits from both male and female. André dedicated the volume to another prominent physiologist, J.F. Blumenbach (1752–1840) of Göttingen, from whom he drew the concept of 'formative force' (*Bildungstrieb* or *nisus formativus*).[6] This was an innate force believed to act upon the embryo to give form to the 'idea of the organism' (i.e. its inherent properties), on the basis of which development could proceed epigenetically.[7] André used Blumenbach's term in the sense of an organised force that would be characteristically different between variants (races) of a species, as well as between different species, although still subject to environmental influences during development. Thus did André consider the vexed subject of 'generation' (*Zeugung*)—the origin of a new living being typical of that particular species or race, with all its characteristics and propensities, at each conception.

In 1798, André moved to Brno to take up a teaching appointment at the city's first evangelical school. Expecting better opportunities in Moravia for developing his scientific interests, particularly in mineralogy, he was not

---

[2] C.C. André 1795      [3] Roe 1981, p. 126*ff*; Wolff 1759
[4] Gaissinovitch 1990, p. 195, i.e. he believed that the ambient temperature affected the uptake of nutrients into the body, which determinent the nature of the following generation, e.g. in rela-tion to skin pigmentation.                                    [5] Roe 1981, p. 131
[6] Blumenbach 1781      [7] Russell 1930, p. 34; Pinto-Correia 1997, pp 4–5

disappointed. Within six years of residence, he had published a textbook *Instructions in the Study of Mineralogy for Beginners.*[8] In the thriving industrial city, he found congenial company with a group of naturalists who had organised themselves into a private Natural Science Society. From the outset, the new member revealed his managerial skill in promoting scientific activity and its practical applications. A visitor to Brno from Bavaria in 1805, J. Röckel, wrote about André: 'He is a passionate mineralogist who now devotes himself less to his teaching than, with painstaking endeavour, to his extensive scientific correspondence and scientific activity.'[9]

A perfect chance to extend his influence came when he gained the interest and co-operation of Count H.F. Salm (1778–1836) (Hugo Franz Altgraf zu Salm-Reifferscheidt), an influential aristocratic innovator and entrepreneur in industrial textile production in Brno and throughout Moravia.[10] In his castle at Rájec (Raitz), not far to the north of Brno, the Count had a fine library of 59,000 volumes, mainly in French but some in English, and an elaborate laboratory where experiments were conducted on dyeing and waterproofing. His library was rich in works on natural science and technical subjects, although his interests also extended to the history of art, alchemy, the occult and freemasonry. In 1811 he appointed André his economic adviser (*Wirtschaftsrath*), thereby allowing him the financial independence to give up teaching. We may notice how André switched his attention to an aspect of industry in which Brno was already beginning to excel.

Salm's various industrial ambitions had been enlarged by a visit to England in 1801, accompanied by one of André's close friends, the pharmacist V. Petke (1758–1805), who acted as consultant on matters chemical to the Brno textile industry. They made the visit 'with scientific aims', to gather information of value to industry.[11] Before setting out, Salm had the opportunity to consult a knowledgeable friend, fellow freemason and seasoned traveller, Count Leopold Berchtold, who had already made several visits to England.[12] Berchtold was eager to advise any 'patriotic traveller' about making influential contacts and asking questions likely to invoke the most useful responses.[13] Salm was thus well prepared for what proved to be a most profitable visit.

He arrived in England hungry for information of advantage to his businesses and to Moravia. Being an unusually well educated and perceptive individual, he was able to offer his hosts valuable information in return. Sir Joseph Banks, President of the Royal Society, gave him and his party a warm reception.[14] Salm had fresh knowledge to share on Merino breeding, which Banks was trying to encourage in Britain. Both men enjoyed the opportunity to build a relationship based on mutual regard and reciprocal advantage.

---

[8] C.C André 1804c    [9] Röckel 1808
[10] d'Elvert 1870, ii, pp. 128–30; Freudenberger 1975, p. 126; Freudenberger 1977, p. 168*ff.*
[11] d'Elvert 1870, ii, pp. 103–12    [12] Pluskal 1816    [13] Berchtold 1789a, 1789b
[14] d'Elvert 1870, ii, p. 111

Salm's principal aim was to introduce English-style industrial textile production into Moravia. His determination led him to engage in a most profitable act of industrial espionage. Unknown to his hosts, he smuggled back to Moravia 'drawings of British spinning machinery and other preparatory textile machines'. Forbidden to take out the plans straightforwardly, he 'secretly purchased a set and calmly folded them around his legs on the voyage across the English Channel'.[15] His connection with Britain was formalised by marriage in 1802 to a Scottish woman, Maria Josepha *née* MacCaffrey Keanmore (1775–1836).[16] At about the same time, Salm had gone into business with an Austrian, Field Marshal O'Brady (von Brady), a native of Ireland[17] who was living in Brno.[18] The British connection was thereby strengthened in multiple ways. Such of Salm's activities that could be publicly revealed were reported in journals founded and edited by André (Chapter 1). Not only was the public informed but André's own horizons were greatly extended.

An important consequence of co-operation between André and Salm was the reorganisation in 1806 of the almost defunct Moravian Agricultural Society. They, André and Salm, acted in response to a decree issued from Vienna, to merge the Society with the active private Natural Science Society in Brno. It was renamed The Moravian Society for the Improvement of Agriculture, Natural History and the Knowledge of the Country (hereafter called the 'Agricultural Society'), with Salm as President (Director) and André as Secretary. In 1811 it was united with the Silesian Agricultural Society.[19] One of the designated responsibilities attached to the old Agricultural Society, founded in 1770, had been to take all necessary steps to improve wool production. This obligation passed without question to the new Moravian and Silesian Society. Sheep, more than all other products of the land, had brought prosperity to Moravia due to the industrialisation of cloth manufacture, to a degree unparalleled in the Hapsburg lands.[20] The Agricultural Society soon grew to a membership of 300–400,[21] becoming a model for other Austrian provinces to follow.

From the outset André had ambitious plans for the new Society, which he intended should combine the functions of both Academy of Sciences (*Academie der Wissenschaften*) and Economics Society (*Economische Societät*). As models André listed the Royal Society and the *Académie française*.[22] This lofty ideal led him, on the one hand, to insist upon the value of theoretical research in mathematics, chemistry and statistics and, on the other, to encourage the application of such knowledge within the economic context of Moravia:[22]

---

[15] d'Elvert 1870, pp. 85–6; Freudenberger 1977, pp. 174–6

[16] Her name and dates are taken from a pamphlet issued by the library at the Castle of Rajec, although elsewhere (Wurzbach 1874, p. 143) her name is given as Macaffry-Keanmors-Maguire

[17] Freudenberger 1975, p. 126     [18] Freudenberger 1977, p. 178

[19] C.C. André, 1815; Freudenberger 1975, p. 126; d'Elvert 1870ii, 16–26; Orel 1983

[20] Freudenberger 1977     [21] Blum 1978 pp. 288–9     [22] C.C. André, 1815

Zoology, Botany, Mineralogy, Physics and above all, Mathematics and Chemistry should teach us to go more quickly, surely and easily along old paths, to make shortcuts, to pursue new ways, better illuminated and richer in profit.

The region was rapidly emerging into the industrial age, making increasing demands on agricultural and horticultural production. Above all, it was consuming ever greater quantities of wool in the factories of the expanding textile industry.

André's sense of cultural history inspired him to emphasise the contemporary significance of great scientific discoveries of the past, like those of Copernicus or Newton. He believed that members of the Agricultural Society could prepare the ground for equally important discoveries in Moravia, which would withstand the test of time.[22] Increasingly André concentrated his journalistic efforts on sheep breeding, to encourage an increase in wool production, both in quantity and quality, by spreading news of the triumphs of Geisslern and his friends. He was hopeful that scientific truth about the production of high quality wool would be revealed through the practical experience of breeders.

## Sheep breeding in Moravia progresses from Art towards Science

André was well informed about new foreign ideas on sheep breeding, particularly those coming out of France and England. Reviews of foreign books and articles were a particular feature of André's journals. In the first volume of *Patriotisches Tagesblatt* (*PTB*) (1800), he published favourable accounts of Lasteyrie's book on sheep breeding in Spain and Parry's work on Merinos in England, both in the same year.[23] Parry's book ('essay'), *Facts and Observation on Fine Wooled Sheep*, published in Bath in 1800, was reviewed most favourably. The anonymous critic, most probably André himself, picked out for comment Parry's endeavour to convince British landowners that Spanish Merinos kept in Britain would give exactly the same quality wool as produced in Spain. Coming from Parry it had to be taken seriously. The widely knowledgeable Englishman obviously created a big impression with his book. It was also known that he had attracted the friendship and admiration of Dr Edward Jenner, who dedicated his treatise on vaccination to him (1798). Jenner was a hero in Brno, where a statue was erected to him, one of only two on the continent of Europe. Profound confidence in Parry's judgement by a German readership was matched by his own admiration of German scholarship, illustrated by his decision to send his son to study zoology at Göttingen with Professor Blumenbach.[24]

---

[23] Anon. ('D. Le') 1800, p. 47
[24] Carter 1979, p. 32: letter from Parry to Banks, April 1800

The timely appearance in 1804 of the German translation of *Observations on Livestock* (second edition), written by Bakewell's closest pupil George Culley, struck a deep chord with C.C. André, stimulating him to publish a major paper about animal breeding in the first volume of *Hesperus*, in 1809.[25] According to a footnote, the article was due to have appeared in 1805 in *PTB*, but publication was stopped in nervous reaction to the dissemination of ideas of enlightenment at a time of political tension in the Empire at war with Napoleon. With the publication of *PTB* suspended, André founded *Oekonomische Neuigkeiten und Verhandlungen* (*ONV*), a weekly journal with a strongly scientific and technical bias, the first issue of which appeared in 1811. Reviews by André of foreign books on sheep breeding were a particular feature of the new publication.

As André developed his assessment of sheep breeding from a Moravian and Silesian perspective, Albrecht Thaer in Saxony was writing his monumental work on English agriculture, issued between 1798 and 1804, the final volume being on animal breeding. He followed it with a textbook on fine-wooled sheep breeding, published in 1811. When introducing the latter work, Thaer stated that he made no claim to have extended scientific knowledge but wished merely to offer what was most important and proven in sheep breeding. Later he was to admit its deficiencies, due to pressure on him to get it written and published quickly.[26] The lack of stimulating new ideas met with disappointment in Moravia, prompting André's son Rudolf to write an unfavourable review in *ONV*,[27] in which he also took the opportunity to heap criticism on two other acknowledged Merino experts, C. Pictét (mispelt as Picket) in Switzerland and H. Tessier in France, who had published books in 1808 and 1811, respectively.[28]

Expressing profound dissatisfaction with much of what all three experts had written, Rudolf contrasted the sketchiness of their information with the richly informative material available for writing a truly comprehensive textbook of practical sheep breeding, based on local expertise in Moravia. Two paragraphs of the impatient young man's review are here reproduced in translation:[28]

> If I were to rewrite my notes and arrange them in a unified account, a new book would result and, without exaggeration, one at least three times thicker than that of Thaer. The pick of foreign writings on sheep breeding, as for example the ones I have before me by Mr Thaer, the one by Tessier, and the writings of Pictét, fall far short of giving me a glorious impression of sheep breeding abroad. With what eagerness to instruct myself did I read these books. As great was my expectation, so was my dissatisfaction when I finally

---

[25] Anon. [probably C.C. André] 1809

[26] Thaer 1825, Introduction to translation of a French text by Jotemps, Jotemps and Girod

[27] R. André 1812      [28] R. André 1813

put them down. I particular wished to find an answer to my doubts concerning the wisdom of crossing fine rams with common ewes—a controversial but common breeding practice. I have read Pictét's original French version; I thought to find everything on the subject, and I found almost nothing, despite the following words being spelt out on the title page: 'and crosses' [*et les croisements*]. It seems often to be the case, especially these days, that the title of a book promises more than the work contains. How could Thaer call his publication a 'textbook'? A textbook ought to be comprehensive and exhaustive. The uninformed should be able to consult the books for advice.

We possess in Moravia, a school for sheep breeding such as the rest of the Empire and foreign countries can scarcely boast of. Professor Weber became convinced of it on his trip through our land—and yet he saw by no means everything that we can show in sheep breeding.

Weber was Professor of Agricultural Science in Breslau in Prussian Silesia (now Wroclaw in Poland), outside Austrian political influence. He could therefore be quoted as a witness unbiased in Austria's favour. André needed to assure his readers that his criticism of major German, Swiss and French experts with international reputations was not simply an expression of petty nationalism. At the time they might well have thought so, although five years later, when André published his own book,[29] he was able to justify his criticism.

The same year that Rudolf's review was published, his father serialised a paper in *ONV* by the Austrian B. Petri (his name was printed backwards as 'Irtep'), in which he gave vivid and informative descriptions of his experiences when visiting sheep breeding centres in Spain, France and England. He assured his readers that his opinions were based on 'practical experience and sound theory' (*Erfahrung und gesunde Theorie*).[30] In England Petri was encouraged to draw parallels between animal and plant breeding, and reached the conclusion that different races (breeds) of sheep were equivalent to different varieties of plants.[30] Extending the argument to crosses, he expressed his conviction that both animal and plant breeders had a serious chance of being successful in selecting new varieties within the progeny of crosses 'to make these accidental varieties endure' (*zufällige Varietäten bleibend zu machen*). As a condition for success, Petri warned that breeders must free themselves from any prejudice against inbreeding. It can be of no surprise that it is in England that he learned to draw parallels between animal and plant breeding in the areas of hybridisation and inbreeding. We may be sure that T.A. Knight, President of the (London) Horticultural Society, was a key influence in this respect.[30]

The matter of inbreeding, whether it exerted a favourable or an unfavourable influence on the continuity between generations, was becoming

---

[29] R. André, 1816      [30] 'Irtep' 1812

a major discussion point within the Agricultural Society. Petri was convinced that the Spanish were in possession of a 'genetically fixed race' (*genetisch befestigte Rasse*) of Merinos, a privilege they had enjoyed since the middle of the fourteenth century. The Spanish claimed that the race was maintained in its uniformity by the climatic conditions in which every individual sheep was reared. Petri interpreted this in terms of an 'internal prototype of organic formation' conditioned to act in such a way that if a reversion (*Rückschläge*) or a freak of nature (*Spielart*) appeared, it corrected itself, regaining its main 'racial form' (*Hauptsgeschlechtsform*) again. However, he was unprepared to believe that the condition of the race was being left entirely to nature. Indeed, it was his firm conviction that the Spanish, far from leaving matters to chance, practised selective inbreeding without speaking or writing about it.

Referring to Merino breeding in France, Petri reported a visit to the National Sheep Breeding Farm at Rambouillet, where he had observed that sheep were still not yet constant in wool quality and that rams were still being imported from Spain. Nothing compared, however, with the low view he had formed about Merino breeding in England, on which subject he wrote: 'In highly enlighted Britain where all refined sciences, as one may refer to them, reach the highest levels, the production of fine wool is still very backward.'[29]

The central issue of breeding was the enigmatic process of generation and what influenced its continuity. Petri summarised his views on generation in an article published by André, entitled 'My ideas and principles on the breeding of domestic animals'.[31] There we can read that 'upon the generative substance hangs responsibility for fertilisation and the progeny produced'. The conclusion he reached was that:

> when pairing two completely homogeneous animals of each sex, under favourable conditions, their inherent properties and superiority appear in the progeny, which are referred to as animals of the pure animal race [*reine Racethiere*].

The problem, of course, was to produce two completely homogeneous animals. Breeders in the Geisslern tradition would say it was impossible, which was why they insisted on maintaining constant and vigilant selection.

In his journals, C.C. André was willing to give a platform to a variety of opinions. Aided by his son he created in Brno an effective means of disseminating the latest information on fine-wool production, of exchanging and evaluating breeding results, of encouraging improvements in sheep husbandry and of speculating on the nature of generation. What was it that induced constancy in the internal organic structure that made races different? Was it 'favourable conditions' that stabilised the organic structure or was it selective breeding? The latter would imply active intervention in breeding, a course

---

[31] Petri 1813

with which the two Andrés were strongly sympathetic. They were convinced that directed artificial selection, coupled with inbreeding, could be a most powerful force for improvement.

As a sequel to his critical comments on Thaer's textbook of sheep breeding, R. André reviewed a handbook for shepherds by C.A. Hubert (1814) published in Berlin.[32] He was appalled at the 'superficial treatment' given to breeding methods and he returned to his earlier critique of the writings by Pictét and Thaer, repeating his comment that the latter's sheep breeding book was 'imperfect'. Continuing to search the international literature for enlightenment, he reviewed the recently published book on sheep husbandry by Petri (1815).[33] He was again disappointed, forced to conclude that, even though this was the best work on sheep he had seen, breeding methods were still not being explained in sufficient detail.

Within a year, R. André had published his own book on sheep breeding, *Introduction to the Improvement of Sheep on Principles Deriving from Nature and Experiments* (1816), published in Prague and dedicated to Geisslern, his patron (Figure 9.2). Stimulated into existence by André's frustration at the inadequacies of other authors' books, it contained extensive description of methods. André underlined the fact that these arose from personal experience, obtained directly from his observations when living and working with Geisslern at Hoštice. Claiming to reveal the intimate details of his patron's breeding strategy for fine-wool production, he wrote:[29]

> Sheep breeding became my favourite study from the first moment I came into contact with practical agriculture, to which I have since devoted myself deeply and passionately. I have been most fortunate to attend the best schools, to study the noblest flocks and their management, and also to play an essential part in sheep breeding myself. I flatter myself, therefore, that in this book I will not be guilty of stating the obvious.

In the next line André continued:[29]

> How very important it is that finally this special part of sheep husbandry, the most important and most difficult in fact, should be treated as far as possible comprehensively. It is necessary because no author dealing with sheep breeding known to me, even Petri, has dealt with the business of improvement [*das Veredelungsgeschäft*] sufficiently deeply or convincingly.

Such words from the enthusiastic André might be judged as little more than personal advertising, except for the fact that we can compare the books side by side and appreciate for ourselves the unusual degree of technical detail in André's account.

Favourably reviewed in the *Journal of the Patriotic and Economic Society of Prague*, in a piece that was reprinted by André senior,[34] the book brought

---

[32] Hubert 1814; R. André 1815a    [33] R. André 1815b    [34] Anon. 1816a, p. 296

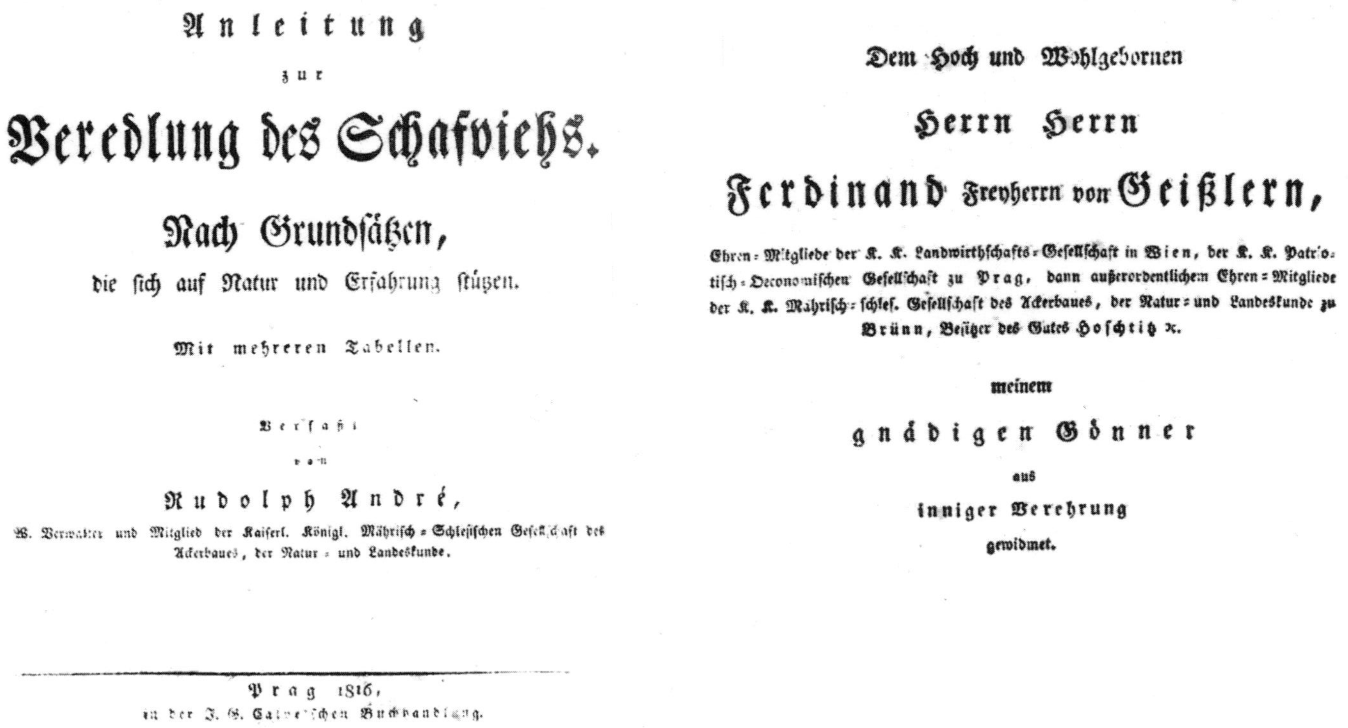

# Anleitung

## zur

# Veredlung des Schafviehs,

## Nach Grundsätzen,

die sich auf Natur und Erfahrung stützen.

Mit mehreren Tabellen.

Verfaßt von

## Rudolph André,

K. Mitglied der Kaiserl. Königl. Mährisch-Schlesischen Gesellschaft des Ackerbaues, der Natur- und Landeskunde.

Prag 1816,
in der J. G. Calve'schen Buchhandlung.

---

Dem Hoch und Wohlgebornen

Herrn Herrn

# Ferdinand Freyherrn von Geißlern,

Ehren-Mitgliede der K. K. Landwirthschafts-Gesellschaft in Wien, der K. K. Patrio-tisch-Oeconomischen Gesellschaft zu Prag, dann außerordentlichem Ehren-Mitgliede der K. K. Mährisch-schlesl. Gesellschaft des Ackerbaues, der Natur- und Landeskunde zu Brünn, Besitzer des Gutes Hoschtitz ꝛc.

meinem

## gnädigen Gönner

aus

inniger Verehrung

gewidmet.

**Fig. 9.2.** Title page of R. André's book *Introduction to the Improvement of Sheepstock According to Principles Drawn from Nature and Experience* (1816), and the second page on which he acknowledges 'his most gracious patron, Ferdinand Geisslern'.

credit to Moravia. The anonymous reviewer was convinced by R. André's claim to originality: 'He describes in detail and to the fullest extent the most important and most difficult aspect—the business of improvement which up to now has not been treated satisfactorily by any other author.'[33]

R. André's uniquely practical book was especially valued by those whose livelihood depended on sheep and who saw the chance of greater prosperity. This is what his father intended. The book would never have existed but for his own dedication to the cause of wool improvement. It represented the crowning achievement of André senior's life's work. When now one examines the contents of the many articles, letters, reviews and reports published by the two Andrés, it is possible quite easily to trace the development of the various ideas that form the background to the book. Reports of observations on the relationship between parents and progeny in traits of economic importance led to the establishment of scientific breeding principles, based on observation and experience, which Rudolf sets down in detail, and which will shortly be considered. The journals edited by André senior were the primary vehicle for spreading Geisslern's ideas and making them known outside Moravia. His son's book added a wealth of practical information. R. André began by describing the aims of sheep breeding and defining its terminology; most of the rest of the book is devoted to a step-by-step account of the technology of sheep breeding by the method of grading. Finally, in a short section of five pages entitled 'Improvement, or even greater perfection of sheep of the noble race', André expanded upon the feasibility of enhancing wool quality by selection, to make pure Merino sheep even better than those raised in Spain.

From an economic standpoint, the most crucial matter exercising breeders in Moravia, as elsewhere, was whether it was necessary to import rams from Spain on a regular basis and, if so, how often? Opinions differed according to the supposed effect of the local environment. R. André, in his book, asserted that the quality of wool produced by Merinos was not dependent on the Spanish climate, and he was not the first in Moravia to make this daring claim. In 1811, André senior had published a paper written by Martin Köller, under the pseudonym 'K in Mähren', who claimed that sheep breeders in Moravia could already offer home-bred pure Merino sheep without any hereditary deficiencies. This gave him good reason for importing no further 'original' sheep from Spain. R. André was prepared to go further than Köller by suggesting in his book that Moravian breeders had the skill not only to *maintain* but also to *improve* the quality of Spanish Merino breeding stock, by increasing the intensity of selective breeding.

The volume of *ONV* containing the paper by 'K in Mahren' introduced an idea that was new to Moravia, namely leasing rams at an appropriately high price, as Bakewell had done. This would allow a breeding ram to be selected not simply on outward form but on the quality of its lambs produced from earlier matings. The author, C.C. André, linked this idea to a proposal dear to his heart, of forming a Society of Breeders like the British Merino Society, established a year earlier. He again stressed the urgency of forming such a Society in a published article.[35] Referring to a recent embargo on Merinos by the King of Spain, which had increased demand for 'original Spanish sheep' all over Europe, he reminded his readers of what had been the outcome. In the years following the ban, only a very few exceptional breeders, those who practised progressive breeding methods, had continued to produce fine-wooled sheep of consistently high quality. The examples he singled out were Petri in Austria and the group of breeders around Hoštice, associated with Geisslern. André also touched on the subject of changing an animal's shape by selective breeding as Bakewell had done with the New Leicester breed of sheep. He listed eight distinct traits to be considered and, for the first time in print, he used the expression 'by artificial breeding' (*durch künstliche Zucht*), reminiscent of the expression 'artificial selection' made famous by Darwin several decades later, when he drew the analogy with natural selection. We can find no earlier use of the term than by C.C. André in 1812.

Moravia's foremost breeders were well aware of the need for a body of theory, a reliable ground plan, to guide them in their selection programmes to improve wool production. André senior would later recall how genuine had been the search for breeding theory in Brno and how essential it was to 'move from words to action' in attempting to formulate the theory and test it.[36] The growing demand for reliable breeding stock encouraged Salm, with André in support, to propose establishing a market in Brno for buying, selling and exchanging rams.[37] André's next action was to issue an invitation 'to friends and patriotic industry, especially those concerned with sheep breeding' to attend a meeting in Brno in May 1814 to create the Sheep Breeders Society.

## The Sheep Breeders' Society

The formation of the Society was closely connected with the entry of C. André's son Rudolf on to the scene. He described what happened in the foreword to his book of 1816:[29]

> When the more clear-sighted [*klügern*] of the agriculturalists recognised, one by one, that their work and efforts would be fruitless without a basis in pure principles, they began to search for them. One stated this, another that, mostly keeping their ideas to themselves. Very seldom was there a critical exchange of views, which can be so useful and beneficial. Most breeders kept secret the art by which they improved their own flocks, and informed nobody of their observations, which, if connected, might have led to useful results

[35] C.C. André, 1812    [36] André's translation of Jotemps *et al.* 1825, p. 57
[37] Salm and André 1814; Anon (C.C. André) 1811a

but were lost. Thus the majority of breeders remained long in darkness, ignorant of generally acceptable, incontrovertible principles to be applied to ensure favourable results in sheep improvement. . .

It was for this reason that the Agricultural Society in Brno, recognising the need, hit upon the splendid idea of organising a society to provide help for our experts and enthusiasts in sheep breeding. Only within such a context would it finally be possible to construct a foundation of pure knowledge for this important branch of agriculture, by open communication, discussion and instruction. Through this excellent and unique institution, sheep breeding in Austria will soon gain the ascendancy over foreign countries, as was the case in Spain some years ago, but without engendering the suffering that was caused in Spain.

When forming the Society, André senior had in mind not only an exchange of knowledge and experience of breeding but also the stimulation of members to carry out properly designed field experiments to investigate the transmission of traits determining wool production. As he was fond of saying, 'To instruct is good but action is better' (*Belehren ist gut, Handeln noch besser*).[38] Determined to make the breeders responsive to the demands of industry, he introduced manufacturers and textile merchants into the membership from the very beginning. It was for this reason that the full name of the Society was 'Society of Friends, Experts and Supporters of Sheep Breeding for the achievement of a more rapid and thorough-going advancement of this branch of the economy and the manufacturing and commercial aspects of the wool industry that is based upon it'. Although the name was usually abbreviated to 'The Sheep Breeders' Society', its wider aspirations were always clearly recognised.

C.C. André needed to exercise considerable managerial skill and tact in bringing together these disparate elements, including leading members of the Brno business community, into what was basically a scientific society. In so doing, he created a unique blend of expertise, perhaps his most progressive venture. Its success may be judged from the reaction of Elsner, a Silesian expert on sheep breeding, who concluded: 'It can be asserted without exaggeration that the establishment of this first Sheep Breeders' Society in Brno on 16 May 1814 marked a new era in German sheep breeding.'[39]

A later appraisal came from K.N. Fraas (1810–75), Professor of Agricultural Science in Munich, who wrote of the Society in his classic work on the history of agriculture published in 1852, that 'its merit for science and sheep breeding is immeasurable'.[40]

At the opening meeting of the new Society held in Brno on 16 May 1814 and attended by 150 participants, Count Salm, President of the Agricultural

---

[38] C.C. André, 1825, p. 57; Orel 1974    [39] Elsner 1828, p. 117; see also Elsner 1826, 1849
[40] Fraas 1852, p. 337

Society, recalled the Government's letters patent of 1811, according to which the Society was charged with taking care of the two most flourishing economic developments of the province: 'superior sheep breeding and the textile industry'.[40a] All over continental Europe, noble (i.e. Merino) flocks had been devastated by war. Only in Moravia was it not so, and Saxony and Prussia were only too grateful to turn to this province for breeding stock. Salm took pleasure in informing this historic meeting that, in tribute to their most renowned breeder Ferdinand Geisslern, it had been decided to award him honorary membership of the Society. Other landowners who had established valuable sheep farms in the province, mainly in direct co-operation with Geisslern, received praise for their good sense and enterprise. Baron Emanuel Bartenstein, who was prominent both in sheep breeding and in commerce, was elected President of the new Society, with André as Secretary. The rest of the committee numbered 13, six from Moravia, three from the province of Austria, three from Hungary and one from Bohemia, a balance of membership that was intended to reflect the dominance of Moravia in sheep breeding and the relative weight of expertise elsewhere. Membership would soon be extended to other Hapsburg provinces and even to Silesia, Brandenburg–Prussia and German territories beyond. The international composition of the Society was a key factor in its influence.

In his presidential address, Bartenstein concentrated on Geisslern, who 'has given practical proof that Spanish sheep can be reproduced on Moravian soil without reversion. His procedures have created enlightenment and eliminated prejudice.' Discussion followed, then talk gave way to practical business when 300 sheep from prominent farms were exhibited and offered for sale. This marked the beginning of a tradition in exhibiting, evaluating and selling breeding stock at every annual meeting.

After electing the Society's committee, the members planned the programme for the next meeting. They took the decision to meet annually in the middle of May in Brno. The minutes of the first meeting were brought to a wider public by C.C. André who printed them in *ONV* (1814). Half a century later, d'Elvert published interesting quotations from a whole series of articles dealing with the activities of the Sheep Breeders' Society, from 1814 onwards, taken from *ONV* and other sources, 'for those who will later write a special history'.[41] Critically important to the central purpose of the Society was the frank exchange of information on the production of the finest quality wool, in open discussions. Methods of trait recording, the matching of selected rams and ewes, progeny testing and consanguineous mating (including 'breeding in-and-in' in the Bakewell manner) would often be discussed. We know this because of extensive reports published by the

---

[40a] Anon (probably C.C. André) 1814      [41] d'Elvert 1870, pp. 140–47

two Andrés in *ONV* and, after 1821, in *Mittheilungen*, the journal of the Agricultural Society.

The second meeting saw participants exchanging views on a variety of aspects of sheep husbandry and the evaluation of wool quality.[42] Martin Köller, Geisslern's long-time associate, summarised the major obstacles to the improvement of fine-wool production. He had discovered that most landowners, their officials and shepherds were fundamentally ignorant about how to keep fine-wooled sheep in a thriving state. He deplored the ruination of good stock when grazed on unsuitable meadows, or wintered in unsatisfactory stables. For written guidance on sheep husbandry, Köller recommended books by seven different authors, as listed in Chapter 8. But he was convinced that little progress would be achieved without a system of formal instruction. Another member recommended the founding of a training school for shepherds, all of whom would be required to pass an examination. As a graphic demonstration of the true principles of sheep improvement, Köller displayed wool samples from the breeding farm at Kvasice, which he was managing at that time, and from Zdounky, the breeding farm of Baron Kaschnitz. Pospíšil, Geisslern's farm manager, then demonstrated the record cards used to register traits of wool quality in individual sheep at these two breeding farms. He also showed the micrometer used for measuring wool fineness. The idea of measuring wool with a micrometer came from the French school of Daubenton and was introduced on to Geisslern's farm some years before the foundation of the Sheep Breeders' Society. Earlier the chief mechanic in Count Salm's factory workshop had produced a specially adapted microscope for this purpose.

By the third annual meeting it was impossible to ignore the bitter truth that the quality of fine wool was on a downward path. This was becoming obvious not only in the Hapsburg provinces but also in neighbouring German countries. The decline was a major topic on the agenda of this meeting and also of the two that followed. The problem to be solved, as had been defined in their very first meeting by a prominent Moravian sheep breeder, J. Petersburg, was how to breed sheep producing abundant wool with the right characteristics of elasticity, firmness, density, length and colour, in accordance with the requirements of the woollen cloth industry. They needed to establish high standards and work to them, i.e. select for specific 'noble' features. The industry had the opportunity to state exactly what was needed through its representatives in the Society. André recommended setting up a 'Wool Institute' in Brno, in co-operation with the manufacturers, to house a permanent exhibition of samples for reference. Study, analysis and comparison of wool samples would be made in order to chart the progress of

---

[42] Anon. [probably C.C. André] 1815, pp. 353–8

improvement or check its decline. Such a centre, intended to serve the interests of all Moravian breeders, was never in fact brought into existence, although an unofficial collection of samples was put on display at the breeding farm managed by Köller at Kvasice.

With C.C. André's encouragement, discussion took place again on the proposal to institutionalise the training of shepherds. Köller and André joined forces in bringing this into effect. Köller established the first training school at Kvasice and R. André established a second one on Salm's estate at Rájec. R. André's course began in 1818, the same year in which he published an instruction manual for his students: 'Concise instruction in sheep husbandry for sheepmasters and their assistants', issued in German, Czech and Polish versions.[43] The course ended with an examination before a commission chosen by the committee of the Society. Shepherds who passed the examination received a pay rise and enjoyed a higher status.

## The enigma of inbreeding

As Secretary, C.C. André was always in the vanguard of developments, putting his mark on all the Society's activities. He arranged the programmes of the annual meetings, organised discussions and enquiries, and published reports. At the meeting held in 1817, the participants were introduced to R. André's major book (1816), after which the author gave a practical demonstration of his method of evaluating sheep. Baron J.M. Ehrenfels (1752–1843), a prominent Austrian breeder, could not overlook the regrettable reduction in wool fineness in Austrian flocks, which demanded an explanation.[44] On the basis of his 'theoretical and practical knowledge', he perceived three main reasons for it, which he termed 'harmful factors': (i) the practice of selecting sheep for breeding mainly according to their physical appearance; (ii) a neglect of wool quality in favour of quantity; and (iii) an overemphasis on matching sheep 'in closest consanguineous relationship'. His remedy for the second harmful factor was to ensure that selection criteria included wool fineness as well as wool length, the aim being, as ever, to raise wool quality above that in Spain. In a footnote, C.C. André reminded members what R. André had written in his book: 'A combination of maximum yield with the greatest degree of fineness has yet to be achieved. Usually when wool is produced in considerable quantity, it is of second grade in fineness.'[43]

Concerning his third 'harmful factor', the pairing of nearest blood relatives, Ehrenfels took a view directly opposite to that which had inspired Geisslern. He claimed that close inbreeding must lead to 'natural climatic degeneration' (*die natürliche klimatische Rückbildung*) by disrupting 'the principle plasma of the animal's organisation' (*Hauptplasma der thierischen*

---

[43] R. André, 1818      [44] Ehrenfels 1817

*Organisation).*[44] We may note how closely Ehrenfels linked constancy of inheritance with climate. On his own farm he avoided close matings because he believed they would automatically be accompanied by a decrease in wool fineness, although he recognised that other members might have a different opinion. He was open to the opinions of the general membership of the Society, realising that 'Not every master of mathematics has to be a Newton, not every sheep breeder a Buffon, nor every collector of herbs, a Linnaeus.'[44]

The Society's judgement on the matter was reserved until the following year's meeting when sheep from Ehrenfel's farm were exhibited in comparison with Geisslern's. Although the Society's experts commended Ehrenfel's wool for its fineness, they classified it overall in the third rank, while Geisslern's wool gained ranks 1 or 2, on the basis of both quantity and uniformity in quality. The anonymous author of the report on this meeting added:[46]

> This is why Moravian sheep have the appearance of being created on a unique model from the excellent school of Geisslern. Here we can see how, through a single man, culture and wealth have been presented to the whole province.

The report was sent to C.C. André for publication in *ONV*, to be printed in some thousands of copies and read throughout the German-speaking world. The anonymous author, a Doctor of Philosophy who had worked four years in agricultural economics (as André explained in a footnote), was clearly fascinated by the theoretical implications of what he had heard in Brno. In response to the textile industry's request for clarification, André added in his footnote 36 questions on wool quality to be examined by the Society's members. According to him, the two most pressing issues were the influence of environmental variation on wool quality and the transmission of individual traits from father or mother. The idea of heredity was constantly alluded to, without using the term explicitly. The reporter noted that, very recently, President Bartenstein (1818) had expressed views on how to combine quantity with quality, and density with fineness. In doing so, he had stated his agreement with Professor Thaer in Berlin, who recommended breeders to select primarily for wool quantity, density and fibre thickness, and only later to cross to stock bearing fine wool. Ehrenfel's reaction was to favour the opposite procedure: to graft wool quantity and density on to a breeding stock already selected for wool fineness, a state which was attainable, he claimed, after only two generations. Count E. Festetics (1769–1847), a Hungarian expert, joined Bartenstein in rejecting this advice.

Count Festetics came from a family well known for agricultural enterprise, supported by an exceptionally rich agricultural library[45] owned by Count

---

[45] Kurucz 1990      [46] Anon. 1818a; Bartenstein 1818; Orel and Wood 1998

Gyrgy Festetics at Keszthely on Lake Balaton. In 1797, Count Gyrgy had established a high-level agricultural teaching establishment, named the Georgikon. The library contained most of the major English agricultural works, including publications by Young, Sinclair, Marshall and Culley and, most importantly, the County Surveys produced by the Board of Agriculture.[45] When a member of this highly informed Hungarian family joined the Brno debate, the meeting witnessed both an extension of the conflict and an attempt to find a compromise solution. First came a restatement of Ehrenfel's opposition to mating very close blood relatives. Bartenstein countered by stressing that close inbreeding need not produce the ill effects that Ehrenfels feared. The answer, he suggested, lay in progeny tests, to avoid incorporating individuals with hereditary malformations into the chosen breeding flock. Festetics deferred to Bartenstein and other members who were agreed on the value of assessing an animal not simply for its own qualities but for those of its parents and descendants. They viewed the selected population as an indivisible whole.[46]

In an extensive footnote to the discussion, André underlined the need for compromise, on the matter of inbreeding, suggesting that a misunderstanding had partly arisen from differences in terminology.[47] Members used the terms 'crossing' (*Kreuzen*) and 'pairing' (*Paaren*) in different senses. Appreciating the confusion that could arise, André recommended restricting the term 'pairing' exclusively to matings within a race, and 'crossing' for matings between races. In stressing the need for consistency of terminology, he drew a parallel with what had been achieved in the science of mineralogy, by Gottlieb Werner (1749–1832) of Freiburg in Saxony, who, in creating a new standard terminology for the subject, had transformed its scientific status. He asked whether the same could not be done for sheep breeding.[48] As to consanguinity, he was sure that it would be found to have predictable consequences, behaving according to some 'physiological natural law'. He was convinced, however, that unconditional inbreeding in closest kinship, if long continued, must result in organic weakness. But to do justice to the problem he would need to write a book about it. 'What subtle problems are here to be solved before we can approach nearer to the truth with confidence. Here we are interfering with the innermost secrets of Nature', he wrote.[47]

André was aware that leading breeders were sometimes prepared to match sons with mothers, or fathers with daughters. But he also knew that they mostly bred from less closely related animals and were even prepared to use a ram from outside their own flocks if it was an equally good example of the same race. The dichotomy of views expressed in the Society about inbreeding had arisen, André continued to stress, largely from a misunderstanding

---

[47] C.C. André 1818; Orel and Wood 1998
[48] C.C. André 1818; Franke and Orel 1983; Orel and Wood 1998

of one another's views. Ehrenfels did not recommend pairing *only* unrelated animals, nor did proponents of inbreeding, such as Festetics, expect to pair only relatives. Nevertheless, the debate had clearly become polarised: Ehrenfels continued to believe that inbreeding disrupted the constancy of inheritance, while Festetics, supported by Bartenstein, defended the practice as beneficial because it could result in more constant inheritance.

The following year, André gave the platform to Festetics to air his views more fully. Festetics began by acknowledging André's primary influence in leading him to 'search for the truth'.[49] He had discussed the question of consanguinity with his esteemed friend Baron Ehrenfels but he still failed to understand André's proposition that unconditional application of close inbreeding must lead to 'organic weakness'. He needed clarification of what André meant by this term. Did 'organic weakness' imply a condition preventing the race from maintaining its identity (*Erhaltung seiner selbst*), by which he meant, did it affect continuity between generations? For the characteristics of a breed to remain stable, Festetics agreed with Ehrenfels in assuming a 'strong constitution' to be necessary, determined partly by the inborn component (*theils angeboren*) and partly by rearing (*durch Erziehung*). When imperfect individuals, with diminished traits, appeared in the progeny of a healthy father, it was because their constitution was weakened. Imperfect progeny born to two healthy parents were considered freaks of nature (*Spiele der Natur*). In addition, grandparents might have traits that were absent in their children but reappeared in their grandchildren. Festetics quoted particular human examples to explain the transfer of parental traits under the influence of inbreeding. In some Hungarian villages colonised by different nationalities, consanguineous marriages were practised. These resulted in both harmful and beneficial effects, spiritual as well as physical. Festetics draw parallels with Bakewell's principle of breeding 'in-and-in', practised most successful by certain German cattle breeders. Whether close inbreeding brought benefit or disadvantage depended on the progeny selected for breeding.

According to Nestler, Festetics obtained his first breeding stock of the noble (Merino) race in 1803, one ram and nine ewes.[50] After that (until 1819), he had maintained a closed race which he would continue to do, being unable 'with full confidence' to buy better rams than his own. In a footnote to Festetics's article, André contrasted the procedure of inbreeding with that of crossing, after which he stated what he believed to be a natural law in favour of crossing, in the following terms: 'It is not in the homogeneous [condition], but in reciprocal reaction, that new products with more significant, stronger action, constitution and forms are generated.'[49]

---

[49]  Festetics 1819a; R. André (1819); see also Nestler 1838b
[50]  Nestler 1839; Buffon 1795

## 'Genetic laws of Nature'

André called upon Festetics to formulate his own rules about when and where to use the inbreeding technique. Festetics responded to his call by summarising 15 years' experience of closed-race breeding, under the heading 'Genetic laws of Nature' *(genetische Gesetze der Natur)*, in four points:

(1) Animals of healthy and robust constitution are able to propagate themselves, and their characteristics are inherited (*vererben sich*) (i.e. he equated pure breeding with robust good health).

(2) Traits of grandparents not reproduced in their progeny may appear again in succeeding generations (i.e. he appreciated that inherited traits could be recessive for at least one generation).

(3) Animals possessing the same suitable traits can have offspring with divergent traits. These are variants, freaks of nature, unsuitable for propagations if heredity is the aim (i.e. he recognised the extent of inherent variation, even in pure breeds).

(4) The precondition for applying inbreeding is scrupulous selection of stock animals. Only animals possessing the required traits in noticeable excess can prove effective in breeding. (We may assume he accepted the point, made by Bartenstein, that looking at progeny already produced by breeding stock could provide extra important information.[51])

In a footnote to the term 'scrupulous selection', André added: 'In my opinion this is decidedly the main point.' Geisslern, and Bakewell before him, would have agreed.

By defining his genetic laws, Festetics alluded to heredity without reference to its physiological basis, although he saw it being connected with health and robustness. He was fully aware that heredity of wool quality involved different traits, but stressed that these had to be integrated into a healthy whole. Based on extensive practical experience, his laws can be said to be empirical. The traits that breeders were concerned with, that they might recognise as having a separate identity, could appear in different combinations.

Festetics' laws have been reviewed by Orel and Wood (1998). They arose from practical questions related to the still controversial topic of consanguinity. Does inbreeding lead to degeneration, i.e. a breakdown of heredity, or does it lead to exactly the opposite effect, to more certain heredity? Festetics' answer was that it depends precisely on whether the parents are carefully paired for the same strongly selected traits, which must include those relating to health, reproductive capacity and a robust constitution as well as those directly concerned with wool quantity and quality, and other

---

[51] Festetics 1819b

economically significant features. Festetics, who shared the views of Geisslern, had come to a similar conclusion as that reached by English sheep breeders, influenced by Bakewell: that inbreeding could be critically valuable to strain improvement and was relatively safe to apply as long as it was prac-tised with discretion and accompanied by stringent selection. Although the idea was not original, it was new to codify it in quite this way. The stress on the word 'genetic' in the context of inbreeding and outbreeding found no direct echo in other breeding literature of the period that we can discover, although wide interest was shown in identifying the laws of nature and trying to obey them. Albrecht Thaer lectured to his students on this very subject in 1812. The lecture was published in 1816 with the title 'On the laws of nature which have been observed by the farmer and must be obeyed in the improve-ment of his domestic animals and the production of new races.'[52] Regarding the transfer of parental traits to progeny, Thaer recognised that both parents made a contribution in terms of the *Anlage* (endowment). But he was unable to explain the laws by which similarities and dissimilarities arose between parents and progeny. The matter remained an enigma for Thaer, even in his later writings.[53]

The validity of Festetics' explanation in terms of inbreeding was confirmed by R. André (1819). His personal experience at Geisslern's farm assured him of the value of inbreeding as a principal instrument for preventing organic weakness. He noted how breeders of Arab horses had reached the same conclusion, experiencing no detriment to their breed. The highest aim of breeding was organic robustness, expressed in terms of homogeneity for desired characteristics. Such animals, the product of inbreeding, which he referred to as *Racethiere*, transmitted their uniform characteristics to their offspring with impressive regularity.

At the meeting in 1819, Rudolf André demonstrated his scheme for judg-ing sheep presented for exhibition. Each breeder was invited to submit a small sample of wool. André also promised to try to answer queries on phenomena related to the efficient selection of sheep. He demonstrated a newly designed micrometer for evaluating wool into seven grades of quality. Reporting on the meeting, Festetics described R. André's latest book as a classic work, appreciating, above all, his 'mathematical' evaluation of wool quality: 'It will be judged as marking the beginning of an epoch in the sci-ence of breeding, that in 1819 the grades of wool fineness were established and defined with mathematical precision.'[54] Such praise was generous from a man who had played a major part himself in transforming breeding from art to science.

---

[52] Klemm and Meyer 1968, p. 214    [53] Steiner 1978
[54] Festetics 1820, p. 33

## The Pomological and Oenological Society

As had been done with sheep, serious attempts were made under C.C. André's leadership to raise the breeding of fruit trees and vines to the status of science. As early as 1806, when André drew up the programme for the reorganised Agricultural Society, he highlighted wine as an obvious area for improvement and expansion, to Moravia's economic advantage. When finally the fruit and wine producers joined together in 1816 to form the Pomological and Oenological Society (hereafter referred to as the Pomological Society) as a section of the Agricultural Society, alongside the Sheep Breeders' Society, André was convinced that animal breeding would provide a pattern for plant breeding, as proposed by T.A.Knight in England.[55] Later, when Petri[56] reported upon a visit to England, he too drew an analogy between animal and plant breeding. He brought Knight's achievements once again to the attention of the scientific community[56] just at the time when André was establishing the Pomological Society.

The new Society was modelled on the Pomological Society of Altenburg (Leipzig), established in 1803. The Secretary of the Altenburg Society, G.C.L. Hempel, maintained a regular correspondence with André, and also with Knight at the Horticultural Society of London. In 1820 the London Society elected Hempel as an overseas member and in 1826 they extended the privilege to André, after he moved to Stuttgart, and to his patron, Count H. Salm.[57] Hempel and André did everything possible to establish a free flow of information between Altenburg, London and Brno.[58]

Undeniably the journal *ONV* was essential reading for members of the Pomological Society. As in the case of sheep breeding, writers sought to give their work a theoretical basis. In the 1820 edition, André included a paper by Hempel, entitled 'On the origin and significance of certain cereal varieties.' Hempel wrote that 'From the grain of seed, formed by artificial fertilisation, a new type of progeny appears with characteristics of both mother and father plant.' In the introduction he wrote of 'scientific pomology' and especially about the 'remarkable experiments of T. A. Knight Esquire' leading to new and excellent varieties of fruit trees created according to a preconceived ideal. In this context Hempel speculated on the personal characteristics required of the future discoverer of a 'law of hybridisation', strangely foreseeing the character of Mendel, as mentioned in Chapter 1.

The Pomological Society owed much to its second President, J. Sedláček von Harkenfeld (1760–1827). A self-made man born to a very poor family, he had risen by sheer talent through all the grades of feudal administration up to the highest, as Director of State Estates. His special interest in wine production made him determined to apply the latest techniques of vine

---

[55] Orel 1978a; Mylechreest 1988 (commenting on Knight 1800; see Chapter 1)
[56] Petri 1812    [57] Anon. 1826    [58] Mylechreest 1988

breeding by hybridisation. In 1818 he used his position in the Pomological Society to set up a fruit tree and vine nursery in Brno for the members. In a published article about inter-varietal crosses, he recorded his expectation that:[59]

> with high probability the new varieties produced would, now and then, unite within themselves many superior characteristics causing the wine pressed from them to be of higher quality; and even occasionally would show such improvement as to be superior to any known varieties, even from abroad.

It can be seen that Sedláček saw the possibility of combining traits by sexual reproduction just as sheep breeders had done during 1816–19.

When Sedláček died in 1827, his successor was expected to continue practical work on pomology and vine growing at an accelerated pace. Progress demanded that the results of experiments and other systematic observations should be disseminated to the up-and-coming generation of growers, through appropriate technical instruction. Who could possibly follow such a lead as Sedláček had set? The clear choice for the Society was Abbot C.F. Napp. He justified his election by promoting pomology and vine breeding, through experiment, observation and instruction, to a degree never experienced before in Moravia, taking full advantage of his influence as head of the Brno monastery.[60] A remarkable skill in producing new varieties of fruit trees and vines was usefully complemented by his knowledge of sheep breeding, which had led to his being elected to membership of that Society two years earlier. Men like Napp, who combined wide scientific expertise with dynamic managerial ability and influence in the highest political and social circles, were hard to find. At the time, none of Brno's breeding experts appreciated more than Napp the parallel between animal and plant breeding, and the need for common scientific principles of breeding to encompass both animal and plant kingdoms.

## Breeding as an academic subject

When the original Moravian Agricultural Society had been founded in 1770, one of its tasks was to institute obligatory examination for managers and clerks of estates.[61] Those successful in passing the examination were deemed capable of improving agricultural production 'through discoveries based on experimentation'. A pressing need for formal instruction to prepare hundreds of clerks annually for the examination encouraged André to propose the first Chair of Agriculture at Olomouc (Olmütz) University in 1808. Teaching of agriculture began there in 1811. By 1816, similar teaching was being given at the Philosophical Institute in Brno, attached to the Bishop's office.[62]

---

[59] Harkenfeld v. Sedláček 1826    [60] Orel 1975b    [61] d'Elvert 1870, ii, p. 43
[62] Orel 1975b, 1978b

The first Professor of real quality to occupy the Olomouc Chair was J.K. Nestler, appointed in 1823. After reading agriculture at Vienna University, he had entered the world of agricultural journalism under André's wing.[63] This was his route to Salm, who proposed him for the Chair at Olomouc. From the onset Nestler took a fresh approach to agriculture by teaching it within the context of natural history.[64] His equivalent in Brno was Franz Diebl (1770–1859), already experienced as an estate manager, whose professorial appointment at the Philosophical Institute was gained through the support of Napp, when newly elected as abbot.[65] Diebl proved himself a born teacher[66] and loyal in his co-operation with Nestler and Bartenstein in ensuring the continued activity of the Sheep Breeders' Society after André's departure to Stuttgart in 1821 (Chapter 10). As President of the Pomological Society, Napp strengthened Diebl's position by making him the Society's Secretary. In the temporary vacuum created by André's exit, Diebl paid particular attention to plant breeding, and Nestler to animal breeding. The two talented men entered upon a period of close co-operation, reinforcing one another as energetic promoters of horticultural and agricultural progress, with Abbot Napp ever in close support. The abbot exerted his influence not only through the powerful position he held within the Agricultural Society but also by ensuring that members of the monastic community attended Diebl's lecture courses and also read the Society's journal (*Mittheilungen*). It was a publication to which Nestler made frequent contributions, sometimes expressing opinions contrary to established academics at Vienna University, which gained Napp's scholarly support on the basis of practical experience.

Nestler was a new type of scholar, whose science was a reflection of practical experience. Concerned to put André's philosophy to the test but prevented by shortage of funds and space from carrying out experiments himself at the university, he co-operated with estate managers in organising field trials and helping to interpret the results. One of the first subjects to attract Nestler's attention was the heredity of horn formation, both in wild and domestic animals, on which he published an anonymous paper in 1827, drawing attention to the new English hornless cattle breeds, which were less likely to injure one another when confined. He recommended selective breeding for this character to produce wholly hornless offspring. He mentioned that the practice was already being followed at Hoštice and neighbouring villages.[67]

Nestler included what he learned from those practical experiments and observations in his teaching. This we know from the fact that he published his lectures in weekly instalments of the 1829 edition of *Mittheilungen* under the title 'The influence of generation on the properties of progeny'. Authorship is confirmed in an editorial footnote which states that the treatise 'is part

---

[63] Orel 1978b     [64] d'Elvert 1870, ii, pp. 203–15     [65] Weiling 1968; Orel 1975b
[66] d'Elvert 1870, ii, pp. 280–89     [67] Anon. [Nestler] 1827

of the model lectures of Dr Nestler, Professor of Agricultural Sciences and Natural History at Olomouc University, in which that worthy author develops the most important aspects of rational breeding". The lectures were divided into 23 parts. In the first he began by considering generation in plants, moving on to make analogies with animals. In most of the following parts he was concerned almost exclusively with animals, especially sheep, although in Part 4 he briefly reviewed Kolreuter's hybridisation experiments in *Nicotiana* and in Part 19 commented on Knight's experiments with various plant species. His concentration on sheep breeding and the transmission of parental traits to progeny was without doubt a reflection of the priority given to wool production by the Agricultural Society in Brno. He gave full credit to[68]

> the famous leaseholder Bakewell and his followers who paired cattle and sheep with the closest degree of inbreeding, simply to produce stock with a special form, size, fattening ability [*Masfähigkeit*] for meat production and slenderness of bones.

The author stated his conviction that animal and plant improvement should be considered in a common framework. He opened his lectures as follows:[69]

> In most plants and animals, generation is the most important and, in many cases, the only way of multiplying [*Vermehrungsart*]. The product of a fruitful generation is offspring, which in their characteristics are more or less similar to the procreators [*Erzeugern*] from which they are descended, depending only on how the essential characteristics, and also the chance characteristics, of the parents are transmitted to the progeny.

Nestler considered 'essential characteristics' first. These are those that differentiate a major sort (*Art*) of plant or animal from related sorts:

> Fruitful generation with heredity [*fruchtbare Zeugung mit Vererbung*] of all essential characteristics is possible only between two sexes which, in the Natural History sense, belong to the same sort. Fruitful generation between sexes of different sorts within the same genus is associated with loss of some essential characteristics.

Nestler's view of heredity was that parental characteristics blend (*verschmelzen sich*) in the progeny. This was a view typical of the period, among both naturalists and breeders. He understood the transmission of parental traits in mechanistic terms, under the influence of the force of generation. Defined like Blumenbach's 'formative force',[6] it was believed to interact with influences coming from the environment, both climatic and nutritional. It was strongest in its effect upon matings of the same sort (*Art*). Some members of the Sheep Breeders' Society called it the 'genetic force'.[70]

---

[68] Nestler 1829, Part 20      [69] Nestler, 1829, Part 1      [70] Ehrenfels 1831

Regarding the appearance of new traits, Nestler believed that 'Predispositions to deviations [*Veranlassungsurchen der Abweichungen*] can be counted upon to be in the stock parents themselves.' Among them would be heritable defects (*vererbbare Fehler*).[71] Irregularities in the transmission of characteristics and random appearance of new traits were observed in the progeny of crosses between different sorts and might also result from a change in environment. Nestler described how, under such circumstances, progeny might inherit only some of the features of the parents. He concluded:[72]

> For lack of relevant, repeated experience, we know little enough about either defects or good characteristics transmitted from parents to progeny. We only know in a general sense that organisational defects, weakness and inadequate activities in blood vessels or in the whole blood vascular system are more easily passed on [*übergehen*] to progeny when both sexes are of the same sort.

When progeny differed from both parents in a particular inherent characteristic, Nestler denoted this a 'sport' or 'saltation' (*Naturspiel, Natursprung*), corresponding with the use of the term *Spiele der Natur*.[73] As an example he gave hornless cattle in the progeny of horned parents. He had noted similar deviations in the progeny of potato plants grown from seed. He stressed the economic potential of such deviations. Hornless cattle normally produced only hornless offspring but, exceptionally, horned animals appeared. This observation, among others, led Nestler to stress that repeated generations of grading would fortify (*befestigt*) the new trait in the progeny 'from within', after which the appearance of the deviation will occur with certainty (*Erblichkeit der Abweichung mit Sicherheit eintritt*). Such a trait with 'fortified heredity' could be used to create a new race. Similar useful deviations occurred in the plant kingdom and could be exploited. We see here an apparent echo of Lamarck's (1809) concept of 'comparative stability' (see Chapter 3), although the two concepts are quite different. While Lamarck believed that a character became comparatively stabilised by repeated exposure to its traditional environment over many generations, Nestler was claiming that stability came from *within*, concentrated by selective breeding, even when the stock was transferred to a new environment. The evidence on which Nestler based his concept had been provided by Geisslern and those who followed him.

In further reflection of Geisslern's practice, Nestler recommended the improvement of local ewes by repeated crossing with imported Spanish rams (grading up). He stressed the economic advantage of purchasing a small noble flock. To follow this course, the breeder would have to pair pure Spanish rams and ewes together, i.e. there would be close inbreeding in the

---

[71] Nestler 1829, Part 9    [72] Nestler 1829; Orel 1977    [73] Anon. 1818b

small noble flock. Most experts on animal breeding of the time rejected such a procedure on principle, sure that it would be followed by degeneration of the race. Referring to the favourable experience of Bakewell in England, as well as prominent sheep breeders in Moravia, Nestler defended inbreeding against such opponents of the day as P. Jordan and I.J. Pesina, Professors at the University of Vienna. According to Nestler, the Spanish breeders themselves relied upon consanguineous matings. As to the English experience, Nestler would later write that 'Without this technique Bakewell could never have existed nor could one exist in the future.'[50]

The dissemination of Nestler's ideas in his published lectures set off a chain reaction among the members of the Sheep Breeders' Society, not to mention the wider readership of the journal *Mittheilungen*, as he encouraged them to believe that the principles of scientific breeding could be revealed by experimental investigation. He was proving himself a worthy pupil and successor to André. He reinforced the breeders' confidence in their selection programmes. At the same time he warned against the idea of racial constancy, in the sense in which it has been proposed in German literature elaborated by Justinus for horse breeding.[74] According to the theory of constancy, Nature has created races with an ineffaceable power of heredity, the characteristics of which stay for ever noble when such animals are pure bred. Practical experience in Moravia made the sheep breeders far from confident in this proposition. Accordingly, they were unwilling to place unquestioning reliance on pedigree and blood relationship. On the contrary, they were convinced that wool of required quality could be obtained only under conditions of careful and permanent selective breeding. Inspired by Geisslern, they welcomed the chance to exploit individual differences, believing that particular inherent characteristics could be improved by selection and that breeds could be changed by crossing. The question of heredity in crosses was for the first time being seriously addressed.

The pages of *ONV* and *Mittheilungen* contained numerous contributions on both sides of the argument. Baron Ehrenfels, who was attracted to the idea of racial constancy, argued that variability and constancy were two sides of the same coin, both arising from the 'genetic force' (*genetische Kraft*), 'the mother of all structural formation in the world which allows existence to that which is capable of living'.[75] In this connection he perceived a difference between *race*, which could be constant, and *variety*. 'But is it not true', asked Bartenstein, 'that by crossing two races or varieties, the breeder can create an entirely new strain [*Stamm*] with completely new traits?'[76] When later R. Löwenfeld defended the idea of race constancy at the Brno meeting of the Agricultural Society in 1835, Nestler disagreed:[77]

---

[74] Justinus 1815    [75] Ehrenfels 1829    [76] Löwenfeld 1835
[77] Nestler 1836

Constancy, i.e. the strict inheritance of a racial trait from parents to offspring, without any deviations, can never be found anywhere, certainly not
in free Nature wherein lies the greatest wisdom of all and where the fallible
human hand plays no part. What does Nature do when deviations appear in
her realm, deviations which are contrary to her intentions? She excludes
them from propagation.

In thus restating the traditional idea of a conservative force in Nature, the
action of which is to eliminate all deviations from the normal,[78] Nestler was
drawing an obvious contrast with what happens on a breeding farm where
the protection afforded by the artificial conditions allows some deviations to
be preserved. If constancy is the breeder's aim, he is obliged to cull his race
artificially in every generation. But if he wishes to create new races, he has the
chance to do so by selective breeding in the direction desired. How different
was this view from a strict interpretation of race constancy by Justinus! Talk
of racial constancy on the one hand and selective breeding of male and
female parents on the other evoked fresh discussion on the underlying
mechanism by which parental characteristics were, or were not, inherited in
the progeny, the central question being the physiological basis of generation.
Ehrenfels was already on record with an attempted explanation of the
enigmatic phenomenon:[79]

Generation, the connection of two beings, gives a living form to the lifeless
chaos of matter. This genetic force, the beginning of life and transformation
of lifeless matter, is an imperfect mechanism acting gradually, modified by
the action of climate and nutrition. An inflexible principle rules in the
arrangement of forms, not the wishes and demands of Man; Nature herself
obeys her own established laws.

Ehrenfels recognised the need to explain why progeny deviate from the
parental pattern, and he returned to the idea of an interaction with the environment during development. He agreed with Nestler that there are two
influences, climate and nutrition, capable of modifing the genetic force. We
may note the continuing adherence to the idea of outside nature impinging
upon inner nature. Following up this point, Ehrenfels made clear his opinion
on the relative strengths of these influences, concluding that the genetic force
was more powerful than the external forces modifying it,[79] as can be seen in
the quotation beneath the heading of this chapter. Another member of the
Society, A. Mayer, reacted immediately to Ehrenfels' conclusion by making a
statement about how Man can deliberately 'remove something or add something, modulate and shape differently' the bodies of animals by selective
breeding.[80] This was the more modern, proactive approach to breeding in the
Bakewell tradition which had been so strongly exemplified by Geisslern and
his friends.

---

[78] Mayr 1982, pp. 488–9    [79] Ehrenfels 1831    [80] Mayer 1831

Views such as these, pronounced at one or other of the Society's Brno meetings, and printed in *Mittheilungen*, show the breeders trying to place the concepts of variation and constancy (the origin of differences and the transmission of similarities) within a consistent theoretical framework. The breeders' ability deliberately to bring about changes in animals by selective breeding was seen to provide evidence for the genetic force being capable of resisting environmental influence. On the other hand, the idea that differences between parents and offspring might originate from the action of climate and nutrition imposing changes during embryonic development could not be ignored.

The capacity of a breed to remain constant, i.e. to generate itself, continued to be a central topic of the Society's discussions. Under Nestler's influence, opinion increasingly reflected the Geisslern tradition that close inbreeding in the Bakewell manner increased constancy. In 1839, Nestler vigorously defended inbreeding against the view of the long-deceased French naturalist G.L. Buffon, whose strong opinion in favour of crossing animals from different environments was still being quoted. Nestler was highly critical of Buffon for having misled the public:[50]

> This popular writer, rich in phantasy about his natural history, has so firmly planted those ideas in the heads of the agriculturalists of the European Continent, who are more gullible than they are independent in thinking, that it will now take endless effort to convince those befuddled people of what is true and what is misleading in the practice of inbreeding.

To discover more about Nestler's answer to Buffon, we have to go back three years. Before the 1836 meeting, the President of the Society gave close consideration to what should be its principal theme. Well aware of the mass of fresh data on sheep breeding that had emerged in the first few years after the foundation of the Society, he reflected that nothing basically new in sheep breeding had been discovered in the recent past. Thus he came to the conclusion that it was necessary to investigate 'the great, mysterious workshop of Almighty Nature'. With this intention he asked Professor Nestler to choose a topic which would be beneficial and appropriate to the occasion. In the Society's report, a group of members, led by Teindl, were agreed in recording Nestler's reaction:[81]

> Professor Nestler believes that the most essential matter of all, as well as the most pressing question at this time, in relation to improved sheep breeding, is the 'inheritance capacity' [*Vererbungsfähigkeit*] of noble stock animals.

On this subject Nestler admitted that most breeders were still stumbling in the dark. By 'inheritance capacity' he implied the strength of the transfer of inherent characteristics, and he believed that enlightenment might come by

---

[81]  Teindl *et al.* 1836; Bartenstein 1837

discovering the principles or rules underlying their transfer. He appreciated the importance of knowing which characteristics were transmitted easily, which with greater difficulty and which were certain to be transmitted, under what conditions. Having no knowledge of the physiological basis of generation or the role of each parent in the origin of the embryo, Nestler proposed to deal with heredity as a separate phenomenon from the still enigmatic concept of generation. Nestler's listeners were greatly stimulated by this appeal.

The following year, 1837, Nestler returned to his aim of finding a theoretical explanation for heredity derived from the breeders' own records when he wrote: 'The question of heredity (*Vererbung*) is very difficult. Its difficulty has been acknowledged by brave sheep breeders, far from ready to overestimate their knowledge of sheep breeding.'[82]

The question of heredity, and more particularly the concept of *Vererbungsfähigkeit*, introduced to the sheep breeders of Brno in 1836, would now become a topic of major concern to them. Given that heredity was a reality, what determined an individual's capacity to transmit its qualities to its progeny? What would their sheep breeding records finally reveal to them about it? We may note with interest their recognition of the *strength* of transmission as the central issue of heredity. Darwin[83] reached the same conclusion in his use of the term 'strength of hereditary tendency' which, according to Bell,[84] led eventually to the modern concept of heritability, a fundamental parameter in the analysis of quantitative traits.

---

[82] Nestler 1837      [83] Darwin 1859      [84] Bell 1977

# 10

# *The sheep breeders' legacy to Mendel*

---

For millenia human beings held no useful answers about inheritance because
they were unable to formulate useful questions. In science, useful questions
are those that are amenable to observation and experiment and, hence, susceptible of being answered.

*J.A. Moore, 1986*

---

We have presented C.C. André as an individual responsible for shaping an
intellectual and organisational environment in which many others were able
to express valuable talents. It was through his vision with its international
scope that breeding began to emerge from an art to a science in Moravia. We
have followed the activities of the Agricultural Society, of which C.C. André
was the first Secretary, and noted the growing influence of Professor Nestler
among other leading figures. As now we begin to ask what, if anything,
Mendel might have owed to this technological transition, we need to bridge
several decades. That is the purpose of this final chapter. At the outset we
describe how André carried his vision from Moravia into Germany, and what
happened to sheep breeding and the Sheep Breeders' Society after he left.
Already it is clear that common ground was being discovered between breeders
of sheep and other stock, to include even honeybees. Cross-connections
were also developing with plant breeders, particularly those with ambitions
to exploit heredity in field crops, fruit trees, vines and, later, ornamental
flowers.

André's motivation for concentrating so much of his attention upon sheep
was, as he admitted, not entirely personal. The subject came up in some of
his correspondence, beginning with a letter he received in 1818 from a honey-
bee specialist in Prague who wrote, 'What a fine thing it is to treat a subject

like sheep breeding with love, as you do . . . will perhaps your next task be apiculture?'

André replied with hope that the apiculturists themselves would respond to this question. In explaining his own motivation towards research, he mentioned an initiative by Baron Ehrenfels to bring about an improvement in apiculture and added,[1]

> The enquirer may find it difficult to believe that it was no special predilection that drew me to sheep breeding rather than apiculture. Much more decisive were questions of patriotism, duty and force of circumstance . . . Ten years ago I recognised sheep breeding to be quite outstandingly important from the viewpoints of economics, technical development and the State, especially in the light of Austrian patriotism. This is why I took responsibility for organising the Sheep Breeders' Society.

André's ideals were shared by many friends and colleagues—landowners, manufacturers and naturalists alike—but not, unfortunately, by those in ultimate power. His residence in Brno witnessed a change in the political climate under Francis I. The emperor made no secret of his rejection of any idea of Enlightenment, including the promotion of science in schools. Teachers at the Laibach *gymnasium* whom he addressed in 1821 were left in no doubt by the terms in which he spoke:[2]

> Hold to the old, for it is good, and our ancestors found it to be good, so why should not we? There are now new ideas about, of which I never can, and never will, approve. Avoid those and keep to what is positive. For I need no savants, but worthy citizens.

In the climate of a police state under the government of Metternich, André's brand of patriotism was not appreciated. Liberal views expressed in his journals *ONV* and *Hesperus* met censorship and tough police control, limiting his activity in Brno and threatening his expulsion. To a degree he was protected by his international reputation, and he surely welcomed any public acknowledgement of the fact. We can guess the satisfaction he found in a long letter from a 'visitor to Moravia' that he published in *ONV*. It described with enthusiasm the free discussions witnessed at the previous annual meeting of the Sheep Breeders' Society, attended by participants 'from Prussian Silesia and even from Mecklenburg' (a distance of more than 1500 km). Referring to André as 'the foremost spirit of the Society', the visitor continued:[3]

> For everything I saw and appraised, the whole of Moravia is indebted to C. C. André, for his stimulation, diligence and perseverance, for his heart and spirit. Without him everything would be discordant, dispersed, without plan, without action. He understands how to unite disparate elements; he deals with science as a citizen of the world, and with his country like a born patriot.

---

[1] C.C. André 1820a    [2] Seton-Watson 1943, p. 165    [3] Anon. 1818a

In an extensive footnote André tried to strengthen support for his case by acknowledging the considerable backing he had received from a number of notable figures, 'splendid men' as he referred to them, whose actions on his behalf had revealed their confidence in his achievement for the benefit of the Province in the previous 20 years. Principal among those he thanked was Count Lažanský (1771–1824), Governor of the Province, especially for his guarantee of mild censorship by the Chief Provinicial Censor, Baron Hager, enabling André's journals to bring beneficial information and instruction not only to Moravia and Austria but also to Germany. He had enjoyed the co-operation of nearly 500 of the 'best minds and talents' of different countries in editing the journals, he claimed. With his residency in Moravia under threat, he needed all the public support he could get. Nevertheless no praise nor action on behalf of this committed teacher, writer, economic adviser and organiser, however justified, was sufficient to prevent the axe from finally falling.

Forced to leave Brno at the end of 1820, *persona non grata* for his radical views, he moved to Stuttgart in the far South-west of Germany where, undeterred by distance, he continued to co-operate with his son Rudolf in editing *ONV*. But because of physical separation, most of his activities for the benefit of his adopted land had to cease, causing many Moravians to view his departure with regret. Even those who did not share his politics could recognise his sterling service to the Agricultural Society and its two lively offshoots.

In Stuttgart, a city with its own reputation for cloth manufacture, André became adviser on agricultural matters at the court of King Wilhelm I. Having ascended the throne in 1816 at a time of serious famine lasting two years, the King was intensely concerned to improve the state of agriculture in his domain.[4] The progressive attitude he showed in meeting the urgent needs of his people would later gain him the title 'King of Agriculture' (announced in Brno in 1840, at the Fourth Congress of German Agriculturalists and Foresters). André, in his new post as scientific secretary of the *Zentralstelle des landwirtschaftliches Vereines* (Co-ordinating Office of Agricultural Societies) and editor of the *Correspondenzblatt des Württembergischen landwirtschaftlichen Vereines*, was strongly instrumental in realising the King's agricultural ambitions.[5]

André's continuing interest in sheep breeding led him to undertake a translation into German, published in 1825, of a French book by Perault de Jotemps, his son Fabry and F. Girod (1824), which he published under the title (now translated into English) *Newest Views on Wool and Sheep Breeding*.[6] Thaer published an independent translation of the same book.[7] In

---

[4] Bayer 1966    [5] Wilhelm 1867, Herrmann 1980    [6] C.C. André 1825
[7] Thaer 1825; Klemm and Meyer 1968, p. 222

extensive footnotes, André recalled the 'extraordinary activities' of the Brno Sheep Breeders' Society. Hardly had the work appeared in the bookshops, than a most tragic event occurred for him. His gifted and beloved son Rudolf died suddenly (probably from pneumonia), far away in Moravia where he was unable to visit. In losing his son, André severed his close connection with Moravian sheep breeding. The notes he wrote for the de Jotemps translation were his final published views on the subject.

In memory of Rudolf, André asked a friend, J.G. Elsner, a Silesian expert on sheep breeding, to produce a new edition of Rudolf's book, which appeared in 1826 published in Prague.[8] In the Introduction, Elsner expressed his conviction that Rudolf would surely have enlarged and revised the new edition. Even though Elsner had tried to update the text, he was unable to add anything extra on breeding methods. He concentrated his attention on the value of 'certain saltations or sports' (*gewisse Sprünge oder Spiele*) as material for selection, not only for animal breeders but also for horticulturalists. What attracted his special attention was a universal property of hybrids, common to both animals and plants, that in their progeny there appeared traits reminiscent of the parental forms (their grandparents), with great variability. This had already been pointed out by Festetics with respect to sheep breeding. It was among such abundance of different forms, in plants as well as animals, that the breeder had the greatest chance of making a favourable selection. Towards the end of the introduction, Elsner put in a word for C.C. André, 'whose merits every unprejudiced German Merino breeder must acknowledge', and, on the last page of the book, he stated his high opinion of the Sheep Breeders' Society of Brno: 'This society attained its objectives to such a marked degree that no other could surpass it.' The Brno Society had provided a model for Professor Thaer to set up a comparable society in Berlin in 1816. Another was instituted in Prague in 1828, and a third in Hungary in 1837.[9] The other societies were transitory, however, fading out within a few years.

André's continuing banishment from Brno seemed to many an irreparable loss. Thaer in Berlin even wrote that the Sheep Breeders' Society in Brno was dead,[10] but this report was unfounded. Soon it was clear that the seeds sown by André had found fertile soil in which to survive. He had left the Society in good hands, conscious of his responsibility of leadership and delegation of responsibility. Before leaving Brno he wrote about leadership as follows: 'It is not easy to find men dedicated to the public good and to unite them, difficult to maintain the Society and even more difficult to enhance its creative humanity.'[11]

---

[8] R. André and Elsner 1826     [9] d'Elvert 1870, p. 146     [10] Anon. 1823
[11] C.C. André 1820b, p. 259

André had found a most distinguished and widely experienced individual to lead the Society as its first President. Baron Bartenstein was a respected local figure who, as André reported, 'knew how to unite disparate elements', showing fairness with judgement, zeal with experience and business acumen with 'noble hearted humanity'.[12] In his youth, Bartenstein had studied law at Vienna University, seeking afterwards to apply his knowledge to the advancement of agriculture, with a particular emphasis on wool production. Losing now the support and advice of André, he turned increasingly to Nestler and to M. Köller, Geisslern's closest associate.[13] Furthermore, at his instigation, the Society welcomed into its leading ranks Dr F.J. Teindl (1768–1859), a lawyer and estate manager. In an address to the annual meeting of 1822, the new member commented favourably on the scientific level of debate at the meetings when treating breeding methods. He found the membership united:[14]

> A common purpose among landowners encourages enthusiasm for progressive improvement of domestic animals as a matter of preference. Unreliable procedures are avoided because correct principles are established, based on experience and precise observations. Such principles warn against mistakes and false steps and raise the culture of sheep to the status of science.

The secretaryship of the Society and of the parent Agricultural Society passed to J.C. Lauer (1788–1869), a journalist who remained Secretary almost up to his death, also assuming the editorship of the newly published journal *AGM* (*Mittheilungen*), the 'mouthpiece' of the Agricultural Society. Among his writings there were papers describing the history of sheep breeding in Moravia, although he did not involve himself directly in the practice.[15]

## Heredity as a subject for research

In the years following André's departure, the sheep breeding records of the Society's members were scoured for evidence of distinguishable patterns of heredity. Nestler took the lead in this search, as described in Chapter 9. In his paper 'On heredity in sheep breeding', published after the meeting of the Sheep Breeders' Society in 1837, he revealed the source of his motivation. He claimed it was a chance question about a prize ram that encouraged him to resurrect this important subject. The topic emerged from a remark by Count E. Wrbna, who bred sheep on his estates in northern Moravia. His best ram was a most beautiful and valuable animal, concerning which he expressed the following dilemma: 'If its superior qualities are inherited by its offspring, I cannot afford to sell it at any price, but if they are not inherited, then it is worth no more than the value of its wool, meat and skin', to which Nestler added 'Thus did Count Wrbna pose the question of heredity to me and,

---

[12] C.C. André 1820b, p. 260    [13] Orel 1977    [14] Teindl 1822
[15] d'Elvert 1870, ii, pp. 338–9; Lauer 1826; 1841

through me, to the Sheep Breeders' Society.' Nestler recorded that many participants had thought the question odd. Some even thought it not worth discussing. 'What a great mistake that would have been', he wrote, for he was convinced that Wrbna had 'thrown the seed of the question into the proper soil in which it can now gradually develop into the luxuriant fruit of science, if the embryo is well cared for'.[16]

Wrbna knew of course that his ram was already fathering good lambs. What concerned him was whether, or the extent to which, it would continue to do so if sold to somebody else. Geisslern and others were claiming that transporting rams into a new environment need not affect their performance or progeny if they were properly looked after. Wrbna wanted assurance on the economics of the matter. He hoped the eminent Nestler could provide him with a reasoned opinion.

Nestler came to the 1836 meeting, travelling the nearly 70 km from Olomouc, at the express invitation of Bartenstein, to outline the most urgent problems in wool production. A marked change was manifest in his attitude to the transmission of traits compared with the line he had taken in his published university lectures of seven years earlier.[17] Then his focus had been on 'generation with heredity'; now he was referring only to heredity. By narrowing debate to the transmission of traits between separated individuals, he aimed to clarify a confusing subject. Ready to leave aside for the moment the enigmatic physiological questions surrounding the process whereby the embryo gains life and complexity, he was now concentrating attention upon patterns of transmission between parents and offspring. The major focus was on the male parent, whose only continuity with the embryo was through the semen. Some of the Society's members were reluctant to follow his lead, stating that it was old-fashioned to revisit this particular problem 'in the sole surviving sheep breeding society in the entire world' because everyone who practised sheep breeding with them must know 'whether and how inherent characteristics are transmitted to progeny'. Nestler dismissed their attitude as complacent and mistaken, although, to his dismay, he then found himself opposed by the Society's respected President, 'our grand master [*Grossmeister*] of sheep breeding' Baron Bartenstein.

It is undeniably difficult to discover a new theoretical principle. But Nestler's admiration for the achievements of Moravian sheep breeders inspired him to hope that this case would prove an exception, that the secret of heredity would be revealed to the world in detail never possible before, and that it would happen through his leadership. Even in 1829, he had written:[17]

> I, as the bell ringer [*Küster-Glöckner und Sakristaner*] in my science, direct
> my enquiry to living practical experience, to men of the present, to sheep

---

[16] Nestler 1837     [17] Nestler 1829

breeders of our province by whom scientific sheep breeding has, without doubt, been vitalised and is being transmitted throughout Europe.

By 1837 Nestler was convinced that the Moravian breeders, with unprecedented access to detailed breeding records, must hold the key to this area of knowledge. After all, was not Moravia a 'motherland' of rational sheep breeding? Had not the Moravian breeders selected sheep systematically according to a stock register for more than 40 years, beginning at a time when rival breeders in foreign countries where still dreaming about it? He was glad to know that at some Moravian breeding farms the long-kept records were still perfectly intact with all their wool sample cards attached. He sought co-operation from the owners of such records, 'from the golden age of sheep breeding', in using them to uncover the true principles of breeding by answering the question, 'What noticeable successes in heredity can be achieved when pairing different races, or stocks of the same race [*Schafstamme*], with equal or unequal traits?' In order to answer this question, Nestler asked each member to pick up his old family stock register and examine critically the 'history of heredity' (*Vererbungsgeschichte*) of his best stock animals in their offspring 'from the top downwards', or their 'developmental history' (*Entwicklungsgeschichte*) in their ancestors 'from the bottom upwards'. From this investigation he expected fresh information, sufficient to create a new breeding system by explaining the patterns of heredity. He was aware that the difficult topic of heredity had already been tackled unsuccessfully by many thinkers and observers in the past, but he was convinced that they had not kept the matter in mind over a sufficiently long time span to provide the necessary insight.[18] The Society had the advantage of a wide spectrum of expertise among its members. Nestler could draw upon academic as well as practical opinion.

The idea of bringing the academic and practical worlds together had been implicit in the foundation of the Sheep Breeders' Society in 1814. It was a view not widely shared. Although agriculture had gradually developed a scientific basis in Moravia, particularly after 1806, the growth in knowledge was often viewed as 'mere empiricism'. M.K. Fraas, who wrote a history of agriculture, published in 1852 at the instigation of Nestler and Napp, described the uneasy relationship existing between agricultural science and natural science as follows:[19]

> The fact that agriculture represents a summary of natural scientific experiments, governed by national economic priorities, is not generally understood. The results of experiments in open fields mean little to the pure naturalist. In contrast, we take it as axiomatic in the history of agriculture that it is fully competent to bring about scientific progress.

[18] Nestler 1837    [19] Fraas 1852, p. 13

Such enlightenment was rare, although at least one naturalist in England would surely have sympathised with Fraas's sentiments. Charles Darwin was intensively preoccupied in the early 1840s in reading 'heaps of agricultural and horticultural books',[20] in support of an analogy he was drawing between artificial selection and what he would come to call natural selection.[21]

In Fraas's *History* we can read of his conviction that systematic breeding of farm animals, crop plants or trees for commercial gain could be as precisely scientific a discipline as anything taking place in a laboratory or experimental greenhouse. Furthermore, because of the sheer numbers involved, it could prove substantially more informative. Whether with cattle or clover, sheep or sugar beet, breeders were creating increasingly productive or more beautiful varieties and thereby adding to the bank of knowledge on heredity. But, to be realistic, whatever answers the sheep breeders came up with, there would be a great problem in convincing the academic community of their true relevance. This was Nestler's task and ambition.

In parallel with the examination of sheep breeding records, Nestler and others also placed fresh emphasis on plant breeding by hybridisation, without using the actual term. It is impossible to follow the development of either animal or plant breeding in Moravia without considering each in relation to the other. The technique of artificial fertilisation in plants had brought together new and valuable combinations of traits in the progenies of fertile hybrids between varieties and races. Fraas's *History*, which highlighted this observation, refers to the achievements of both Nestler and Napp. These two formidable promoters of science strongly reinforced one another in the belief that the empirical results of breeding, whether of animals or plants, were worthy of greater regard and deeper analysis.[22]

Despite his relatively short life, Nestler rose high in the estimation of his contemporaries. In Fraas's view, 'he ranked in first place among those beginning to introduce natural scientific research in agriculture'. His outstanding talent was his unusually perceptive and original interpretation of experimental data.[23] He inspired both admiration and support from the influential Napp, whose ambition was 'to promote science in every respect', as Röhrer a local publisher and amateur botanist, reported.[24] The two talented Moravians, sharing a common interest in the scientific basis of animal and plant breeding, found much to gain from their association. Nestler's conviction that the most essential matter for consideration in respect to sheep breeding was *Vererbungsfähigkeit*, that mysterious property which gave an animal the *capacity* to pass on its particular characteristics, stimulated Napp to pursue

---

[20] Darwin F. 1902, pp. 173–4: letter from Charles Darwin to Joseph Hooker, 1844
[21] Betham-Edwards 1898; Wood 1973; Ruse 1975; Mayr 1982, pp. 485–6
[22] Orel 1975b, 1978a    [23] Fraas 1852, p. 337    [24] Röhrer 1830

the question of why heredity was more certain with respect to some traits than others. In a report of the meeting in1836 we read:[25]

> The highly esteemed Abbot and Prelate of St Thomas's Monastery, Cyrill Napp, asserts that, according to his view, heredity of characteristics from the producer [*Erzeuger*] to the produced [*Erzeugten*] consists above all in the mutual affinity by kinship [*gegenseitige Wahlverwandschaft*] of paired animals. As a result of this, a ram chosen for the ewe should correspond to it in both inner and outer organisation. This process deserves to be the subject of an important physiological study.

We can see in Napp's statement an endorsement of breeding closely related animals, those most likely to share the same traits. The use of the term 'inner and outer organisation' does not appear to be a reference back to traditional ideas about separate influences coming from the two sexes, because the ram and ewe were expected to correspond in both respects. It brings to mind the concept of 'internal organisation' proposed by Lamarck.[26] Closer to home, it seems to have been analogous to the organisational concept of Petri, which he mentioned in a lecture he gave to the 1837 meeting of German Naturalists and Physicians, held in Prague. Speaking on the subject of animal breeding, he attached particular significance to selection from crosses, i.e. blending (*Vermischung*), 'where the inner cohesion of the external formation of the individual in its different varieties has to remain hidden from the eye'.[27] The sense of Napp's statement is then interpreted in terms which view the outer organisation as a manifestation of the inner organisation, just like Petri's concepts of inner cohesion and external formation. As far as heredity is concerned, there was continuing uncertainty in the Petri/Napp era of just how much the 'inner organisation' is influenced by the environment, either during development or after birth. However, by ensuring that animals bred together were chosen for their similarity in particular respects of interest to the breeder, and kept under uniform, high grade conditions, the maximum chance was provided for them to transmit the traits they share to their progeny.

Reactions by Society members to Nestler's and Napp's remarks about heredity, and the discussion which they stimulated, were published in the following year. Ehrenfels commented:[28]

> The discussion on sheep breeding published in issue no. 38 of *Mittheilungen* stimulated by Prof. Nestler's thoughts on heredity of stock animals was of great interest, on account of the direction it took. It proves that advanced sheep breeding, according to principles and rules, is urgently needed, and also that the sciences auxiliary to it, natural history, anatomy and physiology, could be an aid to translating an unrevealed truth into a known one.

---

[25] Teindl *et al.* 1836     [26] Lamarck 1809, Chapter 3     [27] Petri 1838
[28] Ehrenfels 1837

Ehrenfels further reacted by drawing a parallel with Bakewell and his breeding of the ideal type of sheep for mutton production. In his opinion, Bakewell's achievement had come about by harnessing the 'genetic force'. He added, 'this reasoning applied now to fine wool breeding, would raise the following questions. What can be considered the ideal? What, for us scientific investigators of nature, is the attainable ideal?' Old-fashioned breeders would have said that the highest attainable ideal was to retain Spanish quality. The Moravian experts, convinced of the power of selective breeding, had ambitions greater than this. The outstanding questions were how far and in what direction did they want to go?

At the 1837 meeting, Bartenstein tried—with evident difficulty—to define an ideal and true-breeding ram. In his less than clear explanation can be seen emphasis placed on pedigree analysis, trait recording and progeny testing. The contributions by Ehrenfels and Bartenstein are marked in pencil in the copy of *Mittheilungen* preserved in the library of the Augustinian monastery, probably by Napp whose own contribution is left unmarked. Nestler and Napp were both well aware that the question of heredity still seemed strange, even irrelevant, to some participants at the 1837 meeting. Practical breeders needed to be convinced that knowledge of heredity would really give them better wool. Nestler addressed the 'stout-hearted sheepbreeders' in his serialised paper of 1837. First he explained the common framework of animal and plant breeding:[18]

> The terms *species* and *race* in the animal kingdom correspond precisely with the terms *species* and *variety* in the plant kingdom. Only Nature produces, through forces beyond the hand of Man, under constant environmental conditions, natural species with unquestionable constancy. Man, however, produces in the manner of the forces of Nature, in the reproductive process and formation of organic bodies, modified deviations. From the moment of their origin such deviations have the chance of increasing or disappearing in succeeding generations according to their inheritance.

Here Nestler, the professor of natural history, states his understanding of the continuity of species from generation to generation, animal or plant, while Nestler, the professor of agriculture, explains how new animal races or plant varieties may be brought into existence. Selection is not mentioned in either case but is clearly understood as far as races and varieties are concerned, as evidenced by what Nestler writes elsewhere. In this remarkable pre-Darwinian analogy between natural and artificial events, he saw the action of breeders as copying nature.

By this time Napp was becoming less optimistic about the outcome of their discussions in terms of theory. He was obviously impatient by the end of the 1837 Sheep Breeders' meeting, disappointed at the way the discussion had become diverted into breeding technicalities rather than concentrating on the central problem—heredity. The published report says:[29]

---

[29] Bartenstein *et al.* 1837

Prelate Cyrill Napp concluded the course of the continuing debate with rigorous brevity . . . and drew attention to the fact that they had completely deviated from the proper theme of heredity. 'What we have been dealing with', he pointed out, 'is not the theory and process of breeding. But the question should be: what is inherited and how?

Napp's concern was not simply with the inherited traits themselves. Every breeder with experience of Merinos knew that certain characteristics of wool could be inherited. Napp was asking what lies behind an inherited trait. The kind of inductive approach to understanding heredity proposed by Nestler (1837) was proving incapable of solving the problem. The central issue for Napp was the material basis of heredity. His simple question exposed the problem still requiring to be researched in the bluntest possible terms. He had to face the frustrating truth that sheep breeding records could not provide the answer. Nor indeed could enlightenment be gained from any current philo- sophical explanation. Hundreds of thousands of words had been written about the generation of new organisms, by such notable scholars as Buffon, Needham, Wolff, von Haller and Blumenbach, approaching the subject with different theories, either of preformation or of epigenesis. But such erudite theorising did not answer the problems of practical breeders in exploiting heredity. As early as the 1780s, Wolff had asked 'What then is it that is inher- ited, and why, as a consequence, do offspring resemble and differ from their parents?'[30] Napp was asking essentially the same question in 1837. The startling successes in practical breeding since the 1780s had served only to reveal the still undefined complexity of inherited variation in an array of dif- ferent traits. Almost nothing could be said about patterns of heredity for traits of economic importance. The subject was still wide open for investiga- tion if a way could be found. Napp made clear his conviction that heredity was a subject ripe for physiological study.[25] It puzzled him that 'Nothing certain can be said in advance as to why production through artificial fertili- sation remains a lengthy, troublesome and random affair.'[31] Napp was con- cerned about the element of chance in heredity, a matter that would later become of particular interest to his protégé, Mendel.

Why should the opinions of a cleric, even one as influential in the Church's hierarchy as Napp, command respect on such a technical subject? Papers and documents relating to Napp[32] provide the answer, revealing him as a many- sided personality, broadly educated and fascinated by the expanding ideas of science. In his position of influence, he gave wholehearted attention to encouraging the cultivation of knowledge of all kinds through study and research, within the monastery and outside. Charged with the responsibility for his monastic estates, he sought whenever possible to apply scientific

---

[30] Roe 1979 p. 30    [31] Nestler 1841, p. 337
[32] Weiling 1968; Orel 1975b; Czihak and Sládek 1991/1992

principles towards increasing agricultural production, thereby adding to the monastery income. In the early days of his appointment as abbot, sheep breeding was a priority for him, since a large part of the income from the estates arose out of wool production.[33] In his autobiography, written in 1836, he made specific mention of sheep breeding in this context.[34] His special knowledge of the subject, coupled with his elevated social position, made him an obvious choice for nomination as a member of the examination board for master shepherds in Moravia, on which he served for many years. Even though his personal research interest turned increasingly towards fruit trees and vines, his expertise on sheep breeding remained in demand. No reticence could be detected in his willingness to encourage debates within the Agricultural Society or its subsections on any aspect of breeding.

An atmosphere of free discussion characterised the Society when Napp was in a leading position, a state of affairs that received praise from J. Waniek in an article he wrote for the Society's journal in 1845. For Waniek the value of scientific discussion lay in its power to accelerate growth of knowledge far beyond what was possible when communication was restricted to published information, which 'requires a long time to concentrate attention upon a particular problem'. Differences of opinion that were freely expressed in the Society stimulated experiments which opened the way for new reflections and experimentation. For Waniek any action that promoted progress in an atmosphere of free communication must be for the greater good and 'according to natural law cannot be stopped'. He likened science to a growing tree becoming more and more deeply rooted, until finally 'at its full maturity, it produces a golden apple to be plucked by the Fatherland'.[35] Such progress for the ultimate benefit of the state had been exemplified by Geisslern's Hoštice breed of sheep, making its mark all over Germany, even as far as Mecklenberg, which Waniek had seen for himself (Chapter 7).

## Physiology and heredity

Napp's statement of 1836 that heredity 'deserves to be the subject of an important physiological study'[25] seems very significant in the light of further events. How can we explain what prompted it? In November 1835, the foremost Czech physiologist, Jan Evangelista Purkyně (Purkinje) (1787–1869), Professor of Physiology at the University of Breslau (now Wroclaw), visited Napp at the monastery. A year earlier he had published a paper on generation, which touched speculatively on resemblances between parent and offspring.[36] His theory was that parental characteristics are reduced in the germs to 'pure quality' by a process he defined as 'involution'. He located the germs in the egg (ovum) of the female and the semen (or 'farina') of the male.

---

[33] Balcárek 1977    [34] Neumann 1930; Czihak and Sládek 1991/1992
[35] Waniek 1845, pp. 263–4    [36] Purkyně 1834

The meeting of the parental germs then leads to a new process that he referred to as 'evolution', in which the embryo develops to a form revealing the traits of both parents. The use of the term 'evolution' with reference to the sequence of events in embryogenesis was quite normal at the time. Only in the nineteenth century did the term come to include the sequence of fossil forms in the geological record.[37] The word in its original biological usage had been used by the Genevan naturalist Charles Bonnet (1762), a supporter of 'ovism', to mean the development of the embryo from the pre-existing 'germ' in the female.[38] Purkyně modified the idea to allow for the fusion of germs so that both sexes contribute to the progeny. Later, in his manuscript entitled 'Developmental history of the plant and animal organism',[39] he tried to explain the origin of 'rudiments of forms' (*Gestaltungsanlagen*) by drawing a parallel with the inorganic world. Just as in the evolution of the earth, gases were transformed into liquid and later into a solid state, forming crystals, so it was with the germ cells of organisms: the *Gestaltungsanlagen* were a physical reality hidden in liquid, being later revealed in a defined form. A comparison of animal growth with that of crystals goes back to the Swiss naturalist Louis Bourguet (1729), a Professor at Neuschâtel, and correspondent with Buffon on the topic of generation.[40]

Having arrived at his theory by logical deduction, on the basis of natural philosophy, Purkyně was eager to demonstrate its truth by physiological experiments. He hoped to apply exact methods to research on living organisms, to analyse the events occurring during the process of generation and differentiation, in both animals and plants, by histological techniques. In 1832 he obtained an excellent microscope, an achromatic instrument made by S. Plossl of Vienna. He also associated himself with the production of mechanical microtomes to cut thin sections of tissues. His methods of fixation, sectioning and staining were meticulous and he made the nuclei of cells visible by treatment with acetic acid.[41] Purkyně's visit to the academic community of St Thomas in 1835 provided a chance for him to expose his theory and experimental results to Napp and also to another member of the community, Fr. Klácel, a close friend and fellow member of the Bohemian Academy of Science.[42] Klácel had become deeply interested in generation, heredity and hybridisation, from a philosophical point of view.[43] Purkyně made four recorded visits to the monastery between 1835 and 1850.

Towards the end of this period, Purkyně was sharing thoughts on generation and related matters with Rudolf Wagner (1805–64), Blumenbach's successor as Professor of Physiology at the University of Göttingen. As managing editor of the four-volume *Physiological Dictionary, with Special*

---

[37] Bowler 1975; Gliboff 1999     [38] Roger 1997, p. 121; Bonnet 1762
[39] Orel *et al.* 1987; Orel and Janko 1988     [40] Roger 1997, pp. 24 and 124
[41] Priestland 1987     [42] Matalová 1979     [43] Kříženecký 1987

*Regard to Pathological Physiology*, Wagner was well placed to assess current opinion on the generative process. He and his collaborating authors recorded the latest experimental findings relating to cell theory, which was then being formulated. His opinions emerged most clearly in the final volume of the *Dictionary* published in 1853. It included a long chapter entitled 'Generation', written by R. Leuckart, one of his former pupils, who was then Professor of Zoology at the University of Giessen. On receiving and reading Leuckart's 300 pages, containing numerous speculative theories, Wagner had been dismayed by the lack of experimental research. Expressing his critical reaction in an 18 page postscript, he outlined his own and others' microscopical investigations on eggs and semen, and suggested some ideas for future research. He referred to 'the classical investigations of Purkyně and B. Baer on the vertebrate egg', which had led to a new theory that 'It is the nature of the interaction between the semen and the egg, and their relation to the future embryo, that represents the true, central issue of the whole subject known as generation theory.'[44]

Growing evidence, from both animal and human examples, was convincing Wagner, like Nestler and Napp some years earlier, that further progress in research on generation would come from studying heredity. The transmission of physical peculiarities from parents to their children, and onwards, demanded an explanation. For such research, Wagner recommended the crossing of animals showing different traits, followed by statistical analysis of the progeny. 'Only through statistical assembling can a more exact ascertainment of numerical data furnish reliable clues', he wrote.[45] The statistical approach was an advance on anything Nestler or Napp had written, although Napp's concern about the randomness of heredity might be considered a step in that direction.

Underlying Wagner's proposal, as he noted, were 'recently reported' experiments by the Englishman Martin Barry, a young scientific collaborator of Purkyně's, also working for periods of time in the laboratories of J. Müller, G.C. Ehrenberg, T. Schwann and Wagner himself. Barry claimed in 1840 to have demonstrated the presence of a single microscopic spermatozoon within a fertilised egg of a rabbit.[46] The discovery was discussed in a letter from Purkyně to Wagner, dated 27 February 1847.[47] Although Wagner was aware that Barry's observations had been 'until now rejected as unbelievable',[48] growing evidence encouraged him to write a second postscript to Leuckart's chapter, accepting the explanation provided by Barry. While the question of how many spermatozoa (seed fibres) were necessary to achieve fertilisation remained open, Wagner agreed with Purkyně that both parents contributed substances to fertilisation and the formation of the embryo: [48]

---

[44] Wagner 1853, p. 1003; see also Wagner 1837   [45] Wagner 1853, pp. 1006–7
[46] Farley 1982, pp. 56–8; Wood 1989–90; Barry 1840   [47] Orel and Kuptsov 1983
[48] Wagner 1853, p. 1018

Now, if some kind of material is inevitable for the transfer of female peculiarities to the future embryo, why should not one postulate *a priori* the same in relation to the transfer of male peculiarities to the embryo through actual reception of semen elements by the egg. Here, if ever, the proposition of equal causes is valid.

Having accepted the idea of heredity through both parents on an equivalent basis, Wagner was repeating essentially the same arguments for investigating heredity in animals by experimental research that Napp and his colleagues had rehearsed 17 years earlier. He detailed the main topics for research, expressed as a series of questions. These concerned the effect of age, sex, constitution, repeated fertilisations of a mother, and various environmental influences. In fact, there was little of novelty in his questions. André had included most of them in his list of 36 questions published in 1818. Another interesting parallel with the Brno sheep breeders was Wagner's use of the term 'heredity' for the origin of traits. Churchill has pointed out that prior to the 1880s, the word *Vererbung*, used in a biological sense, appears infrequently in German writings.[49] Nestler and Napp were employing the term freely in 1836–7 when Purkyně began his visits.

In the second of his postscripts to Leuckart's monograph (1853), Wagner reported that he had started controlled experimental crosses of his own with frogs and fish. Beginning enthusiastically, he had been quick to realise the extent of the labour and time involved in collecting satisfactory amounts of data over several generations. He had to conclude that there would be a greater opportunity to carry out crossing experiments at institutes for horse or sheep breeding or, in a veterinary context, with dogs. Looking further afield, he speculated about making crosses between animals at London Zoo. Wagner was one of the first physiologists to raise the possibility of investigating heredity directly. It must be said, however, that for all the respect his undoubted talents commanded, he was moving out of his depth in proposing such speculative and unrealistic experiments. He surely had no hope of persuading professional breeders or zoologists to carry out special crosses unrelated to economic priorities. The major interest of his proposals is the extent to which they were out of touch with the reality of animal breeding, both economic and practical, compared with the level of understanding reached in Brno even 25 years earlier.

The contrast between Wagner's unrealistic proposals and the practical breeders' long tradition of experimentation illustrates the gap that existed between theorising academics and knowledgeable practitioners in this area of biology. However, it is interesting that, once Wagner began to consider breeding, he was encouraged to speculate about heredity long before physiologists in general were ready to do so. He expressed his intention to investigate

---

[49] Churchill 1987

heredity in complex organisms like the horse, sheep and dog, or even zoo animals, although he well realised that generation was better understood in plants, being easier to demonstrate and hence potentially more enlightening. He was ready to draw an analogy between plant hybridisation and animal crossing, believing that they rested on a similar physiological basis.[48]

Despite its uneven quality, Wagner's multiauthored *Dictionary* received serious attention from fellow physiologists. Its major significance today lies in his definition of heredity as a problem for research, aimed at answering specific questions in a physiological context, more elaborate questions than those posed earlier by Napp. It must be said, however, that Wagner's proposals for physiological research in this area elicited no obvious response.

## Heredity remains a mystery to animal breeders

Meanwhile, several German animal breeders of Wagner's day were attracting due credit and respect for their practical achievements. One of these in particular deserves our attention, having been described as 'the outstanding representative of pre-Mendelian agricultural science'.[50] August von Weckherlin (1794–1868) made an early reputation by improving the estates of the King of Württemberg, Wilhelm I, who entrusted him with a series of missions to France, Italy, the Netherlands and Britain. Farms for the breeding of cattle, horses and sheep, which he had been charged with setting up on the royal estates, were recognised as models of excellence.[50] With regard to sheep breeding he drew upon the experiences of breeders in Moravia and Silesia.[51] Accounts of Weckherlin's activities appeared in articles in the *Correspondenzblatt des Württembergischen landwirtschaftlichen Vereines*, edited by C.C. André, who had recently arrived from Brno. As a result of these early achievements Weckherlin was appointed in 1829 to membership of the Landwirtschaftliche Zentralstelle des landwirtschaftlichen Vereines, which gave him further contact with André, who was Secretary of this co-ordinating body for agricultural societies. Weckherlin's association with someone so wise in all matters of agriculture, and above all in breeding, cannot have left him unaffected as he strove for progress in animal improvement. By 1837 he had moved to direct the School of Agriculture and Forestry at Hohenheim near Stuttgart, which he transformed into a centre of learning and research, and enlarged the technical workshop. He spent the years 1837–45 reforming this institution and promoting research, before moving to Berlin to accept the post of counsellor on economic matters to the Prince of Hohenzollern.

Weckherlin's English experience led to the book *On English Agriculture* in 1842. A second major work, *Agricultural Animal Production*, contained a

---

[50] Herrmann 1980     [51] Elsner 1827, 1828

volume devoted exclusively to cattle and sheep, which appeared in 1846. Commenting favourably on the achievements of Bakewell, he described how early German horse breeders had tried to determine underlying principles. The prevailing approach to animal breeding in Germany was based on the theory of racial constancy elaborated by Justinus (1815), in which close attention was paid to pedigree recording, and to ensuring an origin from the best ancestors. Conscious of how much horse breeding was limited by a long generation time, Weckherlin became convinced that sheep offered greater potential for elaborating 'certain breeding principles' (*gewisse Züchtungsregeln*). He remarked that 'What can be revealed from horse breeding in a century, can be achieved in sheep breeding within a decade.' Assessing progress made in this direction, he gave special credit to the 'Moravian Association' (i.e. the Agricultural Society) which had brought to public notice 'most valuable contributions' in its journal *Mittheilungen.*[52]

The development of animal breeding in Germany during this period has been reviewed by Berge (1961), who concluded that the pedigree concept, and the doctrine of constancy of race associated with it, grew deep roots in northern and central Europe after 1848. Weckherlin went along with this trend, although careful to draw a distinction between traditional races and newly created races. He believed the former to be produced by long exposure to external influence, generation after generation, mating only among themselves and giving rise to what amounted to a subspecies of similar individuals, constant in the transmission of their characteristics. By contrast, the new races, created by selective breeding, were less stable and more likely to revert to a primitive condition if not carefully checked by selection. He was adding his authority to a distinction made by earlier breeders, which Lamarck (1809) had rationalised with his proposal that an animal's 'inner organisation' acquires increasing stability the longer it interacts with the same environment.

Weckherlin set out his mature view on the theory of animal breeding in his book *Contributions to Opinions on Constancy in Animal Breeding* published in 1860. In the introduction,[53] he mentioned the work of diligent agriculturalists who were already attempting to elaborate 'breeding science' (*Züchtungskunde*) and expressed regret that physiologists failed to support animal breeders in the investigation of this important subject. For his part, he had come to acknowledge 'the victory of individuality over race' and that some individuals have 'a special capacity to transmit inherent features'. He had grown convinced that reliable breeding principles would emerge not from 'natural race constancy' but from defining the production ability of individuals and how their traits became fixed in their progeny, which, according to him, was the true principle of constancy in heredity. Weckherlin claimed

---

[52] Weckherlin 1842, 1846, pp. 20–22    [53] Weckherlin 1860, p. 5

constancy of heredity rather than constancy of race. For him heredity was 'the basis of all breeding methods'.[54] His experience told him that when males and females with the same degree of constancy in their ancestry were mated together, the progeny would show a similar degree of constancy; however, there would still be individuals that deviated. For this reason he emphasised the need to monitor constancy of traits and to evaluate breeding animals according to proven performance.[55] He stressed the importance of inbreeding combined with expert selection of the best individuals to achieve constancy of inheritance, and then to improve average quality beyond the 'already existing satisfactory inherent characteristics'. Weckherlin, like Bakewell and those that followed him in Britain, had finally become convinced of the individuality of heredity, even among animals with a common ancestry.

Weckherlin's strongly held opinion about the production ability of particular individuals (*Leistungsfähigkeit des Individuums*) and the fixation of their traits in their progeny was highly influential in the world of breeding. Like R. André and Nestler years before, Weckherlin was rejecting the old idea that breeding should be aimed at preserving some ideal race, the qualities of which could never be exceeded. All rational breeders, he urged, must take account of inheritance (from parent to progeny) as well as ancestry, which together comprise constancy. This is not, however, to underestimate the significance or the value of inbreeding to produce uniformity.[56] He warned sheep breeders of the danger of falling into the hands of dealers peddling sheep that, while looking good, did not have the required pedigree. This is a precise repetition of advice given by Nestler in his article on inbreeding, published in *Mittheilungen*[56a], in which he had pointed to the financial advantage of inbreeding, which had brought such profit to Bakewell in England and to breeders co-operating with Geisslern in Moravia. In 1838 Weckherlin had supported Nestler publicly when he defended inbreeding on the occasion of the second meeting of German Agriculturalists and Foresters in Carlsruhe, an endorsement which Nestler was pleased to note.[56a] The similarity between Weckherlin's interpretation of breeding and the contents of articles published in the Brno journals, mainly by Nestler in the 1830s, seems too close to be accidental. One suspects a covert influence from André at Hohenheim. A continuing interest in the activities of the Brno Agricultural Society is evidenced by Weckerlin's praise for it.

Despite Weckherlin's great reputation and wide influence, his approach to selective breeding does not seem much in advance of that followed by Bakewell in England or Geisslern in Moravia. If we agree with Herrmann that Weckherlin was 'the outstanding representative of pre-Mendelian

---

[54] Weckherlin 1860, p. 11    [55] Weckherlin 1860, pp. 20–29
[56] Weckherlin 1860, p. 29    [56a] Nestler 1839

agricultural science',[50] we must acknowledge that, as regards animal breeding, he was standing on the shoulders of some highly significant pioneers, including those in Moravia. Unfortunately, memory of the Moravian achievement had faded by 1860. The Sheep Breeders' Society had long since ceased its activities, due to the importation of cheap fine wool from the British colonies. Moravia produced no outstanding champions of scientific animal breeding in the second half of the century to demand acknowledgement of its pioneering past.

Among a new generation of German animal breeding experts, H. Settegast was one who would re-examine the Moravian claims. Educated in the natural sciences as well as agriculture, he completed his training in 1845 with a short stay at the Agricultural School at Hohenheim, at which he distinguished himself sufficiently to be awarded the Royal Württemberg Medal.[57] He was following a tradition established by Weckherlin, to which he would add his own particular flair and significant contribution. His first major appointment was as administrator of the Royal Domain at Proskau-bei-Opeln in Prussian Silesia (1847–56), where he published an account of English agriculture based partly on his personal observations during a visit there (1852). He also wrote a review of sheep husbandry in Silesia, compared with that in Prussia and Saxony, and a report on the Höhere Lehrenstalt (High School) at Proskau where he lectured. A growing reputation led to his appointment as director of the Königliche Höhere Lehrenstalt (Royal High School) at Waldau in East Prussia. From that time onwards he became increasingly interested in sheep breeding and produced a number of articles on the subject. It was while based at Waldau that he personally investigated Geisslern's achievement with a visit to Moravia, which led him to write about the breeder's originality and the special excellence of his breeding stock (Chapter 8). He made no secret of having been overwhelmed by the power of selection after his revealing visit to Hoštice.[58] At the end of 1862 he was invited back to Proskau to direct the Lehrenstalt there, where in both lectures and writings, he continued to argue strongly for the significance of individual potency in heredity, providing examples from the achievements of various practical breeders in bringing about inherited changes by selective breeding. His most important work at this time, and arguably his chief work, was *Animal Breeding*, published in 1868, where he came out in favour of Darwin's theory of natural selection.

Two major barriers to understanding heredity can be identified among German animal breeders of this period. The first was the doctrine of racial constancy, the limitations of which are considered by Berge (1961). Sure that its pervasive influence hindered progress, Berge gives credit to those breeders

---

[57]  Phillips 1989, i, p. 341        [58]  Settegast 1861

willing to recognise the extra reproductive potency of certain individual animals in transmitting valuable characteristics. Of all the German experts of his time, Settegast was the one who promoted the idea of individual potency most strongly. The other barrier, discussed recently by Gayon, was to treat heredity as a physical force.[59] Gayon provides evidence that this concept became a major obstacle to research on heredity by German breeders right up to the end of the century. It affected both Weckherlin and Settegast but perhaps was expressed most influentially in the writings of Herman E. von Nathusius (1809–79), educated in zoology.[60] Practising his animal breeding skills in the region of Magdeburg, he founded an Institute of Higher Learning in Agriculture at the University of Halle with an international reputation. In contrast to Settegast, who came from a family of modest origins and impoverished circumstances, Nathusius enjoyed the advantage of social privilege. His career was unvaryingly successful and his influence correspondingly dominant. Unfortunately he drove animal breeding even more firmly into a theoretical *cul de sac*. He is known for his belief in the constitution of an animal being fixed or variable according to the particular character considered. In his *Lectures on Animal Breeding and Knowledge of Races*, published in 1872, he distinguished two classes of inherited character. First there were 'natural characteristics of animals', that is to say 'foundations of forms', which were believed to be wholly inherited. Secondly there were features that he referred to as 'physiologically limited', believed to be inherited only as a potential (*Anlage*), to be transformed during embryonic development into a form dependent on how the animals were looked after (*die Haltung des Thieres*).

Variability in domestic animals, including in some situations a tendency to revert back to an ancestral type, seemed to deny the orderliness of heredity which he sought. In viewing the constitution of an animal as fixed or less fixed according to the characteristic being considered, Nathusius found himself dismissing the straightforward theme of racial purity and questioning the statement that 'like begets like'. As he wrote in 1872, 'The law of heredity is not yet recognised. The apple from the tree of knowledge has not yet fallen which, according to the story, brought Newton to his discovery of the law of gravitation.'[61]

Living in the shadow of these three major figures of German animal breeding was Ferdinand Stieber (1804–85) in Moravia. As a member of the Agriculture Society, he came into contact with Napp and later also with Mendel. Around 1842, discussions were taking place on a possible relationship between the age of a sheep and its capacity to transmit its production traits. It was Stieber's opinion that the creative force determining heredity, which he called *vis plastica*, acted mechanically and continuously throughout

---

[59] Gayon 1996, pp. 61–75     [60] Berge 1961     [61] Nathusius 1872, pp. 120–21

an animal's life. He saw the creative force as being organic in its action and although it could 'deviate accidentally from its limits', it was 'determined by certain laws of Nature'. He stressed the value of inbreeding in fixing the constancy of traits.[62]

Later, in his book *Cattle Breeding and its Profitability*, published in 1851 and dedicated to Abbot Napp, Stieber tried to explain the origin of the embryo as follows: 'In generation juices, the force of life is so concentrated that the smallest part can bring the new being into existence.' He stated 'that in the fertile mating the male seed inoculates the female organ, and thus an embryo is produced, and in its revival both producers take dynamic and material part'. Experience had led him to believe that the more similar the parents, the more equal their contribution to the progeny, enabling the race to advance step by step to perfection. He sought authority from the ancient classics, beginning with Aristotle who, according to him, had advised animal breeders to match fathers with their daughters. Reviewing the recent history of breeding success, he picked out Bakewell, Sinclair, Nestler and Weckherlin as key contributors. From Weckherlin, Stieber drew the concept of heredity as a force, and a conviction that constancy of inherent traits could be influenced by the environment. Considering the transmission of traits from parents to progeny, he could find no clear pattern. He concluded that 'Creation operates in their formation and transmission, according to certain laws, although in our view, often ambiguously.'[63] Although he gave credit to Nestler, he made no specific reference to the Moravian sheep breeders. He leaves the impression of being more under the influence of Weckherlin's writings on cattle than of the Moravian tradition of sheep breeding. His major interest in relation to breeding in the Moravian context rests in his acknowledged connection with Napp.

Breeders elsewhere were no closer to interpreting the basis of heredity. In the *Origin of Species*, published in 1859, Darwin had commented that 'The laws governing inheritance are for the most part unknown.'[64] He himself had scoured the breeding literature for information. By 1840 he had constructed an eight-page questionnaire to be issued to farmers and breeders.[65] Their answers helped him put together his unpublished *Sketch* in 1842 and *Essay* in 1844 in which he imagined some invisible 'effect' being produced during conception, by which the 'simple cell of ovule' (or the 'primary germinal vesicle') is 'impressed with some power which is wonderfully preserved during the production of infinitely numerous cells in the ever changing tissues'.[66] But his words reveal neither the nature of the invisible 'effect' ('power') nor the rules governing its transmission, which were the two questions that Napp

---

[62] Stieber 1842, pp. 41–4    [63] Stieber 1851, p. 123    [64] Darwin 1859, p. 13
[65] Darwin 1840; Clutton-Brock 1982
[66] Darwin 1842, pp. 76–80; Darwin 1844, pp. 226–7

had asked. The gemmule theory[67] was an ingenious, although vain, attempt to solve the problem. The truth lay in quite another direction, in the discovery of which (it is one of the purposes of this book to argue) Napp played a significant part. Which brings us almost to Mendel.

## Prelude to Mendel

A strong interest in the use of artificial fertilisation in plant breeding preceded Mendel's arrival in Brno by some years. The considerable technical difficulties were overcome by practice and improved instruments. Botanically minded members of the Agricultural Society might have been a little slower off the mark than the sheep breeders, but their enthusiasm for uncovering the rules of transmission and combination of traits was no less. Furthermore, since the membership of the two groups overlapped, there was scope for interaction. The teaching of animal and plant breeding as an academic discipline had been introduced to Brno through courses at the Brno Philosophical Institute, given by Professor Diebl in close collaboration with Abbot Napp. When the 1828 issue of *Mittheilungen* carried an article by Baron von Witten 'On Wheat Varieties', previously published in the Journal of the Prussian Horticultural Society, Diebl, as holder of the Chair in Agriculture, criticised the arrogance of the author.[68] He was eager to offer a comment on 'this very important subject', one which was already covered in his lectures. He noted that, although it was usual for natural scientists to see plant species as being characterised by constant traits passed down from the producer to the produced, such distinctions are not always absolute. Under changed conditions it could be observed that certain plant traits become 'invisible', only to reappear again when the plant is returned to its original environment. A possible explanation for such environmentally dependent traits, acceptable to the breeder, is that they arise through hybrid formation. Knowledge of plant physiology allowed Diebl to accept this explanation although he did not dismiss the possibility of some other force, yet to be determined. In 1834 an anonymous article appeared in *Mittheilungen*, probably written by Diebl again. Arising from the 'important discovery of sex in plants', it pointed to the possibility of pairing parental traits and reproducing the combinations as a new variety. At the end of the article there is a brief description of the technique of artificial pollination of beans and peas.[69]

The official textbook for teaching agriculture, written by Professor L. Trautmann of Vienna University, ran into four editions, the first of which appeared in 1814. Trautmann rejected inbreeding by consanguineous mating in animals and made no mention at all of artificial pollination in plant breeding. For his negative attitude, amounting to a total rejection of inbreeding

---

[67] Darwin 1868a, 1868b      [68] Witten 1829; Diebl 1829      [69] Anon. 1834

expressed even in the third edition of his book,[70] he was criticised by Nestler, who pointed out that careful farmers had always practised some degree of inbreeding.[71] For the growing number of students studying agriculture in Brno, some of whom came from outside Moravia, Diebl began in 1835 to issue a serially published textbook (1835–41) under the title *On the Science of Agriculture for Agriculturalists, Especially Those Devoting Themselves to the Study of this Science.* There were four volumes: *General Agriculture* (1835), *Plant Production* (1835), *Animal Husbandry* (1836) and *Economic Aspects* (1841). The whole series was issued in a second edition in 1844. In his introduction, Diebl stressed that the public teacher has a continuing duty to improve his teaching through reflection, observation and investigation, and to make full use of the latest advances in closely related ('sisterly') disciplines. In the second volume, on plant production, he included an account of improvement through artificial fertilisation, a clear departure from the Trautmann line. He also drew attention to abnormal plant forms appearing for the first time, which naturalists referred to as degeneracies. Diebl proposed that some of these forms might be of practical value in plant improvement. He paid special attention to variability in roots, tubers and fruits, which he believed capable of improvement through hybridisation.

Nestler in Olomouc and Diebl in Brno took a very similar line when lecturing on breeding. For both of them, physiology was the central issue. In 1839 when Nestler summarised the final big discussion on heredity in sheep breeding, he was sensitive to opposition about inbreeding. He recognised it as a 'spectre' that breeders in the past had overcome with difficulty.[72] Successful breeders had learned to master it in order to produce stock with an enhanced capacity to transmit valuable traits. In support of Nestler, Diebl expressed his faith in the power of science and he reminded the Society that Man had placed himself above the animal kingdom, and that natural sciences now offered new opportunities for improving breeding methods in plants and animals alike.[73] He expanded on this theme in a handbook on breeding for fruit, wine and timber production, published in 1844, 'for help in private instruction'. In a section on fruit trees, Diebl emphasised the creation of 'completely new noble varieties through artificial pollination'.[74] Two years later, Mendel was attending his lectures, on which he passed three examinations with top marks.

After 1848, a big influence on the teaching of plant physiology at Vienna University came from Professor F. Unger (1800–1870). One of his pupils, Netolička, teacher of physics and natural history at the Gymnasium in Brno, wrote a textbook called *Elements of Plant Physiology with Basic Principles of Anatomy, Chemistry and Geography of Plants for Schools and Self-instruction,*

---

[70] Trautmann 1823, p. 231 (1st edition 1814)    [71] Nestler 1829, sections 19 and 20
[72] Orel 1997    [73] Diebl 1839    [74] Diebl 1844

published in Brno in 1855. Here was another book that Mendel would have known. Significantly, the author attached high significance to the Italian microscopist Amici's observation of fertilisation in *Orchis morio*, in which he followed the growth of a single pollen tube into one embryo sac. As a member of the Natural Scientific Section of the Agricultural Society, Netolička was in contact with both Napp and Mendel.

While breeders speculated on what lay behind heredity but were unable to resolve its puzzling patterns, either in animals or plants, physiologists were investigating fertilisation and early embryonic development microscopically with increasingly refined instruments. Such 'famous physiologists', as Mendel referred to them in his *Pisum* paper, had come to accept that 'Propagation in phanerogams [flowering plants] is initiated by the union of one germinal and one pollen cell into one single cell.'[75]

The proposition that sexual reproduction in plants takes place by the fusion of the contents of two cells had become generally accepted by 1856, although not without considerable debate.[76] The Italian Amici had a major influence on the discovery, although for Mendel the category of 'famous physiologists' was bound to include also experts nearer at hand. First among them was Purkyně. Mendel publicly acknowledged Purkyně's fame as a physiologist when, as chairman of a monthly meeting of the Brno Natural History Society, he announced his death in 1869. When Mendel came to design his *Pisum* experiments and elaborate the theory of germ cells in the process of fertilisation and transmission of traits, research by his fellow countryman Purkyně could provide much of the theory on which he relied. As a result of his own experiments, he felt able to presume beyond doubt 'a complete union of elements from both fertilising cells' as the physiological basis of a new embryo. It was such inherited elements that provided the potential for each of the contrasting characters he had investigated.[77]

The connection between Mendel and his admired abbot, patron, protector, social superior and role model, whom he succeeded as head of the monastery, must also be considered. The manner in which Napp, in company with Nestler, accepted the sheep breeders' conclusion that the laws of heredity could be defined only when inherited traits were considered separately from one another has added significance in the light of events. Their conclusion, reached on the basis of practical experience, deviated from the mainstream tradition of natural history. As Gasking (1959) has pointed out, an interest in isolated characters rather than holistically conceived species represents a 'way of thinking more like a farmer's than a biologist's'. This way of thinking would lead the plant breeders of Brno to take a similarly

---

[75] Mendel 1866 (translation by Eva Sherwood in Stern and Sherwood 1966, p. 41)
[76] Olby 1985, pp. 33–7 and 159–63
[77] Mendel 1866 in Stern and Sherwood 1966, p. 41

character-based view, a tradition continued by Mendel in the design of his own experiments. This vital element of Mendelian thinking was remarked upon by R.D. Darbishire in his book *Breeding and the Mendelian Discovery*: 'The general realisation of the fact that the unit which has to be handled in experimental breeding is not the individual animal or plant, but the independently heritable unit-character, marks the beginning of a new epoch in the history of the practice of breeding.'[78] If we accept Darbishire's statement at its face value, the epoch had begun before Mendel.

In support of a link between Mendel and the Moravian breeding tradition (as interpreted by Nestler and Napp), it is worth noting some correspondences in the use of technical terminology, although it is difficult to know how strongly they should be weighted. The same word may be used in a rather different sense by individuals writing 30 years apart, particularly in a subject area which is expanding rapidly. Mendel, well versed in the exact science of physics, as well as possessing the latest information on plant physiology, was able to approach heredity with a fresh vision.[79] However, we feel bound to call attention to the term *Entwicklungsgeschichte*, because it was considered by both Mendel and Nestler to be particularly significant in relation to heredity. Mendel (1866) states in the introduction to his paper his wish to reveal a 'generally applicable law of the formation and development [*Entwicklung*] of hybrids', stressing that a numerical approach seems to be 'the only correct way of finally reaching the solution of a *question* whose significance for developmental history [*Entwicklungsgeschichte*] must not be underestimated'. What is this question? It depends on the meaning attached to the word '*Entwicklung*'. Is the stress to be placed on embryonic development or does it refer also, or principally, to development in a historical sense, as many authors, from Bateson onwards, have supposed when opting to translate *Entwicklungsgeschichte* as 'evolutionary history'? In the sense attached to it by Nestler (1837), it clearly referred to the developmental history of domestic varieties or races. He made his position clear by associating the term with *Vererbungsgeschichte* ('hereditary history').[79a]

We suspect that Mendel's use of the word '*Enwicklungsgeschichte*' had the same motivation as Nestler's, being related to organic transformations guided by Man rather than those taking place in Nature. Then, as now, the drive to discovery in genetics would frequently take place under the stimulus of economics, 'by co-operation with the men who control the breeding herds and the plant breeding nurseries'. The words are those of Willet Hughes, a founder member and guiding spirit of the American Breeders' Association. They occur in his opening remarks at the first meeting of the Association in 1903 in which he deplores the biologists' obsession with 'providing [evidence]

---

[78] Darbishire 1911, p. 244      [79] Orel 1996, pp. 160–80      [79a] Orel and Wood 2000b

for the ten thousandth time that Darwin's main contention is true' and urges his audience to concentrate their energies upon solving 'the great economics problem of evolution guided by man'.[80] It is easy to see how an obsession with natural selection could lead to a limited view of *Entwicklungsgeschichte* at the time, soon after 1900, when Mendel's paper was first translated into English.

Another term associated with Mendel, '*Anlage*', was in common use by breeders. R. André and Nestler used the word in a metaphysical sense, to mean a predisposition for the transmission of particular traits to progeny. The concept was at least as old as the Bakewell era in England. Napp progressed the idea by hinting at a distinction between inner and outer organisation, the one determining the other. Mendel explained *Anlagen* as the determinants of discrete traits inherited through the germ cells. He concluded that 'This development proceeds in accord with a constant law based on material composition and arrangement of the elements that attained a viable union in the cell.'[81] His explanation couched in a new scientific language was a very big advance on anything written before, although its roots were still just visible.

It is a matter of record that scant attention was given to heredity by the mainstream of biologists in Mendel's day. According to W. Bateson (1861–1926), this 'manifestly arose from the curious delusion that the laws of heredity were untraceable'.[82] But why should such a delusion persist when others besides Mendel were making similar crosses while failing to discover the laws? L.C. Dunn has discussed the matter:[83]

> The question why Mendel's predecessors who had also observed the off-spring of hybrids did not discover the rules cannot, of course, receive any definite answer. An essential difference between Mendel and them was that he was looking for such rules and they were not.

An expansion of this idea was proposed by I. and L. Sandler when they reached the conclusion that Mendel had explained heredity 'in purely genetic terms and produced a correct and amazingly complete answer but to an as yet unformulated question'.[84] But was the question unformulated? The sequence of events we have described indicates that heredity was a central scientific issue in the Sheep Breeders' Society before Mendel arrived in Brno.

What then was the sheep breeders' legacy to Mendel? Was it their appreci-ation that precisely defined individual traits could be considered separately for the purpose of selective breeding? Certainly they had pioneered the idea in farm animals, and there is good contemporary evidence that some of the sheep breeders active in Moravia in the time of Geisslern were ahead of the

---

[80] Paul and Kimmelman 1988    [81] Stern and Sherwood 1966, p. 42
[82] Bateson 1906, p. 97    [83] Dunn 1965, p. 32    [84] Sandler and Sandler 1985

Moravian plant breeders in this respect and influenced them in the same direction. Sheep breeders had the advantage of powerful economic forces behind them. Even so, we cannot be sure that some plant breeders, working less in the public eye but known to Mendel, had not preceded them. A degree of uncertainty is attached to all aspects of empirical discovery pertaining to heredity that led to the burst of breeding achievements in the first half of the nineteenth century.

Of more certain significance is the *atmosphere of enquiry about heredity* created by the sheep breeders. It had produced a strong impression on several individuals upon whose advice and guidance Mendel was destined to rely. No group of breeders had demonstrated more publicly the reality of heredity and no group had raised more questions about it. And Brno had been the very centre of these questions in a period immediately prior to Mendel's arrival in the monastery. At the time he approached his experiments, he must surely have benefited from all this. His lectures from Diebl would have provided an ideal introduction. The influence of Napp—who recognised his potential and provided the education for his talent to flower and, right up to the end, invested precious monastic funds in his experiments—would have been inescapable. Nothing could have been more natural for Napp in the atmosphere of enquiry about heredity with which his office had been surrounded than to encourage the young monk's plant breeding experiments. After all, was it not Napp who had the vision to ask the question 'What is inherited and how?' It was only after 1900 that biologists in growing numbers finally disclosed that the apple from the tree of knowledge of heredity sought by Napp and many others, and missed by Nathusius in 1872, had already fallen from the hand of Napp's protegé in 1865.

# Overview

'It seems to me remarkable that for the past fifty years the old conflict on inbreeding has been revised approximately once every decade.' With these words, written in 1839, J.K. Nestler, Professor of Natural History and Agriculture at the Moravian University of Olomouc, drew attention to a major controversy in animal improvement that would not go away. On the one hand, proponents of inbreeding pointed to its unique potential for producing homogeneous 'noble' lines with an enhanced capacity to transmit valuable characteristics to progeny. On the other hand, there were those who, with good cause in many cases, feared its unpredictability and the danger of disastrous consequences arising from it, particularly in terms of fertility. The controversy was heightened by religious considerations about incestuous matings transgressing God-given law.

In this book we have considered the controversy in the context of sheep breeding in different European countries and beyond. Experimentation with inbreeding gave skilful breeders a lead over less adventurous colleagues, although others, lacking the necessary degree of skill, might find their stock degenerating before their eyes. Antipathy to inbreeding was typically associated with a belief that an animal's health and ability to reproduce its own type were influenced above all by the conditions of life in which it grew up and the extent to which it continued to exist in harmony with these conditions. Instances in which sheep were transported from one country to another produced opportunity for the environmental theory to be tested. Limitations upon its influences were exposed in Sweden as early as 1723 with the successful naturalisation of Merino sheep transported out of Spain by Jonas Alströmer and his followers.

A central figure in our story has been Robert Bakewell. As a leading exponent of inbreeding in eighteenth century England, Bakewell attracted a wide and enthusiastic following, though he also aroused opposition from traditionalists among some of his own countrymen. Influential French opinion, led by Buffon and Daubenton, stressed the greater value of crosses between races originating from different environments, as a means of breeding improvement. The arguments became polarised to such a degree that both inbreeding and outbreeding were often practised or opposed, more intensively than could possibly be beneficial.

Observing the controversy, sheep owners outside these two countries tended to adopt less extreme breeding schemes, willingly embracing influences from various directions. Their actions were not based only on technical considerations. They also reflected aspects of economics, culture, politics and social organisation. Several German and Austrian breeders applied Bakewell-style techniques to improve wool quality in imported Merino sheep and for creating new breeds out of crosses between Merinos and local varieties. In the Austrian province of Moravia, which enjoyed particular prosperity from its rapidly expanding textile industry centred around Brno, a call for reliable breeding rules led in 1814 to the formation of a Sheep Breeders' Society with a uniquely broad membership. In regular meetings it brought together not only practitioners of the art but also landowners, academics, journalists and representatives of the textile industry. Examination of breeding records and wool samples led to the publication of articles and books written at an increasingly scientific level. Attempts to imitate the Brno Society in other countries were transitory in their influence by comparison. This lively association, active for a quarter of a century in the city that would later see the secret of heredity revealed by Mendel, has been central to our story. We find that the search for rules of heredity by sheep breeders prior to Mendel's time had an impact upon progress of thought about heredity in a wide context, echoed in the actions of commercial plant breeders.

The idea of plant breeders copying the techniques of animal breeding can be traced to an Englishman very familiar with Bakewell's achievement. Thomas Andrew Knight had chosen to concentrate much of his abundant energies away from sheep and cattle towards the production of new plant varieties, particular fruit trees, by artificial pollination. The close analogy he drew between animal and plant improvement, which he defined in an article published in 1799, created a deep impression on Moravian plant breeders who in 1816 formed a Pomological and Oenological Society to put the new ideas into practice.

By 1818, growth in knowledge of sheep breeding in Brno had reached a level that encouraged members of the Society to attempt a generalisation of breeding principles in the form of 'genetic laws'. The next step, taken in the 1820s, was to include 'scientific animal and plant inbreeding' in the teaching of natural history and agricultural science at the Moravian University of Olomouc. Professor Nestler, who was responsible for these lectures, gave the Society access to their content in published form. The debates that followed in the Sheep Breeders' Society led increasingly in the 1830s to discussions on heredity, i.e. transmission of traits between generations, as a separate issue from the enigmatic process of generation itself. Co-operating with Nestler as an active seeker after the truth about heredity in such debates was Prelate C.F. Napp, Abbot of the Augustinian monastery in Brno.

When, in 1843, Napp recruited Mendel into the monastery and supported his studies in natural sciences and, later, his experiments in plant breeding, he was acting in character. Sixteen years earlier he had been elected President of the Pomological and Oenological Society, since when he had worked closely with Franz Diebl (1770–1859), Professor of Natural History and Agriculture at the Brno Philosophical Institute. Together they were encouraging members to apply artificial pollination in breeding new varieties. By 1846 Mendel was attending Diebl's lectures, through which he became acquainted with the latest theory and practice of breeding. Here we may note Diebl's stress upon the value of controlled fertilisation, his recognition of peas and beans as ideal subjects for experimentations, and his conviction of Man's power through science to control the natural world in all its aspects. Fresh from absorbing the latest development in physiology, he wrote of how Man has 'raised himself from his original animal position in nature and placed himself over the animal kingdom'[1]. Deeply impressed with these lectures, Mendel gained top marks in the examinations that followed them.

For his major published study, Mendel chose a self-fertilising plant, one that could not avoid being inbred and which patently suffered no ill effects from it. It permitted him easily to isolate pure-breeding varieties to be crossed with absolute confidence. He was following the advice of his teacher, Diebl. When one looks for Mendel's precursors, it is natural to focus on earlier plant hybridists, famous in the wider world outside Moravia, five of whom Mendel himself acknowledged by name in his 1866 paper. We hope that the evidence of this book will encourage scholars to look more widely for the shoulders on which Mendel stood to extend his view of heredity. In his local background he benefited from a rich tradition of breeding, with spirited and adventurous pioneers striving to elaborate breeding methods for creating more productive animals and plants on similar principles. Often the greatest energies had been directed towards sheep improvement because of the extreme economic significance of wool. Readers must judge for themselves how far this aspect of genetic prehistory adds to knowledge of how it was that Mendel set about his own experiments and achieved the breakthrough that began the era of genetics.

---

[1] Orel 1996, p. 28

# References

## Abbreviations and note

| | |
|---|---|
| *CBA* | *Communications to the Board of Agriculture* |
| *Folia Mendeliana* | published yearly by the Moravian Museum in Brno since 1966. Issues 1–9 appeared as special volumes. Issues 10–20 were published as a supplement to *Acta Musei Moraviae*, sc. nat. with the pagination of these volumes. The part *Folia Mendeliana* were also reprinted separately. *Folia Mendeliana* Vol. 21 and further volumes are being published again as special volumes with their own pagination. |
| *Hesperus* | *Belehrung und Unterhaltung für Bewohner des österreichisches States*—Hesperus, Prag. |
| *Mittheilungen* | *Mittheilungen der k.k. Mährisch-Schlesischen Gesellschaft zur Beförderung des Ackerbaues, der Natur- und Landeskunde in Brünn* |
| *ONV* | *Oekonomische Neuigkeiten und Verhandlungen*, Prag |
| *PTB* | *Patriotisches Tagesblatt oder öffentliches Correspondenz- und Anzeiger-Blatt für sämtliche Bewohner aller kais. köng. Erbländer über wichtige, interessierende, lehrreiche oder vergnügende Gegenstände zur Beförderung des Patriotismus, Brünn* |

'Agricola' [James Anderson?] (1793). Northumberland sheep. *Annals of Agriculture* (London), **19**, 536–41.

Aldcroft, D.H. (1983) Aspects of eighteenth century travelling conditions. In *Der curieuese Passagier, deutsche Englandsreisende des Achzehnten Jahrhunderts* (ed. Spiekermann, M.L.), pp. 27–45. Carl Winter, Heidelberg.

Alström(er), C. (1770). *Tal om den Fin-Ulliga Får-Afveln*. Stockholm.

Alström(er), C. (1772). Discours sur la race des brébis à laine fine. Traduid de Suedois, par M. Albin, Libraire à Stockholm. *Journal de physique*, **1**, 441–56, 534–59.

Alström(er), C. (1773). *Essai sur la race des brébis á laine fine.* Paris. [Translation from original Swedish.]

Alström(er), J. (1727). *Den swänska fåra-herdens trogne wäg-wisare til en god fåra skiötzel, jemte et bihang om potatoes eller jord-päron, såmt almwild-apal- och eketräns planterande.* Stockholm. Translated into French as *Guide fidèle du berger Suédois,* 1727.

Alström(er), J. (1733). *Fåra-herdans hemliga konster uptäkte, . . . hwar til och är bifogat något angående potatoes planterande. . .* Skara. An abstract was published in France under the title *Traité d'élevage du mouton en Suède.*

Anderson, J. (1796). *Letters and Papers on Agriculture, Planting, etc. Selected from the Correspondence of the Bath and West of England Society for the Encouragement of Agriculture, Manufactures, Arts and Commerce* (4th edition), Vol. 8. R. Cruttwell, Bath.

Anderson, J. (1799). An inquiry into the nature of that department of natural history which is called varieties. *Recreations in Natural History,* **1**, 49–100.

André, C.C. (1795). *Der Zoologe, oder Compendiöse Bibliothek des wissenswürdigen aus der Thiergeschichte und allgemeinen Naturkunde.* Eisenach, Halle.

André, C.C. (1802). Herr Hofrath von Geissler zu Hoschtitz, einer unser ersten Landwirte in Mähren. *PTB,* 919–21.

André, C.C. (1804a). Das Gut Hoschtitz in Mähren. *PTB,* 578–81.

André, C.C. (1804b). Cited in 'K in Mähren' [K. Köller] 1811 and republished by d'Elvert 1870, ii, pp. 145–52.

André, C.C. (1804c). *Anleitung zum Studium der Mineralogie für Anfänger.* Vienna.

André, C.C. (1812). Anerbieten, Gutbesitzern auf dem kürzesten und sichersten Wege zur höchsten Veredlung ihrer Schafherden behülflich zu seyn. *ONV,* 181–3.

André, C.C. (1815). Rede bey der ersten Eröffnung der vereinigten Gesellschaft des Ackerbaues, der Natur- und Landeskunde. In *Erster Schematismus der k.k. Mähr. schles. Gesellschaft zur Beförderung des Ackerbaues, der Natur- und Landeskunde* (ed. C.C. André), pp. 93–113. Brno.

André, C.C. (1818). Terminologie für Woll-industrie. *ONV,* 302–43.

André, C.C. (1819). *The Farmer's Magazine* 1818, Nr 75, Auszug 1818. *ONV,* 469–71.

André, C.C. (1820a). Anfragen. Bienenzucht. *ONV,* 182–3.

André, C.C. (1820b). Oekonomische Gesellschaft im Königreiche Sachsen. Commented by C.C. André. *ONV,* 257–63.

André, C.C. (1825). *Neueste Ansichten über Wolle und Schafzucht.* Prague. [German translation from the cited French book by de Jotemps *et al.* 1824 with commentary from André (see de Jotemps *et al.* 1825a)].

André, R. (1812). Review of the book by A. Thaer: *Handbuch der feinwolligen Schafzucht,* Berlin, 1811. *ONV,* 401–2.

André, R. (1813). Review of the German translation of the French book by Tessier from 1811: *Ueber die Schafzucht, insbesondere der Merinos,* Berlin. *ONV,* 281–3, 289–95, 297–300.

André, R. (1815a). Review of the book by C.A. Hubert: *Die Wartung, Zucht und Pflege der Schafzucht, ihre Benutzung und Veredlung,* Berlin. *ONV,* 137–41.

André, R. (1815b). Review of the book by B. Petri: *Das ganze der Schafzucht in Hinsicht auf unser deutsches Klima*, Wien. *ONV*, 297–8.

André, R. (1816). *Anleitung zur Veredelung des Schafviehes. Nach Grundsätzen, die sich auf Natur und Erfahrung stützen*. J.G. Calvé, Prague.

André, R. (1818). *Kurzgefaster Unterricht über die Wartung des Schafviehes, für Schafmeister und ihre Knechte*. Brünn.

André, R. (1819). Meine Ansichten und Bemerkungen über organische Schwäche, besonders bei feinwollignen Schafen; veranlasst durch den Aufsatz des Herrn Grafen Emmerich von Festetics im Jännerheft 1819. *ONV*, 161–2.

André, R. (1823). Bemerkungen über die Elektoral-Schafrace, besonders in Hinsicht auf die mährische Schafzucht. *Mittheilungen*, 74–77.

André, R. and Elsner, J.G. (1826). *Anleitung zur Veredlung des Schafviehes. Nach Grundsätzen, die sich auf Natur und Erfahrung stützen*. Revised edition, Prague.

Anon. (1781) [Report of Bakewell's farm] *Leicester Journal*, 3 February 1781.

Anon. (1795) Robert Bakewell. *Gentleman's Magazine* **65** (pt 2), 969–71.

Anon. [D. Le] (1800). Schafzucht. *PTB*, 47.

Anon. [J.L.] (1800). Robert Bakewell. In *The Annual Necrology for 1797–1798 Including Various Articles of Neglected Biography*, pp. 199–209. R. Phillips, London. [J.L. stands for John Lawrence].

Anon. [probably C.C. André] (1809). Ueber die Veredlung der Hausthiere. *Hesperus*, 94–101.

Anon. [prob. C.C. André] (1811a). Patriotischer Vorschlag zu einem Schafmarkt an unsere Güterbesitzer. *ONV*, 114–15.

Anon. (1811b). Ueber die Schafzucht in Kärnthen. *ONV*, 431–6.

Anon. [M. Köller] (1811c). Müssen immer wieder ächte Merinos nachgekauft werden, um die Feinheit der Wolle zu erhalten? *ONV*, 25–8.

Anon. (1811d) *On the Nature and Origin of the Merino Breed of Sheep*. J. Harding, London. [British Library, Banks Library B514 (14)].

Anon (1814). A memoir of George Culley. *Farmer's Magazine* **114**, 271–5.

Anon. [probably C.C. André] (1814). Bericht aus der ersten Versammlung des Vereines den 16. Mai zu Brünn. *ONV*, 257–61.

Anon. [prob. C.C. André]. (1815). Bericht über die zweite Zusammenkunft des Vereins den 17. May 1815 zu Brünn. *ONV*, 353–58, 367–8, 375–8.

Anon. (1816a). Anempfehlung von Rudolph André's Anleitung zur Veredlung des Schafviehes. *ONV*, 296.

Anon. (1816b). Künstliche Befruchtung des Obstes. *ONV*, 447.

Anon. (1818a). Wirksamkeit der Ackerbaugesellschaft in Brünn. Schreiben an den Herrn Rath André in Brünn. *ONV*, 297–304, 305–10.

Anon. (1818b). Ein neuer Schafzüchter-Congress in Ungarn. *ONV*, 329–30.

Anon. (1823). Mährischer Schafzüchter-Verein im Jahre 1822. *Mittheilungen*, 44–6.

Anon. (1826). Die Gartenbau-Gesellschaft in London. *ONV*, 336–7.

Anon. [J.K. Nestler]. (1827). Ueber die Hörner, über die Hornbildung bei den gehörnten Rassen aus dem Rinder-, Schaf- und Ziegen- Geschlechte, und über ungehörnte Rassen. *Mittheilungen*, 346–50.

Anon. (1828) Eigenes über den gegenwärtigen Zustand eigener der edelsten englischen Rindviehrassen, nebst Bemerkungen über Inzucht. *Mittheilungen*, 340–43.

Anon. [F. Diebl] (1834). Wie muss man die Getreidearten veredeln. *Mittheilungen*, 333–4.

Anon. ['W.T.T.'] (1875). Recent British Agriculture *Encyclopaedia Britannica*, 9th edn, Vol. 1., pp. 303–306, Edinburgh.

Anon (1878) Biographical entry on J.H. Finke in *Allgemeine Deutsche Biographie*, **18–19**, Verlag von Duncker und Humblot, Leipzig.

Atkinson, J. (1828). Remarks on the Saxon sheep farming: the result of observations, made during a tour through that country in the summer of 1826. *Australian Quarterly Journal of Theology, Literature and Science*, **1**, 96–108.

Baert, M. (1800). *Tableau de la Grande Bretagne, de l'Irlande, et des possessions anglaises*. Paris.

Bailey, J. and Culley, G. (1805). *General View of the Agriculture of the County of Northumberland*. Drawn up for consideration by the Board of Agriculture and Internal Improvement, London. Nicol, London.

Baines, E. (1858). *An Account of the Woollen Manufacture of England*. Paper read before the British Association for the Advancement of Science, Leeds 1858. Published by Thomas Baines in *Yorkshire Past and Present 1875*. New edition with introduction by K.G. Ponting, Newton Abbot, 1970.

Bajema, C.J. (1982). *Artificial Selection and the Development of Evolutionary Theory*. Hutchinson Ross Co., Stroudsburg, PA.

Bakewell, R. (1808). *Observations on the Influence of Soil and Climate on Wool*. London.

Balcárek, P. (1977). Die ökonomische Bedeutung der Veredlung des Schafviehs in Mähren in der ersten Hälfte des 19. Jahrhunderts. *Folia Mendeliana*, **12**, 223–8.

Banks, J. (1787). *My pamphlet*, reprinted by Carter (1979), p. 528.

Banks, J. (undated). *Anecdotes of the Revnd. John Smith L.L.B.*, reprinted by Carter (1979), pp. 523–24.

Banks, J. (1800). A project for extending the breed of fine wooled Spanish sheep, now in the possession of his Majesty, into all parts of Great Britain, where the growth of fine clothing wools is found to be profitable. *Philosophical Magazine*, **7** (First Series), 350–55. (Reproduced as Appendix 7 in Carter 1979, pp. 533–5.)

Banks, J. (1808). Some circumstances relative to Merino sheep chiefly collected from a Spanish shepherd. . . etc. *CBA*, **6**, 269–86. (Reprinted in Carter 1979, pp. 548–53.)

Barry, M. (1840). Researches in embryology. *Philosophical Transactions of the Royal Society of London*, **130**, 529–93.

Bartenstein, E. (1818). Bericht des Herrn Präfes Baron Bartensteins an die k.k. Ackerbaugesellschaft. *ONV, ausserordentliche Beilage*, 81–4.

Bartenstein, E. (1837). Ausserungen von Bartenstein über das von dem Herrn Professor Nestler bei dem Schaf-Züchter-Verein im Jahre 1836 aufgestellte und debatirte Thema der Vererbungsfähigkeit edler Stammthiere. *Mittheilungen*, 9–10.

Bartenstein, E., Teindl, F., Hirsch, J. and Lauer, C. (1837). Protokol über die Verhandlungen bei der Schafzüchter-Versammlung in Brünn in 1837. *Mittheilungen*, 201–5, 225–31, 233–8.

Bateson, W. (1906). The progress of genetic research. In *Report of the Third International Conference on Genetics (1906)* (ed. W. Wilks). Royal Horticultural Society, London, pp. 90–97.

Baumann, C. (1783). *Entdeckte Geheimnisse der Land- und Hauswirtschaft, für jedes Land und zum Besten aller Innwohner Deutschlands.* J. Gerold, Vienna.

Baumann, C. (1785). *Nothwendige Anstalten zur Vermehrung, Verbesserung und Verschönerung der Pferd- Rindvieh- Schaf- Geiss- und anderer Thierzuchten ohne Ausartung.* Frankfurt-Leipzig.

Baumann, C. (1803). *Der Kern und das Wesentliche entdeckter Geheimnisse der Land- und Hauswirtschaft, zur bequemern Uebersicht und zum ausgebreitetern Gebrauch, mit der neunsten bewährten Versuchen und Nahrungsquellen, Liebhabern zum Handbuch gewidmet.* F.K. Siedler, Brünn.

Baumgart, H. (1957). Origins and distribution of sheep breeds in Germany. Part 2, the German Merino. *Wool Knowledge*, **4**, 9–13.

Bayer, D. (1966) *O, gib mir Brot. Die Hungerjahre 1816–1817 in Württemberg und Baden.* Ulm.

Bell, A.E. (1977) Heritability in retrospect. *Journal of Heredity*, **68**, 297–300.

'Benda', M. (1800). Robert Bakewell. *The Annual Necrology for 1797–8, including also various articles of neglected biography.* R. Phillips, London, pp. 199–209.

Berchtold, L. (1789a). *An Essay to Direct and Extend the Inquiries of Patriotic Travellers to which is Annexed a List of English and Foreign Works.* Two volumes. London.

Berchtold, L. (1789b) *Essai pour diriger et étendu les recherches tous les états, chez les différentes nations et les différentes gouvernement.* French translation of 1789 by C.P. de Lasteyrie. Du Pont, Paris.

Berchtold, Count de (1790). Some particulars concerning the present state of Spain. *Annals of Agriculture*, **13**, 81–98.

Berge, S. (1961). The historical development of animal breeding. *Max Planck Institut für Thierernährung*, 109–27. (Cited by Gayon 1996.)

Berry, H. (1826). Whether the breed of live stock connected with agriculture be susceptible of the greatest improvement, from the qualities conspicuous in the male or from those conspicuous in the female parent. *British Farmer's Magazine*, **1**, 28–36.

Berry, H. (1829). Prize essay on 'Whether the breed of live stock connected with agriculture be susceptible of the greatest improvement, from the qualities conspicuous in the male or from those conspicuous in the female parent'. *Transactions of the Highland Society of Scotland,* **7** (New Series 1), 39–42. (Shorter, rewritten version of Berry 1826.)

Betham-Edwards, M. (1898) (Editor). *The Autobiography of Arthur Young with Selections from his Correspondence.* Smith, Elder & Co, London.

B-F-S (1843). Alexandre Henri Tessier. *Biographie Universelle*, **41**, 190–92. Paris.

Billingsley, J. (1795). *Letters and Papers on Agriculture, Planting, etc. Selected from the Correspondence of the Bath and West of England Society for the Encouragement of Agriculture, Manufacture, Arts and Commerce* (4th edition), **7**, 352–60. R. Cruttwell, Bath.

Bischoff, J. (1828). *The Wool Question Considered*. J. Richardson, London; J Baines, Leeds.

Bischoff, J. (1842). *A Comprehensive History of the Woollen and Worsted Manufactures*. Two volumes. Smith Elder & Co., Leeds. (Reprinted 1968 by Frank Cass & Co. Ltd, London.)

Bishko, C.J. (1982). Sesenta anos despues: La Mesta de Julius Klein a la investigacion subsiguente. *Historia Instituciones Documentos*, Vol. 8. University of Seville.

Blacklock, A (1838) *A Treatise on Sheep*. London

Blum, J. (1978). *The End of the Old Order in Rural Europe*. Princeton University Press, Princeton, NJ.

Blumenbach, J.F. (1781). *Über den Bildungstrieb und das Zeugungsgeschäft*. Göttingen. (2nd edn, 1789; 3rd edn, 1791.)

Bökönyi, S. (1974). *History of Domestic Mammals in Central and Eastern Europe*. Akadémiai Kiado, Budapest.

Bökönyi, S. (1976) Development of early stock rearing in the near east. *Nature*, **264**, 19–23.

Bonnet, C. (1762). *Considérations sur les corps organisés*. Two volumes. M.-M. Rey, Amsterdam.

Bonwick, J. (1894). *Romance of the Wool Trade*. Sampson Low, Marston & Co., London.

Boswell, J. (1829 (1825)). Prize essay on 'Whether the breed of live stock connected with agriculture be susceptible of the greatest improvement, from the qualities conspicuous in the male or from those conspicuous in the female parent?' *Transactions of the Highland Society of Scotland*, **7** (New Series, Vol. 1), 17–39.

Bourde, A.J. (1953). *The Influence of England on the French Agronomes 1750–1789*. Cambridge University Press, London.

Bourde, A.J. (1967). *Agronomie et agronomes en France au XVIIIe siècle*. Université de Paris, SEVPEN, Paris.

Bowden, P.J. (1956). Wool supply and the woolen industry. *Economic History Review* (Series 2), **9**, 44–8.

Bowden, P.J. (1962). *The Wool Trade in Tudor and Stuart England*. MacMillan, London.

Bowler, P.J. (1975) The changing meaning of 'evolution'. *Journal of the History of Ideas*, **36**, 95–14.

Boys, J. and Ellman, J. (1793). Agricultural minutes, taken during a ride (across 15 counties) in 1792 (July 1 to July 26, 679 miles). *Annals of Agriculture*, **21**, 72–146.

Bradley, R. (1726). *A General Treatise of Husbandry and Gardening*. T. Woodward, London.

Braun, H.J. (1971). Some notes on the Germanic associations of the Society of Arts in the eighteenth century; (ii) relations with German societies; Leipzig (1764), Celle (1764), Karlsruhe (1765), Hamburg (1765). *Journal of the Royal Society of Arts*, **119**, 558–62.

Bright, R. (1818a). *Travels from Vienna through Lower Hungary.* Edinburgh.

Bright, R. (1818b). Notions of the conditions of the peasantry and of the present state of husbandry in lower Hungary. Extract from 'Travels from Vienna through Lower Hungary'. *Farmer's Magazine of Edinburgh*, 8 August 1818, 291–8.

Brightly, J. (1805). On the Spanish cross and fine wool improvement in reply to Pastorius. *Agricultural Magazine*, **12**, 314–17.

Bronowski, J. (1973). *The Ascent of Man.* British Broadcasting Corporation, London.

Brown, R.D. (1987). *Lucretius on Love and Sex.* E.J. Brill, Leiden.

de Buffon, L.G. (1795). *Allgemeine Naturgeschichte* I–IV. Troppau. Translation from French (*Histoire naturelle*), Paris.

Burgess, W.H. (1908). *History of the Loughborough Unitarian Congregation.* Loughborough.

Burns, R.H. and Moody, E.L. (1935). The trek of the golden fleece. *Journal of Heredity*, **26**, 433–43, 505–10.

Carlier, Abbé C. (1762) *Considérations sur les moyens de rétablir en France les bonnes espèces de bêtes à laine.* Paris

Carlier, Abbé C. (1770). *Traité des bêtes à laine ou méthode d'élever et de gouverner les troupeaux aux champs, et à la bergerie.* Paris.

Carter, F.W. (1988) Cracow's transit textile trade. *Textile History*, **19**, 23–60.

Carter, H.B. (1958) The farmer, the gene and the fabric. *Bradford Textile Society Journal* 1958–1959, pp. 22–33. (Includes a review of experimental data published earlier by the author and his collaborators.)

Carter, H.B. (1964). *His Majesty's Spanish flock: Sir Joseph Banks and the Merinos of George III of England.* Angus and Robertson, London.

Carter, H.B. (1969). The historical geography of the fine wooled sheep. *Textile Institute and Industry: the Proceedings of the Textile Institute*, **7**, 15–18, 45–48.

Carter, H.B. (ed.) (1979). *The Sheep and Wool Correspondence of Sir Joseph Banks, 1781–1820.* Library Council of New South Wales and British Museum (Natural History), London.

Carter, H.B. (1988). *Sir Joseph Banks, 1743–1820.* British Museum (Natural History), London.

Cavendish, W. (Duke of Newcastle) (1667). *A New Method and Extraordinary Invention to Dress Horses. . .* London.

Chambers, J.D. and Mingay, G.E. (1966). *The Agricultural Revolution 1750–1780.* B.T. Batsford Ltd, London.

Churchill, F.B. (1987). The life sciences in Germany, from hereditary theory to *Vererbung*. The transmission problem, 1850–1915. *ISIS*, **78**, 337–64.

Clerke, A.M. (1886). Brühl, John Maurice, Count of (1733–1809). In *Dictionary of National Biography*, **21**, 1269–71. Oxford University Press, Oxford.

Clarke, E. (1897). John Smith (fl. 1747). In *Dictionary of National Biography*, **3**, 47. Oxford University Press, Oxford.

Clarke, E. (1900). William Youatt (1776–1847). In *Dictionary of National Biography*, **18**, 486–87. Oxford University Press, Oxford.

Cline, H. (1805) On the form of animals. *CBA*, **4**, 440–46.

Clutton-Brock, J. (1982). Darwin and the domestication of animals. *Biologist*, **29**, 72–6.

Cook, H.W. (1942). Robert Bakewell of Dishley, the Pioneer of English Stock Breeding. Unpublished manuscript. Loughborough Public Library.

Cooper, J. (1799) Change of seed not necessary to prevent degeneracy, naturalisation of plants; important caution to secure permanent food quality of plants. *Memoir of the Philadelphia Society for Promoting Agriculture*, **1** (1808 [publication delayed]), 11–18.

Cooper, T. (1893). William Marshall (1745–1818). In *Dictionary of National Biography*, **12**, 1136–37. Oxford University Press, Oxford.

Cox, E.W. (1936) *The Evolution of the Australian Merino*. Angus and Robertson, Sydney.

Criste, O. (1905). *Feldmarschall Johannes Fürst Liechtenstein*. Vienna.

Cross, A.G. (1980). *By the Banks of the Thames. Russians in Eighteenth Century Britain*. Oriental Research Partners, Newtonville, MA.

Crowe, J. (1793). Experiments on different breeds of sheep. *Annals of Agriculture*, **20**, 71–5.

Crutchley, J. (1794). *General View of the Agriculture of the County of Rutland*. Drawn up for consideration by the Board of Agriculture and Internal Improvement. Nicol, London.

Culley, G. (1771). Journal of a tour by George Culley from Denton to Leicestershire, Midlands, Cambridgeshire etc. in the company of William Charge, 20 June–25 July 1771. Northumberland Record Office catalogue no. ZCU1.

Culley, G. (1784). George Culley, Dishley, letter to his brother Matthew, 18 November 1784. Northumberland Record Office catalogue no. ZCU9.

Culley, G. (1786). *Observations on Livestock; Containing Hints for Chusing and Improving the Best Breeds of the Most Useful Kinds of Domestic Animals*. (2nd edn, 1794). G.G. & J. Robinson, London.

Culley, G. (1804). *Culley George über die Auswahl und Veredlung der vorzüglichen Hausthiere*. (German translation of 2nd edn, 1794). F. Maurer, Berlin.

Culley, G. (1807). *Observations on Livestock; Containing Hints for Chusing and Improving the Best Breeds of the Most Useful Kinds of Domestic Animals*, 4th edn (including appendix). G. Wilkie, J. Robinson, J. Walker and G. Robinson, London.

Cuvier, G.L. (1817). *Le règne animal distribué d'après son organisation, pour servir de base à l'histoire naturelle des animaux et d'introduction à l'anatomie comparée*. Paris. [see also Smith, C.H. 1827].

Czihak, G. (1984). *Johann Gregor Mendel (1822–1884). Dokumentierte Biographie und Katalog zur Gedächtnissausstellung anlässlich des hundertjährigen Todestages mit Facsimile seines Hauptwerkes: 'Versuche über Pflanzenhybriden'*. University of Salzburg.

Czihak, G. and Sladek, P. (1991/2). Die Persönlichkeit des Abtes Cyrill Franz Napp (1792–1867) und die innere Situation des Klosters zu Beginn der Versuche Gregor Mendels. *Folia Mendeliana*, **26**, 29–34.

Darbishire, A.D. (1911). *Breeding and the Mendelian Discovery.* Cassell & Co. Ltd, London.

Darby, H.C. (ed) (1936). *An Historical Geography of England before AD 1800.* Cambridge University Press, London.

Darwin, C. (1840). *Questions about the Breeding of Animals.* Society for the Bibliography of Natural History. Sherborn Foundation. (Facsimile 1968, no. 3.)

Darwin, C. (1842). Sketch of 1842. In *Evolution by Natural Selection* (ed. G. de Beer), pp. 41–83 (1958). Cambridge University Press, Cambridge.

Darwin, C. (1844). Essay of 1844. In *Evolution by Natural Selection* (ed. G. de Beer), pp. 91–254 (1958). Cambridge University Press, Cambridge.

Darwin, C. (1859). *The Origin of Species by Means of Natural Selection.* (6th edn, 1872). Murray, London.

Darwin, C. (1868a). *The Variation of Animals and Plants under Domestication.* Murray, London.

Darwin, C. (1868b) *Das Variiren der Tiere und Pflanzen im Zustande der Domestication.* Stuttgart

Darwin, E. (1794–6). *Zoonomia.* J. Johnson, London.

Darwin, F. (1892) *Charles Darwin.* John Murray, London.

Daubenton, L.J.M. (1782). *Instructions pour les bergers et pour les propriétaires des troupeaux.* Paris, Dijon.

Daubenton, L.J.M (1784). *Katechismus der Schafzucht zum Unterricht für Schäfer und Schäfereiherrn.* (Translated by C. A. Wichmann, with added notes.) Leipzig.

Davenport, E. (1907). *Principles of Breeding.* Ginn & Co., Boston, MA.

Davies, R.T. (1958). *The Golden Century of Spain.* MacMillan, New York.

Dawson, W.R. (ed.) (1958). *The Banks Letters.* The British Museum, London.

Defoe, D. (1724). *A Tour through England and Wales.* Everyman edition, 1928.

Diebl, F. (1829). Bemerkungen über die von Freiherrn v. Witten hinsichtlich der verschiedenen Weizenarten geäusserten Ansichten. *Mittheilungen*, 177–9.

Diebl, F. (1835–41). *Abhandlungen aus der Landwirtschaftskunde für Landwirthe, besonders aber für diejenigen, welche sich der Erlernung dieser Wissenschaft widmen.* Four volumes. (2nd edn, 1844.) Brünn.

Diebl, F. (1839). Ueber den nothwendigen Kampf des Landwirthes mit der Natur, um die Veredlung und Erhaltung seiner Kultur-Gebilde. *Mittheilungen*, 270–371.

Diebl, F. (1844). *Lehre von der Baum-Zucht überhaupt, und von der Obstbaumzucht, dem Weinbaue und der wilden—oder Waldbaumzucht insbesondere.* Brünn

Dixon, H.H. (1868). Rise and progress of the Leicester breed of sheep. *Journal of the Royal Agricultural Society of England* (Series 2), **4**, 340–58.

Duncombe, J. (1805). *General View of the Agriculture of the County of Hereford.* Drawn up for consideration of the Board of Agriculture and Internal Improvement. Nicol, London.

Dunn, L.C. (1965). *A Short History of Genetics. The Development of Some of the Main Lines of Thought, 1864–1939.* McGraw-Hill, New York.

Ehrenfels, J.M. (1817). Ueber die höhere Schafzucht in Bezug auf die bekannte Ehrenfelsische Race. Belegt mit Wollmustern, welche dei dem Herausgeber in Brünn zu sehen sind. *ONV*, 81–5, 89–94.

Ehrenfels, J.M. (1829). Ueber Rasse, Varietät und Konstanz im Thierreiche. *Mittheilungen*, 129–34, 137–42.

Ehrenfels, J.M. (1831). Fortsetzung der Gedanken des Herrn Moritz Beyer über das Merinoschaf. *Mittheilungen*, 137–42.

Ehrenfels, J.M. (1837). Schriftlicher Nachtrag zu den Verhandlungen der Schafzüchter-Versammlunmg in Brünn, am 10. Mai 1836. *Mittheilungen*, 2–4.

Eiseley, L. (1959). *Darwin's Century: Evolution and the Men who Discovered It.* Gollancz, London.

Elliot, H. (1914) *see* Lamarck 1809.

Ellis, W. (1749). *A Complete System of Experimental Improvements Made on Sheep, Grass-lambs and House-lambs, etc.* London.

Ellman, J. (1800). A letter to Lord Somerville, dated 25 April 1799. *CBA*, **2**, 459–60. (2nd edn, 1805). W. Bulmer & Co., London.

Ellman, J. (1834). *The Library of Agriculture and Horticultural Knowledge With a Memoir of Mr Ellman of Glynde* (3rd edn). Sussex Agricultural Press, Lewes.

Elsner, J.G. (1826). *Beschreibung meiner Wirtschaft.* Calvé, Prague.

Elsner, J.G. (1827). *Meine Erfahrungen in der höheren Schafzucht.* J.G. Gotta, Stuttgart, Tübingen.

Elsner, J.G. (1828). *Uebersicht der europäischen veredelten Schafzucht* Vol. I. (Vol. II published in 1829.) Calvé, Prague.

Elsner, J.G. (1849). *Die Rationelle Schafzucht. Ein Handbuch für Landwirthe, Schafzüchter etc. Resultate dreissigjähriger Praxix und Erfahrung.* Leipzig.

Elsner, J.G. (1857). In *Die verschiedenen Phasen der deutschen Merinozucht*, pp. 72–3. G. Boselmann, Berlin.

d'Elvert, C. (1870). *Geschichte der k.k. mähr. schles. Gesellschaft zur Beförderung des Ackerbaues, der Natur- und Landes- kunde, mit Rücksicht auf die bezüglichen Cultur-Verhältnisse Mährens und Oestrr. Schlessiens.* M.R. Rohrer, Brünn.

*Encyclopaedia Britannica* (1797) 3rd edn. Edinburgh.

'Epicurus' (1803). Letter from a farmer, October 1802. *Farmer's Magazine*, **4**, 35–7.

Ernlé, R.E. (Baron Prothero) (1888). *The Pioneers and Progress of English Farming.* Longmans & Co., London.

Ernle, R.E. (1936). English Farming Past and Present, 5th edn (Hall, A.D., ed.) (6th edn, 1961). Heinemann, London.

Farley, J. (1982). *Gametes and Spores. Ideas About Sexual Reproduction 1750–1914.* The Johns Hopkins University Press, Baltimore, MD.

Festetics, E. (1819a). Erklärung des Herrn Grafen Emmerich von Festetics. *ONV, ausserordentliche Beilage*, 9–12, 18–20, 26–7.

Festetics, E. (1819b). Weitere Erklärungen des Herrn Grafen Emerich Festetics über Inzucht. *ONV, ausserordentliche Beilage*, 169–70.

Festetics, E. (1820). Bericht des Herrn Grafen Emerich Festetics als Repräsentaten des Schafzüchter-Vereins in Eisenburger Comitate. *ONV*, 25–8.

Finke, J.H. (1785). *Nachrichten über die Stallfütterung der Schafe auf dem Amte Gröbzig*. Franktfurt/M.

Fink, J.H. (1797). Answers to questions posed by Sir John Sincliar, concerning the breeding of sheep, particularly in upper Saxony and neighbouring provinces. *CBA*, **1**, 276–94. (2nd edn, 1804) W. Bulmer & Co., London.

Finke, J.H. (1798). *Beantwortung der von dem Chevalier J. Sinclair aufgeworfenen Fragen, betreffend die verschiedenen Schafarten in Deutschland*. Halle. (German version of Fink 1797.)

Fink, J.H. (1799a) *Verschiedene Schriften und Beantwortungen betreffend die Schafzucht in Deutschland und Verbesserungen der groben Wolle, aus eigener Erfahrung und Thathandlung, zusammengetragen in Frühjahr 1799*. Halle.

Fink, J.H. (1799b) *Various writings and answers dealing with sheep breeding in Germany and the improvement of coarse wool according to personal practical experience and fact collected in Spring 1799*. Halle. (Translation from the German of Fink 1799a.)

Fraas, K.N. (1852). *Geschichte der Landwirtschaft, oder: Geschichtliche Übersicht der Fortschritte landwirtschaftlicher Kenntnisse in den letzten 100 Jahren*, Vols I and II. F. Tempsky, Prague. 2nd edn: *Geschichte der Landbau- und Forstwirtschaft. Seit dem sechzehnten Jahrhundert bis zur Gegenwart*. Gottasche Buchhandlung, München, 1865.

Franke, H. and Orel, V. (1983). Christian Carl André (1763–1831) as a mineralogist and an organizer of scientific sheep breeding in Moravia. In *Gregor Mendel and the Foundation of Genetics* (ed. V. Orel and A. Matalová), pp. 47–56. Moravian Museum, Brno.

Freudenberger, H. (1975). Progressive Bohemian and Moravian aristocracy. In *Intellectual and Social Developments in the Hapsburg Empire from Maria Theresa to World War I* (ed. S.B. Winters and J. Held), pp. 115–130. East European Monographs, No. 11 (published by *East European Quarterly*) New York and London.

Freudenberger, H. (1977). *The Industrialization of a Central European City. Brno and the Fine Woolen Industry in the 18th Century*. Pasold Research Fund, Edington.

Frolov, I.T. (1991). *Philosophy and History of Genetics: the Inquiry and the Debates*. Macdonald, London.

Fuss, J. (1795). *Anweisung zur Erlernung der Landwirtschaft in Königsreich Böhmen*. Prague.

Fussell, G.E. (1947) *The Old English Farming Books from Fitzherbert to Tull, 1523–1730*. Crosby Lockwood and Son Ltd, London.

Fussell, G.E. (1950) *More Old English Farming books From Tull to the Board of Agriculture 1731–1793*. Crosby Lockwood and Son Ltd, London.

Fussell, G.E. (1972). *The Classical Tradition in West European Farming*. David & Charles, Newton Abbot.

Fussell, G.E. and Goodman, C. (1930). Eighteenth century estimates of British sheep and wool production. *Agricultural History*, **4**, 131–51.

Gaál, L. and Gunst, P. (1977) *Animal Husbandry in Hungary in 19th–20th Centuries.* Akadémici Kiadó, Budapest.

Gaissinovitch, A.E. (1990). C.F. Wolf on variability and heredity. *History and Philosophy of the Life Sciences*, **12**, 179–201.

Garran, J.C. and White, L. (1985). *Merinos, Myths and Macarthurs. Australian Graziers and Their Sheep, 1788–1900.* Australian National University Press, Canberra.

Garrard, G. (1800). *A Description of the Different Varieties of Oxen Common in the British Isles.* London.

Gayon, J. (1996) Entre force et structure, genèse du concept naturaliste de l'hérédité. In *Le paradigme de la filiation* (ed. J. Gayon and J.J. Wunenburger), pp. 61–75. L'Hartmattan, Paris.

Gayon, J. and Wunenburger, J.J. (eds) (1996). *Le paradigme de la filiation.* Editions l'Harmattan, Paris.

Gazley, J.G. (1973). *The Life of Arthur Young 1741–1820.* American Philosophical Society, Philadelphia, PA.

Gent, W.S. (1656). *The Golden Fleece, Wherein is Related the Riches of the English Wools in its Manufactures.* (Parts are reproduced in Smith 1747, Vol. 1, pp. 196–201.)

Glacken, G.L. (1967) *Traces on the Rhodian Shore. Nature and culture in Western Thought from Ancient Times to the End of the Eighteenth Century.* University of California Press, Berkeley, CA.

Glass, H.B. (1959). *Forerunners of Darwin, 1745–1859.* Johns Hopkins Press, Baltimore, MD.

Glass, H.B. (1960). Eighteenth century concepts of the origin of species. *Proceedings of the American Philosophical Society*, **104**, 227–34.

Gliboff, S. (1999), Gregor Mendel and the laws of evolution. *History of Science*, **37**, 217–35.

Gorzny, W. (ed.) (1986). *Deutscher Biographischer Index.* K.G. Saur, Munich.

Grant, V. (1956). The development of a theory of heredity. *American Scientist*, **44**, 158–78.

Guerchy, Marquis de (1785) *Mémoire sur l'amélioration des bêtes à laine dans l'Isle de France.* Paris

Gullers, K.W. and Strandell, B. (1977) *Linnaeus.* Gullers International Inc., Chicago, IL.

Haldane, J.B.S. (1959) Natural selection. In *Darwin's Biological Work, Some Aspects Reconsidered* (ed. P.R. Bell), pp. 101–49. John Wiley and Sons, New York.

Hardly, J. (1953). *Scottish Farming in the Eighteenth Century.* Faber and Faber, London.

Harkenfeld, von Sedláček, S. (1826). Zustand des mährischen Weinbaues und Vorschläge ihn durch Einführung der Rebenschulen von den edelsten Sorten zu Vervollkommen. *Mittheilungen*, **21**, 161–7.

Hart, E. (1989). *Sheep. A Guide to Management.* Crowood Press, Marlborough.

Hartwell, R.M. (1972). *The Industrial Revolution and Economic Growth.* London.

Hartwell, R.M. (1973). A revolution in the character and destiny of British wool. In *Textile History and Economic History: Essays in Memory of Julia de Lacy Mann* (ed. N.B. Harte and K.G. Ponting), pp. 320–38. Manchester University Press, Manchester.

Hastfer, F.W. (1752a). *Ütförlig och omständig underrättelse om fullgoda fårs ans och skjötsel, til det all männas tjänst sammanfaltad af Fried. W. Hastfer.* J. Merckell, Stockholm.

Hastfer, F.W. (1752b). *Unterricht von der Zucht und Wartung der besten Art von Schafe.* Leipzig.

Hastfer, F.W. (1756). *Instruction sur la manière d'élever et de perfectionner les bêtes à laine.* Dijon.

Hastfer, F.W. (1756). *Geldgrube eines Landes in die Verbesserung der Schafzucht, nebst einem zuverlässigen Mittel und Rath gegen die Schaafspocken auf dem Dänischen übers 725 gbd, 1 Taf.* Mumme, Copenhagen.

Havemann, K. and Friedrichs, W. (1979). Zum Wirken von Franz Christian Lorenz Karsten an der Universität Rostock. *Tag.-Ber., Akad. Landwirtsch.-Wiss. DDR,* **173**, 87–92.

Hawkesworth, F.T.C. (1920). *Australian Sheep and Wool. A Practical and Theoretical Treatise,* 5th edn. William Brooks & Co. Ltd, Sydney.

Heath-Agnew, E. (1983). *A History of Hereford Cattle and Their Breeders.* Duckworth, London.

Hempel, G.C.L. (1818). Horticultural Societät in London. *ONV,* 408.

Hempel, G.C.L. (1820). Ueber die Entstehung und Wichtigkeit der verschiedenen Sorten der Getreidearten. *ONV,* 161–5.

Herrmann, K. (1980). August von Weckherlin, 1794–1868. In *Lebensbilder aus Schwaben und Franken,* Vol. 14 (ed. R. Uhland), pp. 192–218. W. Kohlhammer Verlag, Stuttgart.

Hill, F. (1966). *Georgian Lincoln.* Cambridge University Press. London.

Hitschmann, A. (1812). Die ersten Begründer der Schaf-Veredlung in Mähren. *ONV,* 323–4.

Holt, J. (1795). *General View of the Agriculture of the County of Lancaster.* Drawn up for consideration by the Board of Agriculture and Internal Improvement, London. G. Nicol, London.

Home, H. (Lord Kames) (1776). *The Gentleman Farmer.* W. Creech & T. Cadell, Edinburgh.

Housman, W. (1894). Robert Bakewell. *Journal of the Royal Agricultural Society of England* (Series 3), **5**, 1–31.

Howitt, W. (1842). *The Rural and Domestic Life of Germany, with Characteristic Sketches of its Cities and Scenery, Collected in a General Tour, and During a Residence in the Country in 1840, 41 and 42.* Longman, Brown, Green and Longman, London.

Howlett, John (1786) *Cursory Remarks on Inclosures, Showing the Pernicious and Distinctive Consequences of Inclosing Common Fields, by a Country Farmer.* London.

Hubert, C.A. (1814). *Die Wartung, Zucht und Pflege der Schafe, ihre Benutzung und Veredelung, oder Dienstanweisung für meine Schäfer, in allen seinen Geschäften, und Dienst-verhältnissen.* F. Maurer, Berlin.

Huet, Bishop of Auranches (c. 1706). *Memoirs of the Dutch Trade.* (Translated from the French.) T. Warner, London. (Reprinted in Smith 1747, Vol. II, pp. 76–103.)

Humphries, J (1885) Biographical article on Robert Bakewell (1768–1843), geologist. In *Dictionary of National Biography*, **1**, 943–44. Oxford University Press, Oxford.

Hunt, J. (1808). On the perfections and superiority of the Leicestershire breed of sheep. *Agricultural Magazine* (New Series), **3**, 83–92. (August).

Hunt, J. (1812). *Agricultural memoirs or history of the Dishley system, in answer to Sir John Saunders Sebright.* Nottingham, pamphlet.

Hutton, J. (c.1796). Unpublished manuscript, entitled *The Elements of Agriculture* lodged in the library of the Royal Society of Edinburgh.

Iltis, H. (1924). *Gregor Johann Mendel. Leben, Werk und Wirkung.* Springer, Berlin.

Iltis H. (1932). *Life of Mendel.* Translated by Eden and Cedar Paul. George Allen & Unwin Ltd, London; Norton, New York. New impression, 1966, Hafner, New York.

'Irtep' [Petri] (1812). Ansichten über die Schafzucht nach Erfahrung und gesunder Theorie. *ONV*, 1–5, 9–16, 21–3, 27–8, 45–8, 60–61, 81–5, 91–2, 106–7.

Jacob, W. (1820). *A View of the Agriculture, Manufacture, Statistics and State of Society of Germany, and Parts of Holland and France. Taken During a Journey Through Those Countries in 1819.* London.

Janke, H. (1867). *Die Grundsätze der Schafzüchtung mit besonderer Berücksichtigung der deutschen Merinozucht.* J. Springer, Berlin.

Jenkins, D.T. and Ponting, K.G. (1982). *The British Wool Textile Industry, 1770–1914.* Heinemann, London.

Jenner, E. (1798) An inquiry into the causes and effects of the variolae vaccinae (etc). Memoire.

Jobson, G. (1957). A 'prophet without honour' gets his due at last. *Journal and North Mail*, Loughborough, November 9, p. 6.

Johnstone, G. (1846). On the general management of sheep in enclosed or arable land. *The Plough, a Journal of Agricultural and Rural Affairs*, **1**, 142–57.

de Jotemps, P., de Jotemps, F. and Girod, F. (1824). *Nouveau traité sur la laine et sur les moutons.* Huzard, Paris.

de Jotemps, P., de Jotemps, F. and Girod, F. (1825a) *Neuste Ansichten über Wolle und Schafzucht.* (Translated from the French by C.C. André with added comments.) Calvé, Prague.

de Jotemps, P., de Jotemps, F. and Girod, F. (1825b). *Über Wolle und Schaafzucht.* (Translated by A. Thaer with added comments.) Rücker, Berlin.

Justinus, J.C. (1815). *Allgemeine Grundsätze zur Vervollkomnung der Pferdezucht, anwendbar auf die übrigen Hausthierzuchten.* Wien and Trieste.

Kamen, H. (1984). *European Society 1500–1700.* Hutchinson, London.

Kames (1776). *see* Home, H. (1776)

Kaschnitz, A.V. (1805). *Praktische Bemerkungen und Anweisung zur Veredlung* der *Schafzucht*. J.G. Trassler, Krakau and Brünn

Kerridge, E. (1967). *The Agricultural Revolution*. George, Allen & Unwin, London.

King-Hele, D. (1963) *Erasmus Darwin*. MacMillan & Co., London.

Kjellberg, S.T. (1943). *Ull och Ylle bidrag til den Svenska Yllemannfakturens historia*. Lund.

Klein, J. (1920). *The Mesta, a Study of Spanish Economic History 1273–1836*. Harvard Economic Studies. Cambridge, MA.

Klemm, V. and Meyer, G. (1968). *Albrecht Daniel Thaer.* UEB Max Niemeyer, Halle.

Knight,T.A. (1797). Letter to Sir Joseph Banks. Manuscript in British Museum, quoted by Mylechreest (1988).

Knight, T. A. (1799). An account of some experiments on the fecundation of vegetables. *Philosophical Transactions of the Royal Society of London*, **89**, 195–204. (German translation: Versuche über die Befruchtung der Pflanzen. *Oekonomische Hefte*, Leipzig, 1800, pp. 322–340.)

Knight, T.A. (1800). *CBA*, **2**, 86

Köcker, M. (1809). Auszüge aus Briefen des Herrn Oekonom Köcker, auf den fürst. Salmischen Herrschaft Raiz in Herbst 1808 an den Herausgeber des letzten. *Hesperus*, 277–303.

Koerner, L. (1999). *Linnaeus: Nature and Nation*. Harvard University Press, Cambridge, MA.

'K in Mähren' [M. Köller] (1811). Ist es nothwendig, zur Erhaltung einer edlen Schafherde stets fremde Original-Widder nachzuschaffen, und artet sie aus, wenn sich das verwandte Blut vermischet? *ONV*, 294–8.

Köller, M. (1815) Auszug der Äußerung des Herrn Inspektors Köller. *ONV*, 367–8.

Körte, A. (1862). *Das deutsche Merinoschaf. Seine Wolle, Züchtung, Ernährung und Pflege*. J. Urban Kern, Breslau.

Křiženecký, J. (1987). Wie J.E. Purkyně im mährischen Karst philosophierte—Purkyně und Gregor Mendel. *Folia Mendeliana*, **22**, 69–80.

Kurucz, G. (1990). The literature of the new agriculture in the Festetics library [English translation of Hungarian title]. *Magyar Knyuszenle*, **106** (1–2), 32–44.

Lamarck, J.B. (1809). *Philosophie zoologique; ou, exposition des considérations relatives à l'histoire naturelle des animaux*. Two volumes. Paris. English translation (1914) with an introduction by Hugh Elliot: *Zoological Philosophy*. MacMillan & Co., London.

Langrish, J. (1974) The changing relationship between science and technology. *Nature*, **250**, 614–16.

Lasteyrie, C.P. (1799). *Traité sur les bêtes-à-laine d'Espagne*. A.J.Marchant, Paris, published in *Feuille de Cultivateur* 7, pp vii + 356. Reviewed in 1800 in *PTB* p 47. German translation (1800): *Abhandlung über das spanische Schafvieh*. Hamburg.

Lasteyrie, C.P. (1802). *Histoire de l'introduction des moutons à laine fine d'Espagne dans les divers états de l'Europe, et au Cap du Bonne-Espérance*. Paris. German translation (1804): *Geschichte der Einführung der feinwolligen spanischen Schafe in die verschiedenen europäischen Länder.* G. Fleischer, Leipzig. English translation (1810): *An Account of the Introduction of Merino Sheep into the Different States of Europe, and at the Cape of Good Hope.* Benjamin Thompson, London.

Lauer, J.G. (1826). Skizzierte Darstellung der Verhandlung des Schafzuchtvereines in 1826. *Mittheilungen*, 193–8, 209–14, 241–6, 265–70, 281–6.

Lauer, J.G. (1841). Züge aus dem Leben verstorbener Gesellschaftsmitglieder. Dr. Johann Karl Nestler. *Mittheilungen*, 319–21. [Republished by d'Elvert 1870, Vol. II, pp. 203–17.]

de Lavergne, L. (1854). *Essai sur l'économie rurale de l'Angleterre, de l'Ecosse et de l'Irlande.* Paris.

de Lavergne, L. (1855). *The Rural Economy of England, Scotland and Ireland, Translated from the French with Notes by a Scottish Farmer.* Blackwood and Sons, Edinburgh and London.

Lawrence, J. ['A Farmer and Breeder'] (1800). *The New Farmer's Calendar ... Comprehending all the Material Improvements in the New Husbandry with the Management of Livestock.* C. Whittingham, London.

Lawrence, J. (1805). *A General Treatise on Cattle, the Ox, the Sheep and the Swine.* London.

Lawrence, J. (1809). *A General Treatise on Cattle, the Ox, the Sheep and the Swine. Comprehending their Breeding, Management, Improvement and Diseases. Dedicated to the Rt. Hon. Lord Somerville*, 2nd edn. London.

Lawrence, W. (1819). *Lectures on Physiology, Zoology and the Natural History of Man.* London.

Lerner, J. (1992). Science and agricultural progress: quantitative evidence from England, 1660–1780. *Agricultural History*, **66**, 11–27.

Leuckhart, R. (1853). Zeugung. In *Handwörterbuch der Physiologie mit Rücksicht auf physiologische Pathologie* (ed. R. Wagner), Vol. 4, pp. 707–1000. F. Vieweg, Braunschweig.

Lichnowski, E.M. (1811). Ueber die fürstlich Lichnowski Schafzucht, sowohl im Oesterreichischen kaiserl als königl. Preussischen Antheile Schlesiens, um Gegend von Troppau. *ONV*, 40.

Liège, M.D. de (1842). *Les animaux domestiques.* Liège.

'Lincolnshire Grazier' [William Youatt] (1833). *The Complete Grazier*, 6th ed. Baldwin and Craddock, London.

Lisle, E. (1757). *Observations on Husbandry*, 2nd edn. Two volumes. J. Hughs, London.

de Lisle, S. (1975). Robert Bakewell. *Journal of the Royal Agricultural Society*, **136**, 56–62.

de Lisle, S. (1993). Robert Bakewell and his animals. *The Ark*, **20**, 374–5.

Livingston, R.R. (1810). *Essay on Sheep, the Varieties, Account of the Merinos of Spain, France, etc.* New York. First published in *Trans. New York Agric. Society* (Series II), **22**, 66 *et seq.* (1806) A new impression was published in 1813, issued from Concord, NH.

Löbe, (1878). Finke, Johann, Heinrich. *Allgemeine Deutsche Biographie*, Vol. 7, pp. 18–19. von Duncker und Humblot, Leipzig.

Loudon, J.C. (1844). *An Encyclopaedia of Agriculture*, 5th edn. Longman, Brown, Green and Longman, London. (1st edn published 1831.)

Low, D. (1845). *On the Domesticated Animals of the British Islands.* Longman, Brown, Green and Longman, London. (1st edn published 1842).

Löwenfeld, R. (1835). Andeutungen um die Veredlung der Schafe in allen vorzüglichen Eigenschaften nach geregelten, aus der Natur und Erfahrung abgeleiteten Grundsätzen mit dem besten und sichersten Erfolge vollkommen ausüben. *Mittheilungen,* **1**, 1–6; **2**, 9–13; **7**, 54–6.

Luccock, J. (1805). *The Nature and Properties of Wool Illustrated, with a Description of the English Fleece*. Leeds.

Lush, J.A. (1951) Genetics and animal breeding. In *Genetics in the 20ᵗʰ century* (ed. L.C. Dunn), pp. 493–525. MacMillan, New York.

Luzzatto, G. (1961). *An Economic History of Italy, from the Fall of the Roman Empire to the Beginning of the Sixteenth Century.* Routledge & Kegan Paul, London.

Lyell, C. (1834). *Principles of Geology, Being an Attempt to Explain the Former Changes of the Earth's Surface, by Reference to Causes Now in Operation*, 3rd edn. Four volumes. London.

Lyell, C. (1863). *The Geological Evidences of the Antiquity of Man, with Remarks on Theories of the Origin of Species by Variation.* John Murray, London.

Macdonald, S. (1974). The development of agriculture and the diffusion of agricultural innovation in Northumberland 1750–1850. Ph.D. thesis, University of Newcastle-upon-Tyne. Chapter 14 ('Livestock'), pp. 230*ff.*

Mann, J. de L. (1987) *The Cloth Industry in the West of England from 1640 to 1800.* Alan Sutton. (First publisished in 1971 by Oxford University Press.)

Mantoux, P. (1961). *The Industrial Revolution in the Eighteenth Century.* Jonathan Cape, London.

Markham, G. (1631). *Cheape and Good Husbandry*, 5th edn. London. [1648, 7th edition].

Marshall, J.D. (1772). *Travels through Germany, Russia, and Poland in the years 1769 and 1770.* J. Almon, London. (Reprinted in 1971 by Arno Press.)

Marshall, W. (1790). *The Rural Economy of the Midland Counties.* Two volumes. G. Nicol, London.

Marshall, W. (1796a). *The Rural Economy of the Midland Counties*, 2nd edn. G. Nicol, London.

Marshall. W. (1796b). *The Rural Economy of the West of England.* G. Nicol, London.

Marshall, W. (1803). *Agriculture practique des différentes parties de l'Angleterre.* Five volumes and atlas. Gide, Paris.

Marshall, W. (1818) (Volumes published separately between 1808 and 1817, reissued in 1818). *The Review and Abstract of the County Reports to the Board of Agriculture.* Five volumes. T. Wilson, York. (Reprinted by David and Charles, Newton Abbot, c.1968.)

Martin, W.C.L. (undated, c.1849). *The Sheep*, pp. 1–220. In *The Farmer's Library, Animal Economy*, Vol. 2. Charles Knight, London.

Mason, I.L. (1967). *The Sheep Breeds of the Mediterranean*. Food and Agriculture Organisation and Commonwealth Agriculture Bureau, Rome and London.

Massy, C. (1990). *The Australian Merino*. Penguin Books Australia Ltd, Ringwood, Victoria.

Matalová, A. (1979). A monument to F.M. Klácel (1809–1882) in the vicinity of the Mendel statue in Brno. *Folia Mendeliana*, **14**, 251–63.

de Maupertuis, P.L.M. (1745). *Vénus Physique*. Paris. Translation (1966) by S.B. Boas: *The Earthly Venus*. Johnson Reprint Corporation, New York.

Mayer, A. (1831). Nachbemerkungen zu den in Nro. 18 der *Mittheilungen* 1. J. erschienenen Gedanken des Freiherrn J.M. von Ehrenfels über das Merinoschaf. *Mittheilungen*, 145–151.

Mayr, E. (1971). Open problems of Darwin research [essay review]. *Studies in the History and Philosophy of Science,* **2(3)**, 273–80.

Mayr, E. (1972). The nature of the Darwinian revolution. *Science*, **176**, 981–9.

Mayr, E. (1982). *The Growth of Biological Thought. Diversity, Evolution and Inheritance*. Harvard University Press, Cambridge, MA.

Mendel, G. (1866). Versuche über Pflanzen-Hybriden. *Verhandlungen des Natur-forschenden Vereines, Abhandlungen, Brünn*, **4**, 3–47. Editions in different lan-guages were reviewed by Matalová 1973. They include the English translation commissioned by W. Bateson, reproduced in *Classic Papers in Genetics* (ed. J.A. Peters) Prentice Hall, Englewood Cliffs, NJ (1959) and a second English transla-tion published by C. Stern and E.R. Sherwood (eds) in 1964: *The Origin of Genetics, A Mendel Source Book*. W.H. Freeman, San Francisco.

Mitchison, R. (1962). *Agricultural Sir John*. Geoffrey Bles, London.

Monk, J. (1794). *General View of the Agriculture of the County of Leicester*. Drawn up for consideration by Board of Agriculture and Internal Improvement, London. G. Nicol, London.

Montserrat, P. and Fillet, F. (1991). The systems of grassland management in Spain. In *Managed Grasslands* (ed. A. Breymeyer), pp. 37–70. Elsevier, Amsterdam.

Moore, J.A. (1986) Science as a way of knowing—genetics. In *Science as a Way of Knowing*, Vol. III: *Genetics* (ed. J.A. Moore), pp. 583–747. American Society of Zoologists, Baltimore, MD.

Moritz, C.P. (1783). *Reisen eines Deutschen in England im Jahr 1782*. Berlin. An edited English translation was produced by R. Nettel in 1965 under the title *Journeys of a German in England in 1782*. Jonathan Cape, London.

Müller, H.H. (1969). Christopher Brown, an English farmer in Brandenburg-Prussia in the eighteenth century. *Agricultural History Review*, **17**, 120–35.

Mylechreest, M. (1988). Thomas Andrew Knight (1759–1838) and the Altenburg connection in the origin of Mendelism. *Folia Mendeliana*, **23**, 27–32.

Napp, C.F. (1836). Autobiography. In Czihak and Sládek 1991/92, pp. 39–43.

Nathusius, H. (1872). *Vorträge Viehzucht und Rassenkenntniss.* Wiegand-Hempel, Berlin.

Neitzschütz, M. (1869). *Studium zur Entwicklungs-Geschichte des Schafes.* A.W. Kafemann, Danzig.

Nestler, J.K. (1829). Ueber den Einflusss der Zeugung auf die Eigenschaften der Nachkommen. *Mittheilungen*, 369–72, 377–80, 394–8, 401–4.

Nestler, J.K. (1831). Neues aus der alten Zeit. *Mittheilungen*, 71–2.

Nestler, J.K. (1836). Ueber die Andeutung zur Veredelung der Schafe, von Herrn Rudolph v. Löwenfeld. *Mittheilungen* pp. 153–155, 163–164, 173–176, 185–189, 205–208.

Nestler, J.K. (1837). Ueber Vererbung in der Schafzucht. *Mittheilungen*, 265–9, 273–9, 281, 286, 289–300, 300–303, 318–20.

Nestler, J.K. (1838a). Obituary notice on Martin Köller. *Mittheilungen*, 225–30, 233–7. (Republished by d'Elvert 1870, Vol. I, pp. 181–92.)

Nestler, J.K. (1838b). Bastarde des Mouflon-Widders und des Merinoschafes. *Mittheilungen*, 66–8.

Nestler, J.K. (1839) Ueber Innzucht. *Mittheilungen*, 121–8.

Nestler. J.K. (1841). *Amts-Bericht des Vorstandes über die vierte, zu Brünn von 20. bis 28. September 1840 abgehaltene Versammlung der deutschen Land-und Forstwirthe.* A. Skarnitze, Olmütz.

Netolička, E. (1855). *Elemente der Pflanzenphysiologie mit den Grundbegriffen der Anatomie, Chemie und Geographie der Pflanzen für Schulen und zum Selbstunterricht.* Brünn.

Nettel, R. (1965). *Journeys of a German in England in 1782.* Jonathan Cape, London. This is a translation of Moritz 1783.

Neumann, A. (1930). *Acta et epistolae eruditor um monasterii ord. S. Augustini Vetero-Brunae*, Vol. 1, (A) 1819–1850. Brno.

O'Connor, T.P. (1995). Size increase in post-medieval English sheep: the ostological evidence. *Archaeofauna*, **4**, 81–91.

Olby, R.C. (1985). *Origins of Mendelism*, 2nd edn. University of Chicago Press, Chicago, IL.

Olby, R.C. (1990). The emergence of genetics. In *Companion to the History of Modern Science* (ed. R.C. Olby, G.N. Cantor, J.R.R. Christie and M.J.S. Hodge), pp. 521–36. Routledge, London and New York.

Orel, V. (1973). The scientific milieu in Brno during the era of Mendel's research. *Journal of Heredity*, **64**, 314–18.

Orel, V. (1974). The prediction of the laws of hybridization in Brno already in 1820. *Folia Mendeliana*, **9**, 245–54.

Orel, V. (1975a). The building of greenhouses in the monastery garden of Old Brno at the time of Mendel s experiments. *Folia Mendeliana*, **10**, 201–8.

Orel, V. (1975b). Das Interesse F. C. Napps (1792–1867) für den Unterricht der Landwirtschaftslehre und die Forschung der Hybridisation. *Folia Mendeliana*, **10**, 225–40.

Orel, V. (1977). Selection practice and theory of heredity in Moravia before Mendel. *Folia Mendeliana*, **12**, 179–221.

Orel, V. (1978a). The influence of T.A. Knight (1759–1838) on early plant breeding in Moravia. *Folia Mendeliana*, **13**, 241–60.

Orel, V. (1978b). Heredity in the teaching programme of professor J.K. Nestler (1783–1841). *Acta universitatis Palackianae Olomucensis, fac. rer. nat.*, **59**, 79–98.

Orel, V. (1983). Mendel's achievements in the context of the cultural peculiarities of Moravia. In *Gregor Mendel and the Foundation of Genetics* (ed. V. Orel and A. Matalova), pp. 23–46. Moravian Museum, Brno.

Orel, V. (1992). Jaroslav Kříženecký (1896–1964), tragic victim of Lysenkoism in Czechoslovakia. *Quarterly Review of Biology*, **67**, 487–94.

Orel, V. (1996). *Gregor Mendel, the First Geneticist*. Oxford University Press, Oxford.

Orel, V. (1997) The spectre of inbreeding in the early investigation of heredity. *History and Philosophy of Life Science*, **19**, 315–30.

Orel,V. and Fantini, B. (1983). The enthusiasm of the Brno Augustinians for science and their courage in defending it. In *Gregor Mendel and the Foundation of Genetics* (ed. V. Orel and A. Matalová). pp. 105–10. Moravian Museum, Brno.

Orel, V. and Janko, J. (1988). Purkyně's concept of procreation and Mendel's research in heredity. In *Jan Evangelista Purkyně in Science and Culture* (ed. J. Purs), pp. 657–70. Academia, Prague.

Orel, V. and Matalová, A. (1983). *Gregn Mendel and the Foundation of Genetics*, Mozarian Museum, Brno.

Orel, V. and Kuptsov, V.I. (1983). Precondition for Mendel's discovery in the body of knowledge in the middle of nineteenth century. In *Gregor Mendel and the Foundation of Genetics* (ed. V. Orel and A. Matalová), pp. 189–227. Moravian Museum, Brno.

Orel, V. and Verbík, A. (1984). Mendel's involvement in the plea for freedom on teaching in the revolutionary year of 1848. *Folia Mendeliana*, **19**, 223–33.

Orel, V. and Wood, R. (1981). Early development in artificial selection as a background to Mendel's research. *History and Philosophy of the Life Sciences*, **3**, 145–70.

Orel, V. and Wood, R.J. (1998). Empirical genetic laws published in Brno before Mendel was born. *Journal of Heredity*, **89**, 79–82.

Orel, V. and Wood, R.J. (2000a). Scientific animal breeding in Moravia before and after the discovery of Mendel's theory. *Quarterly Review of Biology*, **75**, 149–57.

Orel, V. and Wood, R.J. (2000b). Essence and origin of Mendel's discovery. *C.R. Acad. Sci. Paris/Life Sciences*, **323**, 1037–41.

Orel,V., Janko, J. and Geus, A. (1987). The enigma of generation in connection with heredity in the teaching of J.E. Purkyně (1787–1869). *Folia Mendeliana*, **22**, 7–33.

Owen, J. (1764). *Britannia Depicta. The Direct and Principal Cross Roads of England and Wales*. Carrington Bowles, London.

Paget, G. (1987 (1945)). *Sporting Pictures of England*. Bracken Books, London.

Parkinson, R. (1808). *General View of the Agriculture of the County of Rutland*. Draw up for consideration by the Board of Agriculture and Internal Improvement, London. Nicol, London.

Parkinson, R. (1810). *Treatise on the Breeding and Management of Live Stock.* Two volumes. Cadell & Davies, London.

Parkinson, R. (1811). *General View of the Agriculture of the County of Huntingdon.* London.

Parry, C.H. (1800). *Facts and Observations Tending to Show the Practicability and Advantage to the Individual and the Nation, of Producing in the British Isles Clothing Wool, Equal to that of Spain, Together with some Hints Towards the Management of Fine-woolled Sheep.* Bath. Reviewed in *PTB* (1800), 222–3. Reviewed again in PTB (1801), 907–9 under the title 'Ueber die spanische Schafzucht in England'.

Parry, C.H. (1806). An essay on the nature, produce, origin and extension of the Merino breed of sheep. *CBA,* V, Part 1 (XVIII), 337–541.

Paul, D.B and Kimmelman, A. (1988). Mendel in America: theory and practice, 1900–1919. In *The American Development of Biology* (ed. R. Rainger, K. Benson and J. Maienschein), pp. 281–310. University of Pennsylvania Press, Philadelphia.

Pawson, H.C. (1957). *Robert Bakewell, Pioneer Livestock Breeder.* Crosby Lockwood & Son, London.

Pearce, W. (1794). *General View of the Agriculture of the County of Berkshire.* Drawn up for consideration by the Board of Agriculture and Internal Improvement, London. Nicol, London

Perkins, J.A. (1977). *Sheep Farming in Eighteenth and Nineteenth Century Lincolnshire.* Society for Lincolnshire Natural History and Archaeology, Sleaford.

Petersburg, J. (1815). Veredlung des Schafviehs in der Blutverwandschaft betreffend. *ONV,* 1–4.

Petersburg, J. (1816). Aussärung des Repräsentanten für den Olmützer-Kreis Herrn Wirtschaftsraths Petersburg über die acht ausgestellten Haupt-punkte. *ONV,* 113–15.

Petri, B. (1812). Auszüge aus Briefen auf einer landwirstchaftlichen Reise nach Spanien, hauptsächlich über Schafzucht. *ONV,* 149–51, 162–4, 177–8, 217–20, 224, 228, 233–4, 241–3, 251–2, 259–60, 281–3, 289–92, 298–300, 306–7, 314–16, 322–3, 331–2.

Petri, B. (1813). Meine Ideen und Grunsätze über die Zuzucht der Hausthiere. *ONV,* 193–6.

Petri, B. (1815). *Das ganze der Schafzucht in Hinsicht auf unser deutsches Klima.* Wien.

Petri, B. (1825). *Das ganze der Schafzucht in Hinsicht auf unser deutsches Klima, und der angrenzenden Länder, insbesonedere von der Pflege, Wartung und den Eigenschaften des Merinos und ihre Wolle.* C. Schaumburg, Vienna.

Petri, G. (1838). Ueber die Inzucht. *Versammlung der Naturforscher und Aerzte zu Prag vom 18. bis 26. September 1837. Sektion für Landwirstchaft, Pomologie, Technologie und Mechanik.*

Phillips, I.A. (1989). Concepts and methods in animal breeding, 1770–1870. Ph.D. thesis, University of Manchester Faculty of Technology.

Pictét, C. (1802). *Faits et observations concernant la race des Mérinos d'Espagne à laine superfines et les croisements.* Geneva.

Pictét. C. (1808). *Erfahrungen und Beobachtungen über die spanischen Merinos*

*Schaafe, die Feinheit ihrer Wolle und das Kreuzen derselben mit gemeinen Rassen.* (Translation from the French.) Malper and Beck, Vienna.

Pinto- Correia, C. (1997). *The Ovary of Eve. Egg, Sperm and Preformation.* Chicago University Press, Chicago and London.

Pitt, W. (1809). *General View of the Agriculture of the County of Leicester.* Drawn up for consideration by the Board of Agriculture and Internal Improvement, London. Nicol, London.

Pitt, W. (1813). *General View of the Agriculture of the County of Worcester.* Drawn up for consideration by the Board of Agriculture and Internal Improvement, London. Nicol, London and Sherwood, Neeley and Jones, London.

Pluskal, F.S. (1816). *Leopold Graf von Berchtold, der Menschen Freund.* Brno.

Ponting, K.G. (1980). *Sheep of the World.* Blandford Press, Poole.

Pounds, N.J.G. (c.1986). *An Historical Geography of Europe.* Cambridge University Press, Cambridge.

Power, E. (1941). *The Wool Trade in English Medieval History.* Oxford University Press, Oxford.

Practicus (1802). The answer of Practicus to Dr Parry's reply. *Agricultural Magazine*, **7**, 181–6.

Priestland, N. (1987). Who was ... Jan Evangelista Purkyně? *Biologist.*, **34**(5), 249–51.

Priestly, J. (1797). Letter from Dr Priestly to Sir John Sinclair. CBA, **1**, 363–66.

Prothero, R.E. (1888) *see* Ernlé, Baron.

Pugh, R.B. (ed.) (1954). *The Victoria History of the County of Leicester*, Vol. 2, p. 223.

Pujol, R. (1981) Contribution à l'étude de la race ovine landaise. In *La grande lande. Histoire naturelle et géographie historique*, pp. 529–62. Editions du CNRS et du parc naturel régional des Landes de Gascogne, Bordeaux.

Purkinje, J.E. (1834). Erzeugung (generatio, genesis, procreatio). *Encyclopedisches Wörterbuch der medicinischen Wissenschaften, Berlin*, **11**, 515–49.

Quennell, M. and Quennell, C.H.B. (1938). *A History of Everyday Things in England*, 3rd edn. B.C. Batsford, London.

Ramsay, G.D. (1982). *The English Woollen Industry, 1500–1750.* The Macmillan Press, London.

Randall, H.S. (1882). *Sheep Husbandry with an Account of the Different Breeds.* Orange Judd Co., New York, NY. (Based on an original memoir submitted to the New York State Agricultural Society in 1837, reprinted here as 'Letter X' (pp. 129–52).)

Ray, J. (1691). *The wisdom of God manifested in the works of Creation.* London.

Redhead, W., Laing, R. and Marshall, W. Jr (1792). *Observations on the Different Breeds of Sheep, and the State of Sheep Farming in some of the Principal Counties of England.* Edinburgh.

Redhead, W. and Laing, R. (1793). Observations on a sheep tour. *Annals of Agriculture*, **20**, 1–34.

Rees, A.D.D. (1819). *Cyclopaedia*, Vol. 22 (unpaginated), article on sheep.

Rice, V.A. (1926). *Breeding and Improvement of Farm Animals*. McGraw Hill, New York.

Rice, V.A., Andrews, F.N., Warwick, E.J. and Legates, J.E. (1957). *Breeding and Improvement of Farm Animals*, 5th edn. McGraw Hill, New York.

Riches, N. (1967). *The Agricultural Revolution in Norfolk*, 2nd edn. Frank Cass & Co. London.

Robson, M.J.H. (1987). *Sheep of the Borders*. Ovenshank, Newcastleton.

de la Rochefoucauld, F. (1784). *A Frenchman's Year in Suffolk. French Impressions of Suffolk Life in 1784*. (Translated and edited by Norman Scarfe.) Suffolk Records Society, Vol. 30. The Boydell Press, Woodbridge.

Röckel, J. (1808). *Pedagogische Reise durch Deutschland veranlasst auf allerhöchsten Befehl der bayerischen Regierung in 1805*, p. 406. Dillingen.

Röhrer, R. (1830). Botanische Notitzen. *Mittheilungen*, 120.

Roe, S.A. (1979). Rationalism and embryology, Caspar Friedrich Wolf's theory of epigenesis. *Journal of the History of Biology*, **12**, 1–43.

Roe, S.A. (1981). *Matter, Life and Generation. Eighteenth Century Embryology and the Haller-Wolff Debate*. Cambridge University Press, Cambridge.

Roger, J. (1997). *Buffon*. (English translation of *Buffon, un Philosophe au Jardin du Roi*, 1989, made by Sarah L. Bonnefoi). Cornell University Press, Ithaca and London.

Rudge, T. (1807). *General View of the Agriculture of the County of Gloucester*. Drawn up for consideration by the Board of Agriculture and Internal Improvement, London. Nicol, London.

Ruse, M. (1975). Charles Darwin and artificial selection. *Journal of the History of Ideas*, **36**, 339–50.

Russell, E.S. (1930). *The Interpretation of Development and Heredity*. Clarendon Press, Oxford.

Russell, N. (1986). *Like Engend'ring Like. Heredity and Animal Breeding in Early Modern Europe*. Cambridge University Press, Cambridge.

Ryder, M.L. (1983). *Sheep and Man*. Duckworth, London.

Ryder, M.L. (1987). The evolution of the fleece. *Scientific American*, **256** (1), 100–107.

Salm, A. and André, C.C. (1814). An die Freunde der vaterländischen Industrie und der inländischen Schafzucht insbesondere. *ONV*, 113–14.

Salmon, N. (1739). *Modern History, or the Present State of all Nations*. Three volumes. (Abstract reproduced by Smith 1747, Vol. II, pp. 217–22.)

Sandler, I. and Sandler, L. (1985) The conceptual ambiguity that contributed to the neglect of Mendel's paper. *History and Philosophy of Life Sciences*, **7**, 3–70.

Savary (1742). *Dictionaire Universal de Commerce*. Geneva.

Scarfe, N. (1988). *A Frenchman's Year in Suffolk*. Boydell Press, Woodbridge.

Scarfe, N. (1995). *La Rochefoucauld, François, duc de, 1765–1848. Innocent Espionage: the La Rochefoucauld brothers' tour of England in 1785*. Boydell Press, Woodbridge.

Schiebinger, L. (1993). *Nature's Body. Gender in the Making of Modern Science*. Beacon Press, Boston, MA.

Schrader, G.W. and Hering, E. (1863). *Biographisch-Literarisches lexicon der Thierärzte aller Zeiten und Länder.* Verlag von Ebner & Sembert, Stuttgart.

Schulzenheim, Baron D. Schulz. de (1797). Observations on sheep, particularly those of Sweden. *CBA*, **1**, 306–24. (2nd edn published in 1804.)

Sebright, J. (1809). *The Art of Improving the Breeds of Domestic Animals.* John Harding, London. (Reprinted in Bajema (1982), pp. 93–122.)

Sedláček, H. (1826). *see* Harkenfeld, von Sedláček (1826).

Seton-Watson, R.W. (1943). *A History of the Czechs and Slovaks.* London, p. 165.

Settegast, H. (1861). *Die Zucht des Negrettischafes und die Schäfereien in Mecklenburgs.* G. Bosselmann, Berlin.

Settegast, H. (1868). *Die Züchtungslehre.* Breslau.

Sinclair, J. (1831). *The correspondence of the Rt. Hon. Sir John Sinclair.* Two volumes. London.

Sinclair, J. (1832). *The Code of Agriculture*, 5th edn. Sherwood, Gilbert & Piper, London. The 1st edn (published in 1817) was reviewed in *Möglin Annalen*, and the review was reproduced in *ONV* (1820), 201.

Sinclair, J. (1823). *Grundsätze des Ackerbaues nebst Bemerkungen über Gartenbau, Obstbaumzucht, Forst-Cultur und Holzpflanzung.* Wien. Translation of *The Code of Agriculture* (1817).

Sinclair, J. (1908). *History of Shorthorn cattle.* Vinton and Co, London.

Smith, C.H. (1827). *A Treatise on the Order Ruminata of the Baron Cuvier.* London. [See also Cuvier 1817].

Smith, J. (1747). *Chronicon Rusticum-Commerciale, or Memoirs of Wool.* T. Osborne, London.

Somerville, J. (1803). *Facts and Observations Relative to Sheep, Wools, Ploughs and Oxen: in which the Importance of Improving the Short Wooled Breeds by a Mixture of the Merino Blood is Deduced from Actual Practice. Together with some Remarks on the Advantages, which have been Derived from the Use of Salt.* W. Miller, London. (2nd edn published in 1806 and 3rd edn in 1809.)

Spary, E. (1996). Political, natural and bodily economies. In *Cultures of Natural History* (ed. N. Jardine, J.A. Secord and E.C. Spary), pp. 178–96. Cambridge University Press, Cambridge.

Spiekermann, M.-L. (ed.) (1983). *Der curieuse Passagier, deutsche Englandreisende des achtzehnten Jahrhunderts als Vermittler kultureller und technologischer Anregungen.* Carl Winter, Heidelberg.

Spoettel, W. and Taenzer, E. (1923). *Rassenanalytische Untersuchungen an Schafen unter besonderer Berüchsichtigung von Haut und Haar.* Abdruck aus dem *Archiv für Naturgeschichte* **5**R. Stricken, Berlin.

Spooner, W.C. (1844). *The History, Structure, Economy and Diseases of Sheep*, 2nd edn. Simpkin, Marshall & Co., London.

Stanley, P. (1995). *Robert Bakewell and the Longhorn Breed of Cattle.* Farming Press, Ipswich.

Steiner, E. (1978). Über den Beitrag Albrecht Thaers zur Entwicklung agragwissenschaflicher Disciplinen auf dem Gebiete der Tierproduktion. Dissertation an der Biowissenschaftlichen Fakultät der Humboldt-Universität zu Berlin.

Stern, C. and Sherwood, E.R. (1966). *The Origin of Genetics. A Mendel Source Book.* W.H. Freeman & Co., San Francisco and London.

Stieber, F. (1842). In welchem Alter und unter welchen Lebensverhältnissen zeigt sich die Vererbung des Schafbockes und der Schafmutter am kräftigsten und sichersten? *Mittheilungen*, 41–4.

Stieber, F. (1851). *Die Rindvieh-Zucht und Ihre Nutzung.* Brünn.

Stirling, A.M.W. (1908). *Coke of Norfolk and his Friends.* London.

Stone, T. (1794). *General View of the Agriculture of Lincoln.* London.

Stumpf, J.G. (1785). *Versuch einer pragmatischen Geschichte der Schäfereien in Spanien, und der spanischen in Sachsen, Anhalt-Dessau etc.* Leipzig.

Stumpf, J.G. (1800). *An Essay on the Practical History of Sheep in Spain and on the Spanish sheep in Saxony, Anhalt-Dessau, etc.* Dublin. (Translated by Dr Lanigan.)

Swain, J.T. (1986). *Industry before the Industrial Revolution. North-east Lancashire ca. 1500–1640.* Chetham Society, Manchester.

Teindl, F.J., (1822). Vortrag gehalten bei der am 1822 stattgefundenen Schafzüchter-Vereines-Versammlung. *Mittheilungen*, 345–51, 353–60, 361–8, 369–74, 377–80.

Teindl, F.J., Hirsch, J. and Lauer, J.C. (1836). Protokol über die Verhandlungen bei der Schafzüchter-Versammlung in Brünn am 9. und 10. Mai 1836. *Mittheilungen*, 303–9, 311–17.

Tessier, H.A. (1811). *Instruction sur les bêtes à laine et particulièrement sur la race des mérinos*, 2nd edn (augmented). Paris. German translation (1811) by W. Witte: *Über die Schafzucht, insbesondere über die Race der Merinos.* Berlin.

Thaer, A. (1797). Extract of a letter to Sir John Sinclair, translated from the original German. *CBA*, **1**, 376–86. (2nd edn published 1804).

Thaer, A. (1804). *Einleitung zur Kenntnisse der englischen Landwirtschaft*, Vol. 3. Gebrüder Halm, Hannover. (The final of three volumes on English agriculture, 1798–1804.)

Thaer, A. (1811). *Handbuch für die feinwollige Schafzucht.* Auf Befehl des Königl. Preuss. Ministeriums des Innerns herusgegeben. Berlin. Review by R. André in *ONV* (1812), 401–2.

Thaer, A. (1816). Über die Gesetze der Natur, welche der Landwirt bei der Veredlung seiner Haustiere und Hervorbringing neuer Rassen beobachtet hat und befolgen muss. *Abhandlungen der physikal. Klasse d. Lgl. Press. Akad d. Wiss. physikal. Klasse d. Lgl. Press. Akad. d. Wiss. aus d. Jahren 1812–1813, Berlin*, p. 5. (Cited by Steiner 1978.)

Thaer, A. (1825). *Über Wolle und Schafzucht.* Rücker, Berlin. (Translation from the cited French book by de Jotemps *et al.* 1824, with German commentary added).

Thompson, B. (1811). *The First Report of the Merino Society.* Nottingham.

Throsby, J. (1791). *Select Views in Leicestershire.* J. Throsby, Leicester.

Trautmann, L. (1814). *Versuch einer wissenschaftlichen Einleitung zum Studium der Landwirtschaft.* Vienna. (Third edition 1823).

Trentham-Edgar, I. (1934). Bakewell. *Notes and Queries*, **167**, 455–6.

Trimmer, J.K. (1828). *Practical Observations on the Improvement of British Fine Wool and the National Advantages of the Arable System of Sheep Husbandry.* London.

Trow-Smith, R. (1957). *A History of British Livestock Husbandry to 1700.* Routledge & Kegan Paul, London.

Trow-Smith, R. (1959). *A History of British Livestock Husbandry 1700–1900.* Routledge & Kegan Paul, London.

Uggla, A.H.J. (1957). *Linnaeus.* Swedish Institute, Stockholm.

Vives, J.V. (1969). *An Economic History of Spain.* Princeton University Press, Princeton.

Vorzimmer, J.V. (1972) *Charles Darwin: the Years of Controversy. The Origin of Species and its Critics 1859–82.* University of London Press Ltd, London.

Wagner, R. (1837). Fragmente zur Physiologie der Zeugung. *Abh. Math. Phys. Classe Bayerischen Akad. Wiss.*, **2**, 381–414.

Wagner, R. (1853). Nachtrag zu dem voranstehenden Artikel Zeugung und Nachtrag zum Nachtrag des Artikels Zeugung. In *Handwörterbuch der Physiologie mit Rücksicht auf physiologische Pathologie*, Vol. 4 (ed. Wagner, R.), pp. 1001–18a–d. F. Vieweg, Braunschweig.

Wall, R. (1758). *A Dissertation on the Breeding of Horses upon Philosophical and Experimental Principles.* London.

Walton, J.R. (1983). The diffusion of improved sheep breeds in eighteenth and nineteenth century Oxfordshire. *Journal of Historical Geography*, **9**, 115–55.

Walton, J.R. (1986). Pedigree and the national cattle herd circa 1750–1950. *Agricultural History Review*, **34**, 149–70.

Waniek, J. (1842). Repräsentanten Bericht über die fünfte Versammlung der deutschen Land- und Forstwirthe zu Doberan im September 1841. *Mittheilungen*, 145–50, 153–5.

Waniek,. (1845). Repräsentantenbericht über die achte Versammlung der deutschen Land- und Forstwirthe zu München. *Mittheilungen*, 249–52, 261–4.

Watson, J.A.S. (1928). Bakewell's legacy. *Journal of the Royal Agricultural Society of England*, **89**, 22–32.

Watts, S. J. (1984). *A Social History of Western Europe. Tensions and Solidarities among Rural People.* Hutchinson, London.

von Weckherlin, A. (1842). *Ueber Englishe Landwirtschaft und deren Anwendung auf andere landwirtschaftliche Verhältnisse, insbesodere Deutschlands. Stuttgart and Tübingen.* (Later editions published in 1845 and 1852.)

von Weckherlin, A. (1846). *Die landwirtschaftliche Thierproduktion* Vol. 1. J.G. Gotta, Stuttgart and Tübingen. (Later editions published in 1851, 1857 and 1865.)

Weckherlin, A. (1860). *Beitrag zu den Betrachtungen über Constanz in der Thierzucht.* J.G. Gotta, Stuttgart.

Wedge, J. (1794). *General View of the Agriculture of the County of Warwick.* Drawn up for consideration by the Board of Agriculture and Internal Improvement, London. Nicol, London.

Weiling, F. (1967). J.G. Mendels Wiener Studienaufenthalt 1851–1853. *Sudhoffs Archiv für die Geschichte der Medizin und Naturwissenschaften*, **51**, 260–66.

Weiling, F. (1968). F.C. Napp und J.G. Mendel. Ein Beitrag zur Vorgeschichte der Mendelschen Versuche. *Theoretical and Applied Genetics*, **38**, 144–148.

Weiling, F. (1986). Das Wiener Universtätsstudium 1851–1853 des Entdeckers der Vererbungsregeln Johann Gregor Mendel. *Folia Mendeliana*, **21**, 9–40.

Western, R. ['A country gentleman'] (1767). *The Complete Grazier, or Gentleman and Farmer's Directory.* J. Almon, London. (4th edn published in 1776.)

Whatley, S. (1732). *Acta Regia: being the Account which Mr. Rapin de Thoyras publish'd of the History of England to the 12th year of Charles I.* J., J. & P. Knapton, London.

White, G. (1977 (1789)). *The Natural History of Selborne.* The Gilbert White Museum Edition. Book Club Associates, London.

White, K.D. (1970). *Roman Farming.* Thames and Hudson, London.

Wichmann, C.A. (1795). Introduction to his own German translation of the book by Hastfer (1752) *Katechismus der Schafzucht zum Unterricht für Schäfer.* Leipzig.

Wight, A. (1778) *Present State of Husbandry in Scotland.* Four volumes. Edinburgh.

Wilhelm, G. (1867). Christian Carl André. *Allgemeine land- und forstwirtschaftliche Zeitung, Vienna*, 46–50.

Willey, B. (1940). *The Eighteenth Century Background. Studies on the Idea of Nature in the Thought of the Period.* Chatto & Windus, London.

Wilson, C. (1965). *England's apprenticeship, 1603–1763.* Longman, London.

Wilson, J. (1909). *The Evolution of British Cattle and the Fashioning of Breeds.* Vinton & Co. Ltd, London.

Wilson, J. (1912). *The Principles of Stock Breeding.* Vinton & Co. Ltd, London.

Winters, L.M. (1954). *Animal breeding*, 5th edn. John Wiley & Sons Inc., New York.

Witten, V. (1828) Ueber Weizenarten. *Mittheilungen*, 317–18.

Wolff, C.F. (1759) *Theoria generationes.* Halle.

Wood, R.J. (1973). Robert Bakewell (1725–1795), pioneer animal breeder and his influence on Charles Darwin. *Folia Mendeliana*, **8**, 231–43.

Wood, R.J. (1990). Martin Barry (1802–1855) and his theory of blood corpuscles as determinants of the organism derived from both parents. *Folia Mendeliana, ***24–25**, 45–8.

Wood, R.J. and Orel, V. (1982). The sheep breeders' legacy to Gregor Mendel. In *Gregor Mendel and the Foundation of Genetics* (ed. V. Orel and A. Matalová), pp. 57–70. Moravian Museum, Brno.

Wurzbach, C. (1874). *Biographisches Lexikon des Kaiserthums Oesterreichs.* Vienna.

Wykes, D.L. (undated but c.1995) Robert Bakewell (1725–1795) of Dishley: a reassessment. Unpublished manuscript, personal communication.

Youatt, W. (1834). *Cattle, their Breeds, Management and Diseases.* Robert Baldwin, London.

Youatt, W. (1837). *Sheep: their Breeds, Management and Diseases to which is Added the Mountain Shepherd's Manual.* Baldwin and Craddock, London.

Young, A. (1771a). *The Farmer's Tour through the East of England*. London.

Young, A. (1771b). Useful projects. Great improvements made in the breed of cattle by Mr Bakewell of Dishley in Northamptonshire [*sic*]. *Annual Register*, 104–10.

Young, A. (1775). *Ökonomische Reise durch die östlichen Provinzen von England* (German translation of 1771). Caspar Fritsch, Leipzich.

Young, A. (ed.) (1783–1815) *Annals of Agriculture*. London.

Young, A. ['The Editor'] (1786a). A ten days' tour to Mr Bakewell's. *Annals of Agriculture*, **6**, 452–502.

Young, A. ['The Editor'] (1786b). Observations on the bill for restraining the growers etc. of wool. *Annals of Agriculture*, **6**, 506.

Young, A. (1788). Sheep controversy between messieurs Chaplin and Bakewell. *Annals of Agriculture*, **10**, 560–77.

Young, A. (1790–91, 1802) *Annale des Ackerbaues unter anderer nützlichen Künste* (translation of early parts of *Annals of Agriculture*). John Riem, Leipzig.

Young, A. (1790–1791). *Travels in France, Italy and Spain during the Years 1787, 1788 and 1789*. Two volumes. London.

Young, A. ['The Editor'] (1791a). Experiment on the introduction of South Down sheep into Suffolk. *Annals of Agriculture*, **15**, 286–310.

Young, A. ['The Editor'] (1791b). A month's tour to Northamptonshire, Leicestershire, etc. *Annals of Agriculture*, **16**, 480–607.

Young, A. (1801) Report of experience of Mr Crowe in breeding greyhounds and then sheep. *Annals of Agriculture*, **37**, 438–43 (see p. 439).

Young, A. (1802) Death of Mr Duckett. *Annals of Agriculture*, **38**, 625–630 (see p. 630).

Young, A. (1811). *On the Husbandry of Three Celebrated British Farmers, Messrs Bakewell, Arbuthnot and Duckett: Being a Lecture Read to the Board of Agriculture*. London.

Young, A. (1813). *General View of the Agriculture of the County of Lincoln*. Drawn up for consideration by the Board of Agriculture and Internal Improvement, London. Nicol, London.

Young, A. (1932). *Tours in England and Wales (Selected from the Annals of Agriculture)*. Reprint no. 14, London School of Economics and Political Science, London.

Young, Rev. A. [son of Young A] (1813). *General View of the Agriculture of the County of Sussex*. Drawn up for consideration by the Board of Agriculture and Internal Improvement, London. Nicol, London.

Zirkle, C. (1941). Natural selection before the 'Origin of Species'. *Proceedings of the American Philosophical Society*, **84**, 71–123.

Zirkle, C. (1951). The knowledge of heredity before 1900. In *Genetics in the 20th Century. Essays on the Progress of Genetics during its First 50 Years* (ed. L.C. Dunn), pp. 35–57. MacMillan, New York.

Zirnstein, G. (1979). Variabilität, Vererbung und Züchtung bei A. D. Thaer und Zeitgenossen. *Tag.-Ber., Akad. Landwirtsch.-Wiss. DDR*, **173**, 53–60.

# Index